Willis's Practice and Procedure for the Quantity Surveyor

Willis's Practice and Procedure for the Quantity Surveyor

Fourteenth Edition

Allan Ashworth
University of Salford
York, UK

Catherine Higgs
University College of Estate Management
Reading, UK

This edition first published 2023

Edition History
John Wiley & Sons, Inc. (13e, 2013)

Registered Offices
John Wiley & Sons Ltd, The Atrium, Southern Gate, Chichester, West Sussex, PO19 8SQ, UK

Editorial Office
The Atrium, Southern Gate, Chichester, West Sussex, PO19 8SQ, UK

For details of our global editorial offices, customer services, and more information about Wiley products visit us at www.wiley.com.

Wiley also publishes its books in a variety of electronic formats and by print-on-demand. Some content that appears in standard print versions of this book may not be available in other formats.

Library of Congress Cataloging-in-Publication Data

Names: Ashworth, Allan, 1944– author. | Higgs, Catherine, author.
Title: Willis's practice and procedure for the quantity surveyor / Allan Ashworth, University of Salford, York, UK, Catherine Higgs, University College of Estate Management, UK.
Other titles: Practice and procedure for the quantity surveyor
Description: Fourteenth edition. | Chichester, West Sussex, UK ; Hoboken : Wiley-Blackwell, 2023. | Includes index.
Identifiers: LCCN 2022056919 (print) | LCCN 2022056920 (ebook) | ISBN 9781119832126 (paperback) | ISBN 9781119832133 (adobe pdf) | ISBN 9781119832140 (epub)
Subjects: LCSH: Building–Estimates–Great Britain.
Classification: LCC TH435 .W6853 2023 (print) | LCC TH435 (ebook) | DDC 692/.50941–dc23/eng/20221206
LC record available at https://lccn.loc.gov/2022056919
LC ebook record available at https://lccn.loc.gov/2022056920

Cover Design: Wiley
Cover Image: © Rudy Balasko/Shutterstock

Set in 9.5/12.5pt STIXTwoText by Straive, Pondicherry, India
Printed and bound by CPI Group (UK) Ltd, Croydon, CR0 4YY

C9781119832126_130323

Contents

Preface

This book was first written by Arthur J. Willis, who became very well-known because of this book in particular. Hence, the current description is described as *Willis's Practice and Procedure for the Quantity Surveyor*. I am never quite sure whether the apostrophe is in the correct place since there were three generations of Willis and each at one time or another edited the book.

Looking back over previous editions of this well-known book, it is clear that the world of quantity surveying has evolved and is vastly different today. In 1951, when the book was first published, quantity surveying could be summarised as approximate estimating, bills of quantities, and final accounts. Such changes should not surprise us, since they are common in all professions. Accountants do not just account, nor solicitors just solicit!

Even the name quantity surveying has changed as the profession emphasises different aspects of their work today. Many will argue that this name is out of date and restrictive. Many firms today use a vaguer term of management or construction consultant to attract wider and different commissions. Traditionally, quantity surveyors operated in the United Kingdom and most of the commonwealth since they adopted UK practices. But quantity surveyors have worked extensively, for example, in the Middle East for a long time. More recently many of the household names of some practices have been acquired by international consultancies that have their head offices in Europe and the United States. These obviously recognise the work that they do and the value that they can add to projects.

It was suggested some time ago that if there were no bills (of quantities), then there would be no fees and hence no quantity surveyors. How wrong this prediction was! If anything the profession is now busier than ever across even a fuller range of construction projects. There has always been a distinction between the large and small practices and the services that they could provide. Some small practices, for example, offer bespoke services in a limited but valuable aspect of how quantity surveying can be applied to a range of different problems today.

The introduction of computers and information technology forecasts a similar demise. There is no doubt that these brought about challenging times for quantity surveyors. This technology created a sort of revolution of what quantity surveyors did and how they did it. How to grapple with it and how to get it to work to the best advantage for both themselves and their clients? No one can doubt that much of the routine activities were removed and that the technology has allowed practices to work more efficiently and smarter rather than just working harder. There is a current focus on modern technologies such as building information management (BIM) that will have extensive ramifications on the world of quantity surveying. Whether such technology will ever be fully effective is still a matter of conjecture, certainly when considered across the full range and type of construction projects today.

The preface to the ninth edition that was published in 1987 speaks of gradual changes to the profession, describing some of them as far reaching. All of these changes that were envisaged then are considered minor in our world today. Whilst the core skills of analysis and evaluation are the same to those days, their applications are much more far and wide ranging. Practices earn fees in a variety of different ways by being able to adapt their skills and knowledge for a wide range of applications. By the turn of this century a much greater emphasis was already being placed on cost and value management.

The previous edition of this book was published 10 years ago. Vast changes have occurred over these intervening years most notably in the areas of sustainability issues and the wider uses of information technology not only to analyse and evaluate building performance just for today but also to examine the implications and impact on project life cycles in attempting to future-proof design and construction.

Allan Ashworth
University of Salford
York, UK
Cathy Higgs
University College of Estate Management
Reading, UK

1

The Work of the Quantity Surveyor

KEY CONCEPTS

- The role of the quantity surveyor (QS)
- The Royal Institution of Chartered Surveyors (RICS)
- Skills, knowledge and behaviours of the QS
  - Ethical decision making
- QS education
- Life long learning

LEARNING OUTCOMES

After reading this chapter you should be able to:

- Understand the role of the quantity surveyor
- Understand the role of the RICS
- Appreciate the knowledge, skills, and behaviours of a QS
- Appreciate the need for life long learning to continual enhance knowledge and skills to meet the needs of industry.

COMPETENCIES

Competencies covered in this chapter:

- Ethics, Rules of Conduct and professionalism

Willis's Practice and Procedure for the Quantity Surveyor, Fourteenth Edition.
Allan Ashworth and Catherine Higgs.

Introduction

In 1971, the Royal Institution of Chartered Surveyors (RICS) published a report titled *The Future Role of the Quantity Surveyor*, which defined the work of the quantity surveyor as:

> '...ensuring that the resources of the construction industry are utilised to the best advantage of society by providing, *inter alia*, the financial management for projects and a cost consultancy service to the client and designer during the whole construction process'.

The report sought to identify the distinctive competencies or skills of the quantity surveyor associated with measurement and valuation in the wider aspects of the construction industry. This provides the basis for the proper cost management of the construction project in the context of forecasting, analysing, planning, controlling and accounting. Many reading this will reflect that this is no longer an adequate description of the work of the quantity surveyor.

Since the report there have been major drivers for change across the construction sector and quantity surveyors now balance the traditional skills of cost expertise with responding to the changing demands in the sector. The needs of clients have changed markedly over the last 50 years. The large regular-procuring clients of the construction industry are increasingly pursuing innovative approaches to the way in which their projects are planned, designed and delivered to facilitate their business strategies. They tend to work more closely with a smaller number of organisations and more closely with their supply chains to maximise value and achieve continuous improvement in performance both of their construction processes and buildings when in use. Advances in digital technologies have had a profound impact on how quantity surveyors operate, their function and the scope and breath of the services they provide. Large practices have responded to the needs of a global market and, over the last decade, there has been an increase in both multidisciplinary and multinational surveying organisations. Quantity surveying practices have diversified in response to government strategies; the most influential being those that address reducing greenhouse gases and improving efficiency within the industry. Pre-2020 these drivers of change were relatively steady, but the global pandemic has accelerated these changes. This chapter seeks both to show how the quantity surveyor role has evolved and the need for continuous enhancement of knowledge and skills.

Characteristics of the construction industry

The total value of the construction new work output in the UK is in the region of 5% of GDP or £116 bn per annum of expenditure (Office for National Statistics 2021). The industry offers direct employment to around two million people and to others in supporting occupations. In addition, many UK firms and practices, including quantity surveyors, have an international perspective through offices overseas or through associations with firms abroad. There has, for example, been an increasing and expanding role of activities on mainland Europe. Approximately 80% of the UK workload is on building projects as distinct from engineering and infrastructure works. New construction projects account for about 64% of the workload of the industry (2022). The repair and maintenance sector will remain an important component for the foreseeable future as clients place greater emphasis upon the improved long-term management of such major capital assets.

The industry is characterised by the following:

- The physical nature of the product
- The product is normally manufactured on the client's premises, i.e. the construction site
- Many of its projects are one-off designs in the absence of a prototype model
- The traditional arrangement separates design from manufacture
- It produces investment rather than consumer goods
- It is subject to wider swings of activity than most other industries
- Its activities are affected by the vagaries of the weather
- Its processes include a complex mixture of different materials, skills and trades
- Typically, throughout the world, it includes a small number of relatively large construction firms and a very large number of small firms

Construction sectors

Quantity surveying offers a diverse range of employment opportunities, within the construction industry, both within the UK and globally. Quantity surveyors are involved in the following four main areas of work.

Building work

The employment of the quantity surveyor on building projects today is well established. The introduction of new forms of contract and changes in procedures continue to alter the way in which quantity surveyors carry out their duties and responsibilities. They also occupy a much more influential position than in the past, particularly when they are involved at the outset of a project.

Quantity surveyors are the cost and value experts of the construction industry. Their responsibilities include advising clients on the cost and value implication of design decisions and the controlling of construction costs. Great importance is now attached to the management of costs in relation to whole life costing. Work within this sector not only relates to new work but to refurbishment of the existing building stock.

Building engineering services

Whilst building services installations are very much a part of the building project, it has tended to become a specialist function for the M & E quantity surveyor, especially on large complex projects. As greater consideration is given to the energy efficiency of systems and alternative sustainable technologies the professional advice from quantity surveyors in this sector will become increasingly influential in project design decisions. Quantity surveyors employed in this discipline have had to become more conversant with the science, technology and terminology of engineering services in order to interpret engineering drawings correctly.

Civil engineering

It is difficult to define the line of demarcation between building and civil engineering works. The nature of civil engineering works often requires a design solution to take into account physical and geological problems that can be very complex. The scope, size and extent of civil engineering works are also frequently considerable. The problems encountered can have a major impact on the cost of the solution, and the engineer must be able to provide an acceptable one within the limits of an agreed budget, in a similar way that buildings are cost planned within cost limits. However, because of their nature, civil engineering works can involve a large amount of uncertainty and temporary works can be considerable, representing a significant part of the budget.

Civil engineering projects use different methods of measurement and different forms and conditions of contract are also used. These to some extent represent the different perception of civil engineering works. The work is more method-related than building works, with a much more intensive use of mechanical plant and temporary works. Bills of quantities, for example, comprise large quantities of comparatively few items. Because much of the work involved is at or below ground level, the quantities are normally approximate, with a full remeasurement of the work that is actually carried out.

Quantity surveyors working in the civil engineering industry provide similar services to those of their counterparts working on building projects.

Heavy and industrial engineering

This work includes such areas as onshore and offshore oil and gas, petrochemicals, nuclear reprocessing and production facilities, process engineering, power stations, steel plants and other similar industrial engineering complexes. Quantity surveyors have been involved in this type of work for a great number of years, and as a result of changing circumstances within these industries a greater emphasis is also being placed on value for money. In an industry that employs a large number of specialists, quantity surveyors, with their practical background, commercial sense, cost knowledge and legal understanding, have much to offer.

Private and contractors quantity surveyors

As well as specialised by project type, quantity surveyor's role can have either a client focus or work for a contracting organisation. Those working in the public or private sector on behalf of clients are known as Private Quantity Surveyors, referred to as a 'PQS' or the clients QS and those working for a main contractor or subcontractor are Contractors Quantity Surveyors, a 'CQS'. The role of the PQS is primarily covered in this book. The role of the CQS is somewhat different from that of the client's quantity surveyor with a focus on commercial management and the supply side of the sector, in that they consider costs from the contractor's perceptive maximising cash flow and ensuring the project stays within budget These activities could include estimation, financial management, site costing and bonusing, contract management, negotiation with suppliers and subcontractors, interim certificates and payments, contractual matters and the preparation and agreement of claims. Further consideration of the role is given within the relevant chapters.

The quantity surveying profession

The origins of the role

The origins of quantity surveying as a distinct activity are hard to trace back in time much further than the Great Fire of London. However, in the New Testament, there is a story about counting the cost before you build (Luke's Gospel, chapter 14). Perhaps quantity surveyors can trace their roots back to more than 2000 years ago! The building activity that followed the Great Fire of London in 1666 encouraged the emergence of the architect and the growth of the single trades, contracting for their own part of the building work.

The measurers had to be invented if they did not already exist. There was a real need for someone to ensure impartiality between the proprietor and the workmen. The rest is history. In 1834, the fire that destroyed the Palace of Westminster was partially responsible for the use of the quantity surveyor on a major scale. Charles Barry won the competition to replace it and was asked to prepare an estimate of cost. Although detailed drawings were not yet prepared, a quantity surveyor, Henry Hunt, came up with an estimated cost of £724,984. Whilst this figure was basically accurate, changes made by Parliament resulted in a final cost closer to £1.5 m.

The name 'quantity surveyor' conjures up a variety of different images in people's imaginations. For some, the term 'quantity surveyor' is an outmoded title from the past. It certainly no longer *accurately* describes the sole duties that are performed as will be discussed later in the chapter. When the term was first applied to the profession, the work of the quantity surveyor was vastly different from that now being carried out and anticipated in the twenty-first century. New titles for the role, over time, have been debated, and it is common to find those offering current quantity surveying services describing themselves as cost consultants.

The Royal Institution of Chartered Surveyors (RICS)

The Royal Institution of Chartered Surveyors was formed in 1868 and offices were leased at 12 Great George Street, which is still part of the RICS Headquarters building today. The Institution of Surveyors, which later became the Royal Institution of Chartered Surveyors (RICS), has evolved into a renowned international organisation with approximately 134,000 members working in 146 countries. It was granted its Royal Charter in 1881 and in 1922 the Quantity Surveyors Association amalgamated with it. In 1930, the then Institution of Surveyors became the Institution of Chartered Surveyors. In 1946, it was granted the title Royal to become the Royal Institution of Chartered Surveyors. The RICS coat-of-arms with its motto *Est modus in rebus* (There is measure in all things) was adopted.

The RICS is organised around 18 professional groups of which one is designated as Quantity Surveying and Construction. There are a number of other groups to which quantity surveyors are also likely to belong. These include Dispute Resolution, Facilities, Management Consultancy and Project Management.

The RICS's objective as a professional body, as outlined in the Charter, is to secure the advancement and facilitate the acquisition of that knowledge, which constitutes the profession of a surveyor, and to maintain and promote the usefulness of the profession for the public advantage in the United Kingdom – and in any other part of the world (RICS 2022c).

Working with key stakeholders of UK governments, the RICS as a leading expert within the sector informs government policy. Through the promotion of the expertise of chartered surveyors, the RICS's professional standards are adopted within the industry. Key activities in 2021 were advice provided for the UK's recovery from the pandemic, and the long-term value of sustainable development and management of the built and natural environments (RICS 2021a). In addition, the principle purposes of the RICS are:

- a global professional, standards and regulatory body
- existing to secure the advancement and usefulness of the profession for the public advantage
- focused on setting standards and assuring these standards are in the public interest delivering support that is valued by RICS members and their employers
- developing members' professional skills and knowledge and
- expanding opportunities for members to apply those professional skills. (RICS 2022c)

Other quantity surveying bodies

Whilst the RICS remains the premier institution for quantity surveyors, they may also be members of other industry bodies. This may be influenced by the sector they work in, for example quantity surveyors employed by contractors are likely to be members of the CIOB, or by the country in which they are located. The Pacific Association of Quantity Surveyors (PAQS) is an international association of quantity surveyor organisations located in the Asia and Western Pacific region. The membership of PAQS is shown in Fig. 1.1.

The Quantity Surveying International (QSi) was formed in 2004 as a professional body solely for those operating in the commercial aspects of construction and the built environment. Its objectives are very similar to the RICS, except they are fully focussed on quantity surveying.

The role of the quantity surveyor

Traditional role

The traditional role of the quantity surveyor has been described elsewhere and in previous editions of *Willis's Practice and Procedure for the Quantity Surveyor*. This traditional role, still practised by some and especially on small- to medium-sized projects, can be briefly

Full Members

- Australian Institute of Quantity Surveying (AIQS)
- China Cost Engineering Association (CCEA)
- The Hong Kong Institute of Surveyors (HKIS)
- Royal Institution of Surveyors Malaysia (RISM)
- Institution of Surveyors, Engineers and Architects, Brunei (PUJA)
- The Building Surveyors Institute of Japan (BSIJ)
- Canadian Institute of Quantity Surveyors (CIQS)
- New Zealand Institute of Quantity Surveyors (NZIQS)
- Singapore Institute of Surveyors and Valuers (SISV)
- Institute of Quantity Surveyors Sri Lanka (IQSL)
- Philippine Institute of Certified Quantity Surveyors (PICQS)
- Ikatan Quantity Surveyor Indonesia (IQSI)

Associate Members

- Fiji Institute of Quantity Surveyors (FIQS)
- Korea Institution of Quantity Surveyors (KIQS)

Observer Member

- Association of South African Quantity Surveyors (ASAQS)

Fig. 1.1 Members of the pacific association of quantity surveyors.

described as a measure and value system. Approximate estimates of the initial costs of building are prepared using a single price method of estimating (see Chapter 5), and where this cost was acceptable to the client then the design was developed by the architect. Subsequently, the quantity surveyor would produce bills of quantities for tendering purposes, the work would be measured for progress payments and a final account prepared on the basis of the tender documentation (see Fig. 1.2). The process was largely reactive, but necessary and important. During the 1960s, to avoid tenders being received that were over budget, cost planning services were added to the repertoire of the duties performed by the quantity surveyor employed in private practice (PQS). The contractor's surveyor was responsible for looking after the financial interests of the contractor and worked in conjunction with the PQS on the preparation of interim payments and final accounts. On occasions, contractors felt that they were not being adequately reimbursed under the terms of the contract and submitted claims for extra payments. This procedure was more prevalent on civil engineering projects than on building projects, although the adversarial nature of construction was increasing all the time.

The distinctive competence found in quantity surveyors relies heavily on their analytical approach to buildings and this stems directly from their ability to measure construction works. Furthermore, the detailed analysis of drawings leads to a deep understanding of the design and construction which enables them to contribute fully to the process. This intimate knowledge of projects is at the root of the contribution made by the quantity surveyor to the value of the client's business through the provision of the services shown in Fig. 1.2.

- Single rate approximate estimates
- Cost planning
- Procurement advice
- Measurement and quantification
- Document preparation, especially bills of quantities
- Cost control during construction
- Interim valuations and payments
- Financial statements
- Final account preparation and agreement
- Settlement of contractual claims

Fig. 1.2 Traditional quantity surveying activities (circa 1960).

Evolved role

In response to the potential demise of bills of quantities, quantity surveyors began exploring new potential roles for their services. Procurement, a term not used until the 1980s, became an important area of activity, largely because of the increasing array of options that were available. Increased importance and emphasis were also being placed upon design cost planning as a tool that was effective in meeting the client's objectives. Whole life costing (Chapter 6), value management (Chapter 7) and risk analysis and management (Chapter 8) were other tools being used to add value for the client. As buildings became more engineering services orientated, increased emphasis was being placed on the measurement, costs and value of such services. Quantity surveyors had historically dealt with this work through prime cost and provisional sums, but in today's modern buildings to describe the work in this context is inadequate. Other evolved roles have included project and construction management and facilities management (see Fig. 1.3). Because of the inherent adversarial nature of the construction industry they are also involved in contractual disputes and litigation.

The current role of the quantity surveyor reflects a more outcomes led approach by clients. Today's construction is not just about the provision of a building, but the performance of an asset in terms of the client's business outcomes. Client's environmental, social and digital agendas, influenced by the wider changing external political drivers, are having a greater influence on the design and use of buildings. The quantity surveyor's expertise in cost, value and managing risk has therefore increased importance in the 'whole life' success of a project. Whilst many of the services related to consultancy and project delivery offered, a decade ago, are similar, the use of technologies has meant that these services have become more integrated internally within the quantity surveying organisation, but crucially more integrated with the services provided by other built environment professionals using collaborative information platforms. The ability to automate some services has led to enhanced practices in benchmarking, scenario testing and risk analysis. The last decade has seen many quantity surveying mergers with other professions to offer more comprehensive and integrated services and an increase in the offer of post occupancy services. Such services include auditing, benchmarking, and information modelling to advise

- Investment appraisal
- Advice on cost limits and budgets
- Whole life costing
- Value management
- Risk analysis
- Insolvency services
- Cost engineering services
- Subcontract administration
- Environmental services measurement and costing
- Technical auditing
- Planning and supervision
- Valuation for insurance purposes
- Project management
- Facilities management
- Administering maintenance programmes
- Advice on contractual disputes
- Planning supervisor
- Employer's agent
- Programme management
- Cost modelling
- Sustainability Advisor

Fig. 1.3 Evolved role (circa 2012).

on business success. Quantity surveyors also offer specialist skills such as Capital Allowances and Alternative Dispute Resolution services.

Global strategies to support a sustainable future, such as *2030 Agenda for Sustainable Development* and UK legislation to support the sector's trajectory towards meeting net zero, are key drivers informing the client's sustainability objectives and organisational practices. As a result, clients are becoming increasingly aware of the need to consider whole life costs, environmental impact assessments and evaluation of carbon emissions. The impact on the quantity surveying services is discussed in Chapter 18.

Skills, knowledge and understanding

In 1992, the Royal Institution of Chartered Surveyors published a report titled *The Core Skills and Knowledge Base of the Quantity Surveyor*. The report developed earlier themes from reports published by the RICS and others. These included *The Future Role of the Quantity Surveyor* (RICS 1971), *The Future Role of the Chartered Quantity Surveyor* (RICS 1983), *Quantity Surveying 2000* (Davis, Langdon and Everest 1991) and *Quantity Surveying Techniques: New Directions* (Brandon 1992). The *Core Skills* report examined the needs of quantity surveyors in respect of their education, training and continuing professional development. This reflected the requirements in the context of increasing changes and uncertainties in the construction industry and, more importantly, within the profession. The RICS report identified a range of skills that the profession would need to continue to develop if it wished to maintain its role within the construction industry. The report identified a knowledge base that includes:

- Construction technology
- Measurement rules and conventions
- Construction economics
- Financial management
- Business administration
- Construction law

and a skill base that includes:

- Management
- Documentation
- Analysis
- Appraisal
- Quantification
- Synthesis
- Communication.

All of these remain valid requirements 30 years later, indeed they are the core of many quantity surveying courses, although their relative importance has changed to suit changing needs and aspirations, as evidenced when comparing the QS activities listed in Figs 1.2 and 1.3 above.

Quantity surveying, like each specific surveying discipline, has developed its own repertoire of techniques. Skills occur in respect of the levels of ability required to apply these techniques in an expert way. The different array of skills is assimilated with the knowledge base through education, training and practice. Whilst there is a general agreement about the skills and knowledge base required, different surveyors will place different emphases upon the relative importance in practice. Skills and knowledge requirements are also not static but must be updated to reflect an ever changing environment.

The RICS Futures Report 2015 *Our Changing World: let's be ready* identified that both the skills needed by surveyors and the work roles were changing, due to the growing complexity of the sector, major skills gap in sustainability and technological advances (Fig. 1.4).

Current Skills	Future Skills
Outcome focus	Sustainability
Communication	Data analysis
Integrated programme and cost management	Maximising resource productivity
Skills for greater complexity	Risk management
Interdisciplinary working	Leadership
Advisory services	Client focus
Understanding technology	Collaboration
	Ethical behaviour

Fig. 1.4 2015 Skills gap analysis (*Source:* Gray et al. 2016).

An outcome of this report was the development of the World Build Environment Forum (www.rics.org/uk/wbef/) to support continued enhancement of the knowledge and skills of its members.

The second Futures report, *Future of the profession*, identified that new business models and innovative technologies were driving the need for new skillsets within the profession. Skills identified were increased proficiency in the application of technology and non-technical skills such as resilience, emotional intelligence and ability to collaborate (RICS 2019a).

The third Futures report, *Futures Report 2020*, reinforced the need for lifelong learning to maintain, develop and enhance competencies throughout individual's careers to adapt to the changing environments in which they work (RICS 2020). Never was this more relevant than how surveyors adapted their working practices during the COVID-19 pandemic to ensure they could carry out their services competently in a virtual environment.

Though dated Powell's 1998 *Challenge of Change* is still relevant in emphasising the importance of the skills required of the chartered quantity surveyor to be prepared for the requirements of the future, with a need to:

- Develop a greater understanding of business and business culture
- Develop strong communications and ICT skills
- Challenge authoritatively the contributions of other team members
- Understand that value can be added only by managing and improving the client's customers and employer's performance
- Develop skills to promote themselves effectively
- See qualifications only as the starting point
- Recognise the need to take action now
- Become champions of finance and good propriety.

It has long been recognised that the development of skills has a much longer life span than knowledge since the latter is changing and frequently being updated. The Skills Plus Project (William 2003) identified a comprehensive listing of 39 skills organised under three headings of personal qualities, core skills and process skills. The personal qualities included, for example:

- Independence
- Adaptability
- Initiative taking
- A willingness to learn
- An ability to reflect on what has and what has not been achieved.

Core skills included the obvious three 'Rs' but also:

- The ability to present clear information when in a group
- Self-management
- Critical analysis
- The ability to listen to others.

The last of these, whilst obvious, is a skill that is frequently much under-developed in people generally. Common listening styles include the juggler (distracted listening), the

QS Attributes	Approach to Activities
Analytical thinker Highly numerate Able to absorb and distil complex information Methodical and meticulous attention to detail	Work on own initiative Strong work ethic Organised Able to prioritise Thrives under pressure Meets deadlines

Fig. 1.5 Skills requirement of Entry Level Graduates and Assistant Quantity Surveyors.

pretender (pretend listening), the hurry-uper (impatient listening), the rehearser (switched-off listening) and the fixer (fix-it is listening). Listening, by allowing others to speak, is a very powerful tool.

The process skills include:

- Computer literacy
- Commercial awareness
- Prioritising
- Acting morally and ethically
- Coping with ambiguity and complexity
- Negotiating.

Analysis of the skills stated within job descriptions for graduate and assistant quantity surveyors show in addition quantity surveyors need to be highly numerate and analytical in their approach (Fig. 1.5). Analytical skills must be considered in the context of an increasing technology-based role. It is essential that quantity surveyors have both the digital and ethical reasoning skills to select the right data, of the required quality to effectively use technology to efficiency provide answers to inform the professional advice provided (Arrow 2020).

Quantity surveying competences will be discussed later in the chapter.

Acting morally and ethically

Acting morally and ethically is central to a quantity surveyor carrying out their role. Ethical behaviour initially and intrinsically stems from a person's own value system. This can be simply described as knowing the difference between right and wrong and having the courage to do what is right. There is never a right way to do the wrong thing. Simply put, that's what ethics is all about. Personal ethics tell us that if we are going to get along with one another we should not lie, steal and cheat. In business, the same principles apply. Business ethics are based upon a willingness to live up to our word and provide all the necessary information so that the other party can fulfil their obligations in a fair manner. James (2003) notes two types of integrity: personal and process. Personal integrity is when a person's words and actions are the same, and those same things are the right things. Process integrity is what companies trust to produce the right results.

Quantity surveyors in carrying out their responsibilities must therefore act with personal integrity; to do so it is important to be aware of the unethical practices that exist within the construction industry to avoid them. Watts et al. (2021) identified examples of such practices as bribes and inducements, conflicts of interest, fraudulent practices, deceit and trickery, presenting unrealistic promises, exaggerating one's own expertise, concealing design and construction errors and overcharging. Further discussion on the quantity surveyor's compliance with all relevant requirements under the Bribery Act 2010 is discussed in Chapter 4.

In order to minimise the chances of unethical or illegal behaviour in the construction industry, professional bodies such as the RICS have codes of conduct to promote and enforce the highest ethical standards. These codes of behaviours provide guidelines and support individuals to make the right decisions when facing ethical dilemmas. The RICS *Rules of Conduct* are based on the ethical principles of honesty, integrity, competence, service, respect and responsibility. Members are required to:

- be honest, act with integrity and comply with professional obligations
- maintain professional competence
- provide good-quality and diligent service
- treat others with respect and encourage diversity and inclusion
- act in the public interest, take responsibility for actions and act to prevent harm and maintain public confidence in the profession. (RICS 2021b)

The RICS *Ethics Decision Tree* is a useful guide to support professional judgement in ethical decision making. Serious ethical breaches will result in disciplinary action as described in Chapter 4.

Quantity surveying education

Courses in quantity surveying are offered at various levels and modes in universities and colleges in the United Kingdom. These courses are also replicated in many of the former Commonwealth countries around the world. They are typically undergraduate courses of three (full-time), four (sandwich) or five (part-time) years' duration, but there are also postgraduate courses of one- and two-years' duration. Degree apprenticeships, introduced in 2015, are also offered at both undergraduate and postgraduate level. Many courses offer accreditation from institutions such as RICS, CIOB, ICES and CABE. For students intending to become an RICS Chartered Quantity Surveyor, a course must be one that is accredited by the Royal Institution of Chartered Surveyors. These courses allow students who obtain such a degree to become eligible for professional membership (MRICS) upon passing the Assessment of Professional Competence (APC).

RICS accreditation

Quantity surveying programmes accredited by the RICS can be found in universities throughout the United Kingdom and in many parts of the world (Fig. 1.6).

Countries with RICS Accredited Courses	
Australia	Oman
Greece	Singapore
Hong Kong	South Africa
India	Sri Lanka
Ireland	United Arab Emirates
Malaysia	

Fig. 1.6 Countries with quantity surveying courses (*Source:* Adapted from RICS 2022a).

Accreditation is a recognition of the appropriateness and relevance of the quantity surveying course offered by the university. The accreditation process consists of a written application and a visit by an accreditation panel of both RICS staff and professionals from industry. This process addresses all aspects of the programme and the students learning experience, which ensures the course demonstrates:

- Curriculum content meets the level 1 competencies of the quantity surveying and construction RICS pathway.
- Curriculum content that embeds inclusion and diversity, ethics and the relevant professional standards.
- Appropriately qualified staff that effectively deliver the programme.
- Assessment methods that develop critical thinking, problem solving and professional communication skills.
- Provision of a high-quality student experience.
- Robust quality processes in place to ensure continued validity and relevance of the programme. (RICS 2019b)

Assessment of professional competence (APC)

The different professional bodies all test an individual's professional capabilities and competence to practise before a chartered designation can be used. In the RICS, this is the Assessment of Professional Competence (APC), which requires candidates to demonstrate that through both their education and professional experience that they have achieved a set of technical and professional practice, interpersonal, business and management skills. This is demonstrated by recording the experience gained in the workplace in the APC Diary templates. For an individual with less than five years' experience, this is for a period of 24 months. In October 2000, the RICS Practice Qualification Group introduced the need for all APC candidates to be employed in an organisation having a structured training framework. Structured training ensures the experience offered gives the trainee an opportunity to develop professional practical skills, put theory into practice, make decisions, solve problems, and develop the competencies required of their profession. For the quantity surveying and construction pathway, the mandatory and core and competencies are listed in Fig. 1.7. Reflecting the diversity of the quantity surveying profession, there are also

Mandatory	Core: Assessed at Level 3
Ethics rules of conduct and professionalism Client care Communication and negotiation Health and safety Accounting principles and procedures Business planning Conflict avoidance, management, and dispute resolution procedures Data management Diversity, inclusion and teamworking Inclusive environments Sustainability	Commercial management of construction or design economics and cost planning Contract practice Construction technology and environmental services Procurement and tendering Project financial control and reporting Quantification and costing of construction work

Fig. 1.7 Quantity surveying and construction competencies (*Source:* Adapted from RICS 2018).

specialist optional competencies covering a wide range of activities including capital allowances, insurance, litigation, programming and risk management. To support the readers understanding of the knowledge and skills required for each of the competences, a list of competencies is included at the start of each chapter within this book.

Each competency is assessed at three levels

- Level 1 – knowledge and understanding (mandatory and core will be covered within university studies)
- Level 2 – application of knowledge
- Level 3 – depth and synthesis of technical knowledge and implementation.

A good spread of Level 1 competencies with some at Level 2 can be achieved through sandwich work placements, part time working and apprenticeships. Competencies at Level 3 are probably achievable for students having considerable relevant experience post-graduation.

The RICS *Assessment of Professional Competence Candidate guide* provides comprehensive guidance on the requirements of the APC. An APC candidate is required to submit a Summary of Experience; this is a reflective piece providing evidence that they have developed each mandatory and technical competency to the required level. A Case Study submission provides further evidence that they can work to the level of competency required by a quantity surveyor. Prior to the final assessment, candidates are also required to successfully past the RICS Professionalism module. The final assessment takes the form of a one-hour interview with normally three RICS professionals. In deciding whether an individual meets the requirements for professional membership, the following are assessed:

- Understanding of knowledge gained and competencies supported by reference to the written submissions
- Understanding of the role and responsibilities of a chartered surveyor
- Application of professional and technical skills

- Understanding of ethics, RICS rules of conduct and current issues of concern to the profession
- Ability to communicate clearly

Success at the final assessment will enable you to call yourself a Chartered Quantity Surveyor. In 2022/21 for AssocRICS and MRICS assessments combined the success rate was approximately 80% (RICS 2022b).

It should be noted at the time of writing that the RICS' Governing Council's Entry and Assessment Review is reviewing the entry requirements to RICS and RICS' qualification assessment methods to ensure they remain relevant to the sector.

Continuing professional development (CPD)

With the accelerating pace of change in the industry, many professional bodies have recognised the importance of CPD and lifelong learning. Since 1984, the RICS has recognised that for CPD to be effective for the profession it should be mandatory for professional membership and that some form of registration is necessary. Rule 2.5 of the Rules of Conduct states

> 'Members maintain and develop their knowledge and skills throughout their careers. They identify development needs, plan and undertake continuing professional development (CPD) activities to address them and are able to demonstrate they have done so'. (RICS 2021b)

There is also an obligation for firms to ensure that all staff are properly trained. CPD can be either formal, i.e. structured training, or informal, which is self-directed and linked to private study. From a firm's perspective the benefits claimed for CPD include higher productivity and profitability, lower staff turnover and absenteeism, innovation, improved client services, higher quality and improved job satisfaction.

References

Arrow, J. *Construction Technology: Building the Future in a Mirror World*. RICS News & Opinion. 2020. Available at https://www.rics.org/uk/news-insight/future-of-surveying/talent-and-skills/construction-technology-building-the-future-in-a-mirror-world/ (accessed 19/10/2022).

Brandon P.S. (Ed.) *Quantity Surveying Techniques: New Directions*. Blackwell Science. 1992.

Davis, Langdon and Everest. *Quantity Surveying* 2000. Royal Institution of Chartered Surveyors. 1991.

Gray, A., Smith, P. and Shah, N. *RICS Futures Report: Our Changing World*. Real Estate Advisory and Leadership Community Presentation. 2016. Available at https://real.ifma.org/wp-content/uploads/2016/06/RICS-Futures-for-IFMA.pdf (accessed 10/10/2022).

James, R. *The Integrity Chain*. FMI Consulting. 2003.

Office for National Statistics. *Construction statistics, Great Britain:2021*. ONS. Available at https://www.ons.gov.uk/businessindustryandtrade/constructionindustry/articles/constructionstatistics/2021#main-points (accessed 09/12/2022).

Powell, C. *The Challenge of Change*. Royal Institution of Chartered Surveyors. 1998.

RICS. *The Future Role of the Quantity Surveyor*. Royal Institution of Chartered Surveyors. 1971.

RICS. *The Future Role of the Chartered Quantity Surveyor*. Royal Institution of Chartered Surveyors. 1983.

RICS. *The Core Skills and Knowledge Base of the Quantity Surveyor*. Royal Institution of Chartered Surveyors. 1992.

RICS. *Quantity Surveying and Construction Pathway*. RICS. 2018. Available at https://www.rics.org/globalassets/rics-website/media/assessment/qs-and-construction.pdf (accessed 20/10/2022).

RICS. *The Future of the Profession*. Royal Institution of Chartered Surveyors. 2019a. Available at https://www.rics.org/globalassets/rics-website/media/news/future-of-the-profession-post-consultation-report.pdf (accessed 10/10/2022).

RICS. *Global Accreditation – Policy and Process*. Royal Institution of Chartered Surveyors. 2019b. Available at https://www.rics.org/uk/upholding-professional-standards/standards-of-qualification/accreditation/ (accessed 19/10/2022).

RICS. *Futures Report 2020*. Royal Institution of Chartered Surveyors. 2020. Available at https://www.rics.org/globalassets/rics-website/media/news/news--opinion/rics-future-report-2.pdf (accessed 10/10/2021).

RICS. *Impact Report 2021: RICS Influence, Advocacy and Thought Leadership Across the UK*. Royal Institution of Chartered Surveyors. 2021a. Available at https://www.rics.org/globalassets/rics-website/media/knowledge/research/research-reports/rics-uk-advocacy-impact-report-2021-dec.pdf (accessed 21/10/2022).

RICS. *Rules of Conduct*. RICS. 2021b.

RICS. *Course Finder*. RICS. 2022a. Available at https://www.ricscourses.org/ (accessed 18/10/2022).

RICS *Building for the Future*. RICS. 2022b. Available at https://ww3.rics.org/uk/en/annual-review/building-for-the-future.html (accessed 21/10/2022).

RICS. *The Purpose of RICS as a Professional Body*. Royal Institution of Chartered Surveyors. 2022c. Available at https://www.rics.org/globalassets/rics-website/media/governance/1.-purpose-as-a-professional-body_final.pdf (accessed 21/10/2022).

Watts, G., Challender, J., Higham, A. and McDermott, P. *Professional Ethics in Construction and Surveying*. New York. Routledge. 2021.

William A. *Skills Plus: Tuning the Undergraduate Construction Curriculum*. CIB W89 International Conference on Building Education and Research (BEAR). 2003.

2

Digital Technologies

KEY CONCEPTS

- Construction 4.0
- Use of digital technologies in the construction sector
- Building Information Modelling
- Digital Twins.

LEARNING OUTCOMES

After reading this chapter you should be able to:

- Appreciate the range of digital technologies used in the sector, to store, analysis and manage project information during the project lifecycle.
- Understand BIM as both a technology and a process for managing information, and how it is implemented at project level.
- Understand the concepts of a digital twin and its potential use in the sector.

COMPETENCIES

Competencies covered in this chapter:

- Communication and negotiation
- Data management

Introduction

Digital technologies are the enablers to create the design, construction and use of our building assets that add societal and environmental value. Digitalisation is transforming the capability to model scenarios in real time to better understand the time, cost, quality, safety, and productivity implications.

Willis's Practice and Procedure for the Quantity Surveyor, Fourteenth Edition.
Allan Ashworth and Catherine Higgs.

Construction 4.0 is challenging how the sector designs, constructs and manages assets over the project life cycle. Both Building Information Modelling (BIM) and Common Data Environments (CDE) are core to the digital transformation of the sector, with digital maturity linked to BIM usage. As will be demonstrated BIM can play a strategic role. This chapter therefore provides an overview of digital technologies use within the sector, but most importantly focuses on BIM, as it will continue to greatly influence both how quantity surveyors work and the integration of cost management services throughout the project lifecycle.

Construction 4.0

Construction 4.0 is the sectors adaption of the principles of Industry 4.0; the digital revolution defined by emerging technologies in the manufacturing sector. The Construction 4.0 framework consist of a physical and data layer, supported by a digital tool layer that contains technologies and tools to translate and transfer information between the two (Ankan and Venkata 2021). Such digital tools include Virtual Reality (VR), Augmented Reality (AR), Mixed Reality (MR), Blockchain technologies, Big data analytics, Artificial intelligence (AI), Machine learning (ML) and Cloud-based project management

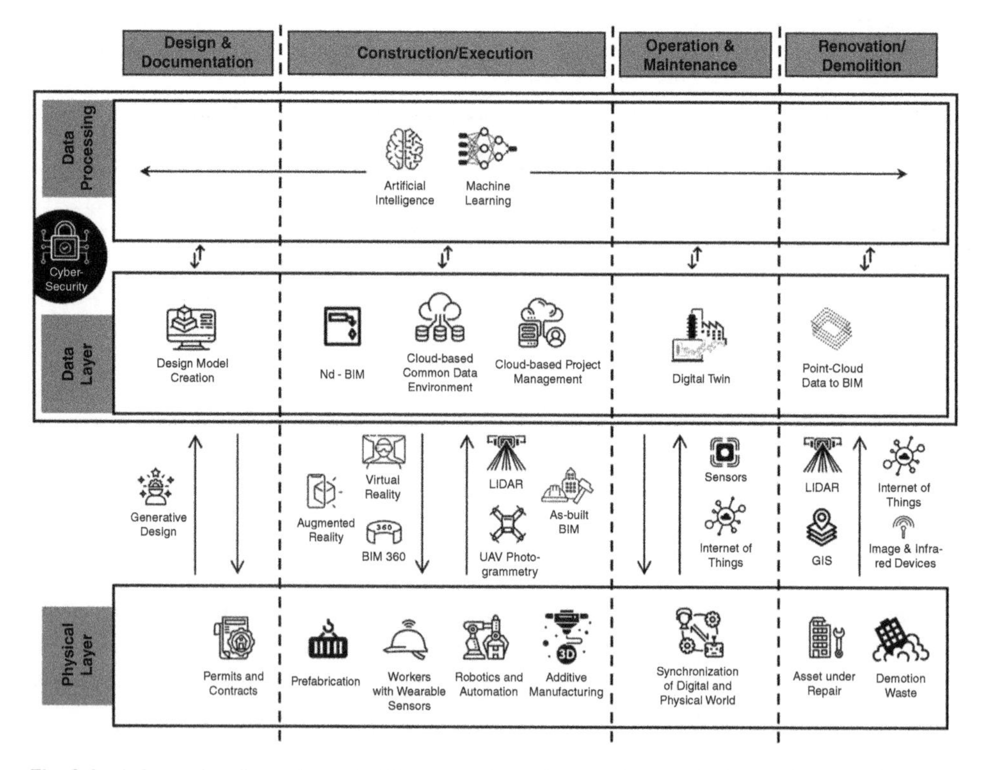

Fig. 2.1 Information flows through different project lifecycle phases illustrating intra- and inter-layer interactions (*Source:* Ankan and Venkata 2021/Journal of Information Technology in Construction/CC BY-4.0).

technologies. Figure 2.1 summarises the information flow across the different layers and between technologies during different phases of the project life cycle.

As shown in the figure, BIM and a cloud-based Common Data Environment are central to the Construction 4.0 Framework, providing a single platform that helps promote:

- Integration of all phases of the project lifecycle, all members of the project team and inter-project learning and knowledge management.
- Linkage between the physical and digital tool layer over the entire project life cycle. (Sawhney et al. 2020)

A detailed explanation of the technologies within the framework, is outside the scope of this textbook; it is recommended that readers refer to *Construction 4.0: An Innovative Platform for the Built Environment* which includes many studies relating to the Construction 4.0 paradigm.

Use of digital technologies within the sector

The *NBS 10th Annual BIM Report 2020* provides an informative insight into the use of digital technologies within the sector and their relative use (Fig. 2.2). Whilst the subsequent survey in 2021 reports on a number of the same technologies there is little that can be inferred from a comparison between the two.

Cloud computing, in both studies has the highest implementation, and has seen the greatest growth over the past years, reflecting that it is an enabler for other digital technologies such as BIM, Internet of Things, AR and Big Data. Microsoft Office 365 and common data environments, like Autodesk's Construction Cloud (previously BIM 360), Aconex, Asite and Viewpoint, are the most common examples of cloud computing (NBS 2021).

The use of immersive technology, such as Virtual Reality (VR) and Augmented Reality (AR), is highest in larger organisations. The strengths of the use of this technology are:

- Enhanced knowledge and understanding
- Effective information delivery and learning through simulation
- A tool to aid and improve communication, collaboration and problem solving. (Gontier et al. 2021)

The NBS Survey provides examples of its use, including the use of VR supporting clients to visualise projects at design stage and the application of AR to overlay instructions and maintenance data to assist those making maintenance designs (NBS 2021).

Technology	**Current Use** (2020)
Cloud computing	42%
VR/AR and Mixed Reality	38%
Drones	32%
3D printing for building components	11%
Analytics and Big data	10%
Digital twins	9%
AI and Machine Learning	6%

Fig. 2.2 Use of technology within the sector: 2020 (*Source:* NBS 2020/NBS Enterprises Ltd).

As cited in the *Future of BIM* the impact of drones on the built environment cannot be underestimated, with predictions of 76,000 drones being in use by 2035. (RICS 2020). Drones offer the opportunity to capture site images with accuracy, so can support any site-based data gathering activities. The ability to survey inaccessible locations has obvious health and safety benefits.

The use of BIM and Digital Twins will be discussed in more depth later in this chapter.

In summary the NBS *10th Annual BIM Report 2020* reported that responses agreed that digitalisation will transpose the way of work and has the potential to improve; productivity, speed of delivery, health and safety and support the construction industry in meeting the challenges of sustainability (NBS 2020).

Building information modelling (BIM)

BIM definition

Before considering the implications of BIM on quantity surveying procedures and practices, it is essential to consider its definition. This definition has evolved as the sector's digital capabilities have matured, from the NBS definition stated in the 13th edition,

> '...a rich information model, consisting of potentially multiple data sources, elements of which can be shared and be maintained across the life of a building from inception to recycling (cradle to cradle). The information model can include contract, specification properties, personnel, programming, quantities, cost, space and geometry'. (NBS 2012)

to 'use of a shared digital representation of a built asset to facilitate design, construction and operation processes to form a reliable basis for decisions'. (BSI 2019)

Both definitions stress collaboration but whilst the NBS definition is focussed on technology, the later definition reflects the evolution of BIM within the industry with a focus on information management. Hence, there is now an established framework of processes, protocols, and standards to support the effective sharing of project information to support decision making over the whole life of an asset.

Traditionally, data relating to a project was not provided for the end user at the handing-over of the building. BIM models offer this capability and, if structured with the intention of use by the facilities manager, can ensure optimum asset performance during the life of the building. BIM will also provide the capacity to better understand and analyse user outcomes and, if captured and feedback given, will provide a more comprehensive understanding of client value for future projects.

BIM stages of maturity

To reflect the evolution of BIM from a strategy to support the implementation of a technology to a fully collaborative approach to managing data across the complete project life cycle, the sector has seen a development from referencing BIM maturity by levels, as defined in the 2011 BIM *Working Party Strategy Paper* and associated with the widely adopted BIM maturity model developed by Bew and Richards in 2008, to information management stages as defined in BS EN 19650-1:2018. For reference the four levels are:

Level 0: The use of 2D CAD files for production of information, but information exchange is predominantly paper-based.

Level 1: The use of both 2D and 3D project information. The 3D information is predominantly used for conceptual design by architects and visualisation. The BIM model is not used collaboratively and there are separate cost management systems.

Level 2: A managed 3D model with separate 'BIM' tools with attached data. Cost information is integrated on the basis of a proprietary interface.

Level 3: A fully open process and data integration enabled by a 'web service', managed by a collaborative model server.

In 2015 the Digital Built Britain Level 3 Strategy redefined level 3 to reflect new digital technology capabilities and better enabled collaborative working in the design, construction and operations of assets. This in turn, enabling best use of capability in the supply chain to deliver value to clients (HM Government 2015).

The information management majority stages introduced in BS EN 19650-1:2018 overlap with the previous levels, in that stage 2 overlaps with the previous levels 1 and 2. Figure 9a and b in the RICS *The Future of BIM; Digital transformation in the UK construction and infrastructure sector* provides a useful schematic of both. Each stage is defined by four layers that describe the extent to which digital as opposed to analogue information management is implemented. The layers are as follows:

Business layer: the benefits realised to the company from the integration of the standards, technology and information layers over the life cycle of the asset.

Information layer: the data that will be captured, interrogated and analysed

Technology layer: the technology used to support storage and management of information, such as the use of a common data environment (CDE)

Standards layer: the processes and policies used, e.g. national standards, such as ISO 19650-1 and -2 (RICS 2020).

BIM as a technology

Unlike 3D CAD, which is a visual representation of a building based on geometric entities represented by points, edges, boundaries and domains, BIM represents the underlying object information, which can be used throughout the project's lifecycle. This object-oriented representation enables properties to be attributed to each of the objects. Objects can be either component objects, those with fixed geometrical shapes such as a door, or layered objects, those that can vary in size such as a wall. Objects in the early stages of design are generic – placeholders to provide a visual representation, and then as design progresses and manufacturer products are specified, the generic objects are replaced by specific objects. BIM objects can be copied directly from BIM object libraries such as those provided by manufacturers or the NBS National BIM Library. (www.nationalBIMlibrary.com).

Linking parametric characteristics to each object, means when design changes are made, the influencing changes on each of the objects is automated in accordance with their defined relationships. Hence, BIM enables holistic design making in relation to the whole life of the building and a more rigorous analysis of proposed design solutions by all members of the design team. The benefits to quantity surveyors of engaging with BIM will be discussed later in the chapter.

It is, however, important to establish at an early stage of the project the level of information to be attributed to each object, based on the client's needs in terms of the operational phase of the asset and the project team in delivering the project objectives. The level of information should be the minimum amount needed to be of value, anything beyond this is considered ineffective and a waste (BSI 2019). Whilst the international standards advocate this approach, many still in the industry refer to the following dimensions of BIM:

- design (3D)
- time (4D)
- cost (5D)
- sustainability (6D)
- facilities management (7D) and
- health and safety (8D).

Specific software has been developed to support each of the dimensions of BIM. Those of most relevance to quantity surveyors are 5D and 6D. As previously stated, it is important to establish exactly what costs are required, i.e. capital costs, operational costs, maintenance costs etc.

BIM as a process

BIM strategy

To implement an effective BIM strategy on the project, it is important to establish the client's requirements in relation to their asset management strategy as this will influence the information requirements that need to be captured at the delivery phase. These will inform the Exchange Information Requirements (EIR) included in the information protocol. As discussed in Chapter 17 the aim of the strategy should, within the constraints of the available time and resource, aim to maximise the value of use of BIM technology for both the client and the project team (Shepherd 2015). At project level the design team should assess the capabilities of each party for each BIM use and identify additional value and risks associated. This evaluation can inform the decision making in whether to implement each potential BIM use (Lu et al. 2019).

BS EN ISO 19650-2:2018 provides a framework to establish the project information requirements to ensure that the collaborative environment enables effective and efficient storing, analysis and sharing of the project information during the design and construction process. Figure 2.3 provides an overview of establishing the specific project needs of the information management process.

The information requirements will be influenced by factors such as the nature and size of the project, the procurement route and the client's post occupancy requirements. The *RIBA Plan of Work 2020 Template* provides a framework for the information exchanges at each of the project stages (from stage 0 – strategic decision to stage 7 – use) and a Design Responsibility Matrix to co-ordinate the production and use of information by cross-disciplined teams (RIBA 2020).

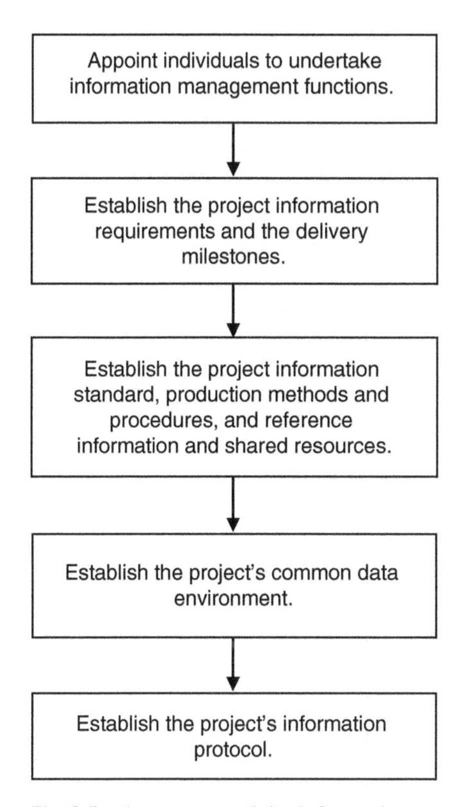

Fig. 2.3 Assessment of the information management process (*Source:* Adapted from BSI 2021).

BIM standards

The UK BIM Framework sets out the approach for implementing BIM in the UK. This framework will evolve to align to the sectors digital transformation, so it is recommended that you visit the website for the most current guidance (www.ukbimframework.org). The framework is based on concepts and principles of the ISO 19650 international standards for managing information over the whole life cycle of a built asset using building information modelling. This focus of each standard in the series is as detailed in Fig. 2.4. These international standards replace the previous UK standards, BS1192 and PAS 1192, which both defined the processes and procedures to support BIM level 2.

Standard	Focus
BS EN ISO 19650-1:2018	Defines the concepts and principles
BS EN ISO 19650-2:2018	Delivery phase of the assets
BS EN ISO 19650-3:2020	Operational phase of the assets
BS EN ISO 19650-4:2022	Information exchange
BS EN ISO 19650-5:2020	Information security

Fig. 2.4 BS EN ISO 19650 series.

Common Data Environment (CDE)

A common data environment (CDE) will be used to enable the collaborative production of the project information. A CDE is, firstly, a technological solution, in terms of how BIM manages both structured and unstructured data. Examples of structured data are the geometric models and databases, whilst unstructured data refers to documentation. This information is contained in 'information containers'. Each container will have a unique ID and, in the UK, will be classified in accordance with UNICLASS 2015 to support a consistent structure and framework for information production and exchange between different software used by each of the design team (see Chapter 10). Secondly the CDE is a workflow process, providing the framework in which the information containers move between different states as the design progresses. They will start as 'work in progress', and then they will be shared before being published. Finally, an information container may be archived (Hamil 2019).

BIM execution plan (BEP)

A BIM Execution Plan (BEP) sets out the proposed approach to the management of information aligned to the assessment described in Fig. 2.2. BS EN ISO 19650–2 2018 specifies a 'Pre-Contract' BEP and a 'Post Contract' BEP. The pre-contract BEP focuses on the design teams' approach to management of information in response to the Exchange Information Requirements (EIR) and also the capability and capacity requirements of design team members to manage information. The post-contract BEP details the project delivery BIM requirements and will inform the tender's evaluation criteria to ensure the contractor's capacity and capability to meet the information management requirements (see Chapter 9).

Information protocol

The information protocol, known as a BIM Protocol, will specify the management, level, use and transfer of information within the model and reflect the requirements of BS EN ISO 19650-1:2018 and BS EN ISO 19650-2:2018.

A BIM Protocol, used at the construction phase of a project will typically:

- set out the obligations of project team members in relation to information management and other BIM processes; and
- contain appendices setting out or referring to the technical documents detailing the technical aspects of those obligations and processes. (JCT 2019)

The indicative content of the BIM Protocol is shown in Fig. 2.5.

Co-ordination of information
Resolution of omission, ambiguity, conflict or inconsistency in information
Obligations of parties
CDE solution and workflow
Management of information
Level of information need
Use of information including licences
Transfer of information – interoperability between software
Liability
Remedies – security
Termination of the protocol

Fig. 2.5 Contents of a BIM protocol (*Source:* Adapted from ISO 19650-2 delivery phase).

It is likely that the *Exchange Information Requirements* (EIR), *BIM Execution Plan* (BEP), *Responsibility Matrix* and a *Master Information Delivery Plan* will be included as appendices to the BIM protocol.

The BIM protocol is a supplementary legal agreement as it will specify the parties' rights, duties, and risks allocation for the BIM process. For a BIM Protocol to apply, this must be identified in the relevant entry in the Contract Particulars (see Chapter 10). As well as the JCT the CIC have also produced a Building Information Modelling (BIM) Protocol for use on projects.

BIM use within the industry

BIM adoption

BIM use varies from country to country. Ullah et al.'s study (2019) found that the USA, UK, Scandinavian countries, and Singapore led BIM adoption within the global construction sector. Since 2011 the NBS have surveyed the construction sector, predominantly UK based in the early years, to establish the use of BIM. Typical of the adoption of new innovations, in this longitudinal study, they have reported a quite rapid growth from very low levels in 2011 (13%) to 63% in 2017, and then a steadier increase in adoption to 71% in 2021 (NBS 2021). This early rise in adoption was further influenced by the UK's government's BIM strategy, in that all public works contracts were required to use fully collaborative BIM Level 2 (with all project asset information, documentation and data being electronic) as a minimum by 2016. The survey also provides evidence of the interpretation of BIM within the sector, with 29% of respondents still using BIM predominantly as a technology only (3D models) and 63% recognising BIM as an information management approach, with 33% following BS1192 and PAS1192 and 30% following ISO 19650 standards.

In relation to tasks relating to BIM processes carried out in accordance with the ISO 19650 standards, as described earlier in this chapter, 63% of professionals surveyed had been involved in BIM execution plans and 60% had collaborated using CDEs over the preceding 12 months. For those that were UK based, 47% were involved with information protocols and 43% with detailed responsibility matrixes (NBS 2021).

With a focus on surveying the *RICS Digitalisation in Construction Report 2022* surveyed how data and information was shared for the following five functions: life cycle carbon emissions, production and fabrication, handover and commissioning, health and safety and wellbeing and quantity take-off and cost estimating. Of note for quantity surveyors was the highest level of data and information sharing was found to be for quantity take-off and cost estimating (RICS 2022a). Figure 2.6 details the flow of information between the quantity surveyor and other members of the project team in the UK.

Quantity Take-Off and Cost Estimating Data and Information Sharing	
Providing and receiving	40%
Receiving only	15%
Providing only	12%
No sharing	33%

Fig. 2.6 Information sharing in the UK (*Source:* Adapted from RICS 2022a).

Barriers to implementation

The most influential barriers to the implementation of BIM on a project were identified as the lack of client demand, cited by 64% of respondents, followed by lack of expertise (56%) and training (48%) (NBS 2021). This reinforces Taylor et al.'s observations that BIM implementation will only be as strong as the weakest link in the supply chain and that training is critical, particularly where BIM competencies are still maturing (Taylor et al. 2016). This work further stressed that cultural barriers to implementing BIM should not be underestimated, as there is a requirement to make significant changes to well-embedded work processes. A further challenge for embedding new technologies, such as BIM, is the selection of the appropriate hardware and software allowing for compatibility and interoperability with the rest of the industry (RICS 2020).

Benefits of BIM

Whilst client demand has been identified as a major barrier to implementation of BIM, the client case for using BIM is that it will create more value. This can be achieved by improving the quality and function of building design, better reasoned design using simulation and BIM analytics, improved accuracy of costs and improved achievement of project milestones (Saxon 2016).

Rethinking Construction (Egan 1998) identified that the construction industry needed to improve performance and deliver better value for money. As BIM is an integrated data environment supporting collaborative working between design team members, it has the potential to support this improvement. A study by Stanford University Centre of Integrated Facilities Engineering (SUCIFE) of 32 projects using BIM in the US observed that the following benefits could be realised when projects were procured using a BIM platform (Azhlar et al. 2011):

- 40% elimination of unbudgeted changes
- Cost estimating accuracy within 3%
- Up to 80% reduction in time taken to generate cost estimates
- Savings of up to 10% of the value through clash detection
- 7% reduction in project time.

Other key benefits relating to information management have been identified as:

- Information once captured can be revised and repurposed
- Information can be reviewed and revised, corrected and controlled
- Information can be checked and validated. (BSI 2010)

BIM as an integrated data environment facilitates collaboration more effectively; faster processes can be established, resulting in information being more easily shared amongst design team members. Sharing rather than duplicating information reduces the likelihood of error due to duplicate data handling but, more importantly allows a design team member to focus on providing innovative solutions within the context of their professional expertise. Visualisation enables the design team to rehearse any complex construction procedures associated with the build so that they might be managed effectively and risks mitigated. Communication between all parties can be supported by the 3D visualisation; this means that proposals are better understood than through other forms of communications such as 2D drawings where there is a higher risk of interpretation error which often later results in design changes. From a client's perspective, projects can be visualised at an early stage, giving them a much clearer idea of the design intent. They can therefore make any modifications at this stage, avoiding 'client led' changes made during the construction phases which impacts on performance.

A, lack of compatibility in project information and missing information result in significant causes of disruption of building operations on site. Using BIM, information is integrated so these issues are reduced. BIM also offers more flexibility in terms of document output, which means information used is 'fit for purpose', resulting in a better quality product. Digitally produced data can also be more easily exported to be used in automated fabrication of equipment and components, thereby enabling more efficient materials handling and reduction in waste.

As identified in the SUCIFE study, 10% of savings can be made through clash detection. As BIM provides an accurate geometric representation of the building, major systems can be checked for interferences that can be resolved at an early stage in the design. For example, the installation of air conditioning pipework can be checked to ensure that it doesn't intersect with structural beams. Interfaces between new and old can also be examined. Identifying clash detection in a 3D environment is more easily achieved than using 2D drawings and resolving issues during design negates costly variations and probable delays during construction.

Faster and more effective processes also result in buildings that better meet the client's objectives, as BIM enables more rigorous analysis of proposed design solutions by all members of the design team. Alternative design proposals can be simulated and performed quickly to support value management and value engineering processes Benchmarking of performance can also be more easily achieved. This ability to 'engage' more readily with the proposed design facilitates development of more efficient, cost-effective sustainable solutions that will better meet the client's objectives and provide value for money.

BIM can be structured to support all information relating to a building which includes not only the physical attributes but functional characteristics and product lifecycle information, enabling environmental performance to be predicted and lifecycle costings to be better understood. For example, an air conditioning unit within BIM can also include information about the supplier and operating and maintenance procedures. This will enable the facilities manager to effectively manage the company's property assets to support business success.

BIM has the greatest potential for improving the whole life costs of the building and, in so doing, the sustainability of the building. Through simulations, energy and thermal analysis can be carried out early in the design stage; changes can be made so design achieves optimum efficiency. As buildings become more energy efficient, the importance of and emphasis on embodied carbon will increase; carbon estimating software can also be linked to the model to support embodied carbon calculations, therefore supporting the client's sustainability agenda as discussed above. But perhaps the major influence of using BIM is that it cultivates working relationships, ensuring that all parties are focused on achieving best value for the client.

Ullah et al. (2019) in their literature review to support the status of BIM adoption in the industry further identifies the BIM benefits at different phases of the building lifecycle (Fig. 2.7).

Benefits of BIM to the QS

The use of BIM enables the quantity surveyor to be more proactive during the design phase, providing cost and value analysis. Benefits of using BIM to support quantity surveying activities include:

- Visualisation of the project will aid the scoping of information gaps when costing and tendering
- Real time costing will allow immediate decisions, removing abortive design and reducing the design/cost programme timeline
- A reduction of time in producing quantities will enable more time to assess the proposal to support value for money and other client key objectives
- Clash detection and improved coordination will produce high-quality project information, therefore reducing the number of instructions and limiting unforeseen costs and delays
- Carbon estimating software can be linked to the model to support embodied carbon calculations, therefore supporting the client's sustainability agenda as discussed above

Phases	Benefits of BIM Use
Pre-construction	• Better concept and feasibility • Effective site analysis to understand environmental and resource-related • Problem identification • Improve effectiveness and accuracy of existing conditions' documentation • Effective design reviews leading to sustainable design • Enhancement of energy efficiency • Resolve design clashes earlier through visualizing the model • Enables faster and more accurate cost estimation
Construction	• Evaluation of the construction of complex building systems to improve planning of resources and sequencing alternatives • Effective management of the storage and procurement of project resources • Efficient fabrication of various building components offsite using design model as the basis • BIM allows better site utilization • Reduce site congestion and improve health and safety
Post-construction	• BIM record model can help in decision-making about operations, maintenance, repair and replacement of a facility • Makes asset management faster, more accurate and with more information • Ability to schedule maintenance and easy access to information during maintenance

Fig. 2.7 BIM benefits through the building life cycle (*Source:* Ullah et al. 2019/Emerald Publishing Limited/CC BY-4.0).

- Programme analysis will support the assessment of preliminaries, valuations and claims for delays
- Increased opportunity to create savings over the life of the building as BIM data can be structured to support asset management.

An additional benefit of efficient extraction of quantity information in a variety of levels of information is that the quantity surveyor can develop a deeper understanding of cost influencing factors over the project lifecycle (RICS 2011). The textbook does not claim to provide detailed discussions of BIM and its function as there are specialised resources and RICS guidance literature that provides this. However, Fig. 2.8 identifies where further discussions on how BIM supports the quantity surveying function is referred to in more depth.

BIM Function	Chapter
BIM and cost estimating	5. Cost Control. Page 85 and 86
BIM and Life Cycle Costing	6. Whole Life Costing. Page 106
BIM and value management	7. Value Management. Page 121 and 122
BIM and risk management	8. Risk Management. Page 145
BIM and tender selection	9. Procurement. Page 156
BIM and contracts	10. Contract Documentation. Page 183
BIM and the measurement process	11. Preparation of Contract Documentation. Page 186, 188 and 193
BIM and valuing and managing change	12. Cost Management. Page 222
BIM and the project manager	16. Project Management. Page 284 and 285
BIM and Facilities management	17. Facilities Management. Page 308
BIM and sustainability	18. Sustainability. Page 324

Fig. 2.8 Further BIM discussion within the textbook.

Digital twins

A digital twin is a digital model connected in real time to a real physical asset, replacing 'as built' drawings with 'as is' (RICS 2020). The RICS *Digital twins from design to handover of construction assets* provides a more comprehensive analysis of current sector definitions and concludes that each has the following components to describe a digital twin:

- virtual or digital representation
- real-world entities, processes or physical assets, or elements
- realistic representation
- synchronisation, including synchronisation frequency and fidelity
- monitoring performance
- insights and intervention. (RICS 2022b)

Whilst Industry 5.0 technologies play an essential role in all aspects of a digital twin system (RICS 2022c), BIM and CDE, are identified within this report, as being core to digital twins. Together they provide the information framework that can be used throughout an asset's life cycle. Whilst BIM and CDE are central to the development of the 'static' virtual representation of the physical asset, it is the synchronisation that distinguishes a digital twin from a digital model. The ability to synchronise between the virtual and the physical asset to produce real-time data gives insights into an asset's performance. Hence the digital twin enables improved data-driven decision making and interventions on the physical twin. This improves strategic design, operational and maintenance decision making (RICS 2022b).

Currently within the sector the use of digital twins is focussed on the operational phase of an asset. However, the RICS survey does provides an insight into the top three current uses or possible uses of digital twins across all phases, as shown in Fig. 2.9.

Design and Construction Phases	Operation/Asset Management Phases
• Facilitating data sharing to deliver performance efficiencies for all stakeholders	• Gathering real-time asset operation and maintenance data
• Gathering real-time site data for decision making and collaboration	• Making better operation, maintenance, and renewal decisions
• Progress monitoring and project controls	• Improving asset performance and owner's bottom line

Fig. 2.9 Top three potential use of digital twins (*Source:* Adapted from RICS 2022b).

In *Digital twin: Towards a meaningful framework* Arup proposes a working evaluation framework to describe a digital twin and its capabilities by categorising the various levels of sophistication. Currently most digital twins are at level 2, in that they focus on the monitoring of the physical environment using sensors. The levels are:

Level 1: A digital model linked to the real-world system but lacking intelligence, learning or autonomy

Level 2: A digital model with some capacity for feedback and control, often limited to the modelling of small-scale systems

Level 3: A digital model able to provide predictive maintenance, analytics and insights.

Level 4: A digital model with the capacity to learn efficiently from various sources of data, including the surrounding environment.

Level 5: A digital model with a wider range of capacities and responsibilities, ultimately approaching the ability to autonomously reason and to act on behalf of users (Arup 2019).

Such level descriptors, give an indication of the potential of digital twins, and in doing so Arup proposes a definition that encompasses the ability to incorporate AI, in that

> 'A digital twin is the combination of a computational model and a real-world system, designed to monitor, control and optimise its functionality. Through data and feedback, both simulated and real, a digital twin can develop capacities for autonomy and to learn from and reason about its environment'. (Arup 2019)

The ability to offer real time data of an asset's components as influenced by occupational factors via the Internet of Things will support Smart Buildings.

To support the UK in its development of Digital Twins, the Digital Twin Hub was set up which operates in collaboration with the new National Digital Twin Programme run by the Government's Department for Business, Energy and Industrial Strategy. More detailed information can be found at www.digitaltwinhub.co.uk.

A continuous digital transformation

RICS (2020) in the analysis of emergent digital technologies for each of the 17 surveying groups, identified the technologies listed in Fig. 2.10 as having potential value to the role of surveyors within the quantity surveying and construction group.

The sector is in a transformative phase, with increased implementation of the use of digital technologies to support the project life cycle. Excluded from this chapter are discussions on Internet of Things, Smart buildings, Blockchain and the Metaverse. The only certainty is the need for quantity surveyors to be aware of the impact of technologies on their expertise and the need for continuous enhancements of technological and data analytical driven knowledge and skills, so their role can evolve in alignment with Construction 4.0. The RICS Tech Partner Programme is a collaborative platform for innovation data and technology, providing a useful repository of links to technology currently in use in the industry.

High Value	Medium Value
Drones Virtual Reality Smart City	Machine learning Blockchain Smart Building Data analytics/Big data 3D printing

Fig. 2.10 The impact of digital technologies on the QS (*Source:* Adapted from RICS 2020).

References

Ankan, K. and Venkata, S.K.D. Construction 4.0: what we know and where we are headed? *Journal of Information Technology in Construction (ITcon), Special issue: 'Next Generation ICT – How distant is ubiquitous computing?'* 26 526–545. 2021.

ARUP. *Digital Twin: Towards a Meaningful Framework*. 2019. Available at https://www.arup.com/perspectives/publications/research/section/digital-twin-towards-a-meaningful-framework (accessed 30/10/2022).

Azhlar, S., Hein, M. and Sketo, B. *Building Information Modeling BIM Benefits, Risks and Challenges*. 2011. Available from https://ascelibrary.org/doi/10.1061/%28ASCE%29LM.1943-5630.0000127 (accessed 29/10/2022).

BSI. *Constructing the Business Case: Building Information Modelling*. BSi Institute. 2010.

BSI. *BS EN ISO 19650-1:2018 Organization and Digitization of Information About Buildings and Civil Engineering Works, Including Building Information Modelling (BIM) – Information Management Using Building Information Modelling: Part 1: Concepts and Principles*. British Standards Institution. 2019.

BSI. *BS EN ISO 19650-2:2018 Organization and Digitization of Information About Buildings And Civil Engineering Works, Including Building Information Modelling (BIM) – Information Management Using Building Information Modelling: Part 2: Delivery Phase of the Assets* British Standards Institution. 2021.

Egan, J., (1998), "Rethinking construction: the report of the construction task force", Department of Trade and Industry, UK

Gontier, J.C., Wong, P.S.P and Teo, P. Towards the implementation of immersive technology in construction – a SWOT analysis. *Journal of Information Technology in Construction (ITcon)*, 26 366–380. 2021.

Hamil, S. *Common Data Environments*. 2019. Available at https://www.thenbs.com/knowledge/common-data-environments (accessed 28/10/2022).

HM Government. *Digital Built Britain Level 3 Building Information Modelling – Strategic Plan*. Crown. 2015. Available at https://assets.publishing.service.gov.uk/government/uploads/system/uploads/attachment_data/file/410096/bis-15-155-digital-built-britain-level-3-strategy.pdf (accessed 24/10/2022).

JCT. *Practice Note – BIM and JCT Contracts*. Sweet and Maxwell. 2019.

Lu, W. Lai, C.C. and Tae, A. *BIM and Big Data for Construction Cost Management*. London and New York. Routledge. 2019.

NBS. *National BIM Report 2012*. NBS Enterprise Ltd. 2012.

NBS. *10th Annual BIM Report 2020*. NBS Enterprise Ltd. 2020.

NBS. Digital *Construction Report 2021 Incorporating the BIM Report*. NBS Enterprise Ltd. 2021.

RIBA. *RIBAPlanofWorktoolboxFeb2020xlsx.RIBA*. RIBA. 2020. Available at https://www.architecture.com/knowledge-and-resources/resources-landing-page/riba-plan-of-work (accessed 29/10/2022).

RICS. *Building Information Modelling Survey*. BCIS. 2011.

RICS. *The Future of BIM: Digital Transformation in the UK Construction and Infrastructure Sector*. RICS. 2020. Available at https://www.rics.org/globalassets/rics-website/media/upholding-professional-standards/sector-standards/construction/future-of-bim_1st-edition.pdf (accessed 24/10/2022).

RICS. *RICS Digitalisation in Construction Report 2022*. RICS. 2022a. Available at https://www.rics.org/globalassets/rics-website/media/knowledge/research/research-reports/rics0112-digitalisation-in-construction-report-2022-web.pdf (accessed 29/10/2022).

RICS. *Digital Twins from Design to Handover of Construction Assets*. RICS. 2022b. Available at https://www.rics.org/globalassets/wbef-website/reports-and-research/digital-twins-from-design-to-handover-of-constructed-assets.pdf (accessed 29/10/2022).

RICS. *Digital Twins from Design to Handover of Construction Assets*. RICS. 2022c. Available at https://www.rics.org/uk/wbef/home/reports-and-research/digital-twins-from-design-to-handover-of-constructed-assets/ (accessed 29/10/2022).

Sawhney, A. Riley, M. and Irizarry, J. Construction 4.0: introduction and overview. In: *Construction 4.0: An Innovative Platform for the Built Environment* (edited by A. Sawhney M. Riley and J. Irizarry). London and New York. Routledge. 2020.

Saxon, R.G. *BIM for Construction Clients; Driving Strategic Value Through Digital Information Management*. RIBA Enterprises. 2016.

Shepperd, D. *BIM Management Handbook*. RIBA Publishing. 2015.

Taylor, M., Sell, P. and Ahmad, T. Information management. In: *BIM and Quantity Surveying* (edited by S. Pittard and P. Sell) London and New York. Routledge. 2016.

Ullah, K., Lill, I. and Witt, E. An overview of BIM adoption in the construction industry: benefits and barriers. In: *10th Nordic Conference on Construction Economics and Organization (Emerald Reach Proceedings Series, Vol. 2)* (edited by I. Lill and E. Witt). Bingley. Emerald Publishing Limited. 297–303. 2019.

3

Organisations and Management

KEY CONCEPTS

- Business structures
- Employer's responsibilities
  - Health and safety
  - Fire safety
  - Data protection
  - Insurance
- Management systems
- Finance and accounts

LEARNING OUTCOMES

After reading this chapter you should be able to:

- Understand the range of different quantity surveying business structures.
- Outline the legal requirements of employers.
- Appreciate the management systems used within an organisation to maintain quality processes to support the business function.
- Understand the principle accounting statements that show a company's financial position and performance.

COMPETENCIES

Competencies covered in this chapter:

- Accounting principles and procedures
- Data management
- Diversity, inclusion and teamwork
- Ethics, rules of conduct and professionalism

Willis's Practice and Procedure for the Quantity Surveyor, Fourteenth Edition.
Allan Ashworth and Catherine Higgs.

Introduction

Setting up and expanding a practice or business, together with general management and finance, are all subjects adequately dealt with in the many textbooks on business management. However, there are certain specific aspects of organisation and management relating to the work of the quantity surveyor that are worthy of consideration in some detail. They include:

- Staffing
- Office organisation
- Employer responsibilities
- Marketing
- Management systems
- Education and training
- Finance and accounts.

Business structures

There are four main types of business structures in the United Kingdom; sole trader, partnership, limited liability partnership and limited company.

Sole trader

Sole traders are self-employed quantity surveyors who run an unincorporated business which is registered with HM Revenue and Customs (HMRC). As they are providing a professional service they are normally referred to as a sole practitioner. This is the simplest way to run a business; however, whilst the quantity surveyor will recoup all the profits generated, they will personally be responsible for any business debts.

Partnership

Where two or more persons enter into a partnership for the purpose of providing surveying services they are jointly and severally responsible for the acts of the partnership under the Partnership Act 1890. Further, they are each liable to the full extent of their personal wealth for the debts of the business. There is no limit to their liability, as there are directors of a limited company for whom failure may only mean the loss of their shares in the company. All partners are bound by the individual acts carried out by each partner in the course of business. They are not, however, bound in respect of private transactions of individual partners. Partnerships come into being, expand and contract for a variety of reasons:

- As a business expands there is a need to divide the responsibility for management and the securing of work
- Through pooling of resources and accommodation, economy in expenditure can be achieved

- The introduction of work or the need to raise additional capital may result in new partners being introduced.

The detailed consideration of partnerships falls outside the scope of this book. Suffice it to say that while it is not a legal necessity, a formal partnership agreement should be in place, which legally sets down how the partnership will operate and covers such details as partners' capital and profit share ratios. The provisions set within the Partnership Act 1890 will apply if no partnership agreement exists.

Limited liability partnerships (LLP)

Limited liability partnerships are recognised under the Limited Liability Partnership Act 2000 and limit the liability of partners within the firm. The key features of LLPs are:

- They will be separate legal entities distinct from the owners (members)
- Members of LLPs will not be jointly and severally liable in the normal course of business
- They will be treated as partnerships for the purposes of UK income tax and capital gains tax
- A partnership that evolves into an LLP will not undergo a 'deemed cessation' for income tax purposes
- They will have to file audited public accounts similar to a limited company
- They will require two or more designated members who will carry out tasks similar to those of a company secretary, for example, signing the annual return.

Therefore, the LLPs combine certain crucial structural features of both companies and partnerships, the general intention being that the LLP will have the internal flexibility of a partnership but external obligations equivalent to those of a limited company. Hence, LLPs should be registered with both the HMRC and Companies House. In common with partnerships, the members of an LLP can adopt whatever form of internal organisation they choose. However, they are similar to limited companies in that the members' liability for the debts of the business will be limited to their stakes in it, and they will be required to publish and file their Annual Accounts at Companies House.

Limited company

Limited companies are registered at Companies House as either a company 'limited by shares' or 'limited by guarantee'. Quantity surveying firms will be the former as they are businesses that are set up to make a profit rather than 'not for profit'. As LLPs, limited companies are separate legal entities, so for the directors, who manage the company, liability is limited to the extent of their shareholding. However, directors still owe a duty of care to carry out their responsibilities and can also be liable for negligence. The company must also publish and file their Annual Accounts.

Regulated by the RICS

Many firms offering surveying services obtain 'Regulated by the RICS' status; this will be a requirement if 50% of the firm's principals are RICS members. This designation is recognised and respected in the sector and therefore communicates to clients the firm's commitment to providing a high standard of service. The RICS *Rules for the regulation of firms* (2022b) set out the criteria and obligations of registration. To enable clients to identify that the firm is regulated, it is mandatory to display the designation on all business materials and state it within any terms of engagement (RICS 2022c).

Staffing

The staffing structure within a quantity surveying practice will comprise fee earning members of staff carrying out quantity surveying services or other specialist service, such as programme and project planning, taxation and dispute advice and those, non-fee earning staff providing the necessary support services such as administration, accounts, human resources, facilities management and information technology. The overall staffing structure will differ from one organisation to another, but certain general principles will be similar.

Practice structure

The way an office organises and carries out its surveying work will to some extent be determined by the size of the practice and the nature of its commissions. The larger the practice, the more specialised the duties of the individual surveyor are likely to be. There are essentially two modes of operation, as shown in Fig. 3.1. Type (a) separates the core quantity surveying work on a project into the cost planning, contract documentation, final account work and specialist services, e.g. value management, taxation advice. Some coordination will be necessary between the phases of a project, as different personnel will carry out the work in each. This type of organisation gives staff the opportunity to develop their expertise in depth and with the growing range of specialist services in larger practices, this may be essential.

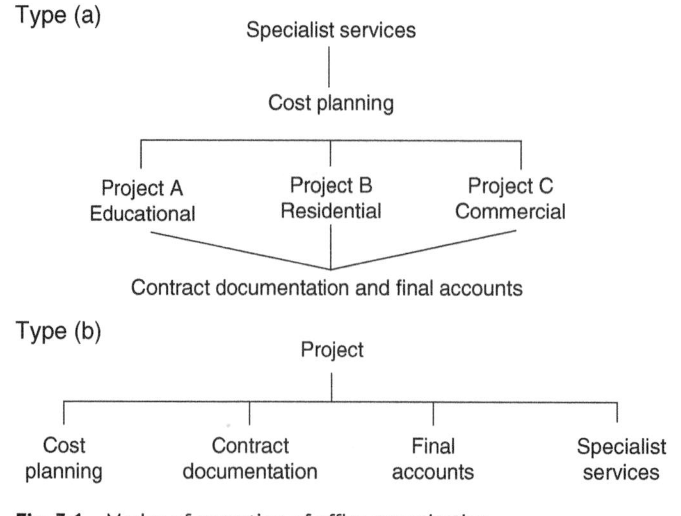

Fig. 3.1 Modes of operation of office organisation.

Type (b) allows the surveyors to undertake all aspects of the project from inception to final completion and provides the surveyor with a clearer understanding of the project. This alternative will often allocate the staff to teams, who then become specialists in certain types of project, depending of course on the type of work undertaken and the workload at any one time. This model does not easily accommodate the development of specialist service expertise, which requires concentration of activity rather than a more generalist approach.

Most practices today also carry out other work that may be incidental to their main work, possibly because of the interest or particular skills and knowledge of members of staff. In other instances, specialist work represents a major source of work for the practice.

Whichever procedure is used by the office for carrying out its work, some coordination and planning of staff time will be necessary if the services are to be carried out efficiently to meet the project requirements and deadlines. A management aid that may be used to plan and control the work is a resource and allocation chart. This, supported by the company's current and anticipated job list can be used to allocate staff according to their expertise and capacity. For any specific project a bar chart might be used, which breaks down the project into its various parts; to be complete it must take into account all aspects of the project, including the preparation of interim valuations and the final account. It may be desirable in the first instance to prepare the chart on the basis of pre-contract work alone, but for overall planning purposes some account must also be taken of the other duties that have to be performed. The type, size and complexity of the project will need to be taken into account in planning the work that has to be done.

Each project will normally be the responsibility of an allocated associate partner/director, or at least one of the senior surveying staff. Having identified the work that needs to be carried out within a particular time, they will determine the resource level required to perform the various tasks, either using staff within the office or calling on freelance staff as necessary. This is likely to be a necessary aspect of calculating the fee when bidding for work or negotiating with clients. It is also essential to pre-plan resources in order to budget costs against anticipated fee income. A major problem often facing the practice is the restricted time available to produce tender documentation and the like. Being at the end of the design process, they have to rely upon the other members of the design team to supply their information in good time. This emphasises the importance that should be attached to an adequate programme of work, by identifying the work that has to be done, who will do it, and the date by which it should be completed.

General organisation costs

Quantity surveyors, whether in private practice or in a support department within a contracting organisation, provide a service, and therefore the direct staff costs and indirect costs (such as employers' National Insurance contributions and pension contributions) form the largest part of the cost of running the practice or department.

The general establishment, expanding and equipping of an office are outside the scope of this book. However, in addition to staffing costs there are significant further costs associated with the running of an office, and it is worth noting the main items that constitute the general office overhead.

Major components of overhead expenditure after salaries is likely to be the cost of premises, utilities costs and technology-related costs necessary for the office to function efficiently. Such costs include computer hardware, software licences required to support the business operations, web hosting, email hosting, data storage and cybersecurity. Other items of overhead expenditure are likely to include vehicle costs, office supplies, marketing and other professional costs, such as accountants and lawyers.

Corporate responsibilities

Quantity surveying company performance is no longer just measured on financial indicators but by demonstrating responsible business practices, such as being socially accountable and addressing the environmental and ethical impact of its operations and management approach. Companies therefore have developed Corporate Responsibility (CR) strategies. Such strategies set out how the business will act socially responsible in the context of 'the triple bottom line' – people, planet and profit.

Stakeholders, such as investors and clients, are increasingly considering an organisations' sustainability and ethical practices. A company's ESG (Environmental, Social, Governance) plan will specify how the organisation embeds environmental, social and governance values in all that it does. At the time of writing there are no standard ESG reporting mechanisms. Figure 3.2 identifies what might be considered under each of the three ESG pillars: Environmental, Social and Governance.

Environmental	**Social**	**Governance**
Use of fossil fuels	Employee health and safety	Boardroom diversity
Water and resource management	Human rights	Fair operating practices
Waste management	Working conditions	Transparency and anti-corruption
Climate action	Wellbeing	Tax strategy and accounting standards
Carbon footprint	Diversity and Inclusion	Ethics and values
	Supply chain transparency	
	Training and development	

Fig. 3.2 The three pillars of an ESG plan.

Responsible businesses, within the quantity surveying sector, have committed to *The Ten Principles* of the United Nations Global Compact. These principles address human rights, labour, the environment and anti-corruption. In embedding these principles into policies and practices not only is the company being responsible in terms of protecting the planet and its people, it will also be beneficial to business success (United Nations 2015).

Inclusive working environments

Strategies to improve the equity, diversity and inclusivity of the surveying profession and its organisations are outside the scope of this book. However, many organisations within the sector are seeking to promote and improve the diversity of their workforces and create more inclusive working environments. At an organisational level, an inclusive workplace is one that values difference and enables all employees the opportunities to use their voice to effect change, irrespective of their background (CIPD 2019). *The Memorandum of Understanding (2022)* supported by the RICS and other key built environment professional bodies sets out an action plan to improve the Equity, Diversity, and Inclusion (EDI) Standards across the construction and property sector. Its purpose is to create a sector that is both inclusive and as diverse as the communities it represents (Joint Institutes 2022). The *RICS Inclusive Employer Quality Mark* is a means by which organisations can self-assess their policies and procedures. As identified by Clack and Gabler (2019) areas that need to be considered in detail in implementing diversity and inclusivity across an organisation are leadership and vision, recruitment, staff development, staff retention, staff engagement and continuous enhancement mechanisms.

As well as organisational culture, both the physical and virtual office should promote an inclusive environment that recognises and accommodates, without discrimination, diverse user needs.

Employer's responsibilities

There are certain legal requirements to be addressed by employers.

Health and safety

The need for any business organisation to correctly attend to health and safety matters cannot be overstated. The Health and Safety at Work Act etc. 1974 is the primary legislation covering occupational health and safety in the United Kingdom. It is important that employers establish the current legislation and ensure adherence. The Management of Health and Safety at Work Regulations 1999 (the Management Regulations) makes more explicit what employers are required to do. In general terms, there is responsibility on the employer in respect of the health, safety and welfare of employees and this incorporates a wide range of obligations including the following:

- The employer must ensure that any plant or equipment is properly maintained
- The employer must ensure that the systems of work are safe
- The employer must ensure that entrances and exits are easily accessible

- Relevant instruction or supervision should be provided where required to avoid misuse of equipment etc.
- In addition to employees, the general public visiting the premises must also be properly safeguarded from any risks
- The provision of adequate heating, lighting, ventilation, sanitary conveniences, washing facilities and drinking water

Under The Health and Safety (Display Screen Equipment) Regulations, employers must protect employees from health and safety risks associated with the poor use of display screen equipment (DSE), such as computers, tablets and smart phones. Employees are at risk of health issues relating to poor posture and infrequent breaks as well as eye strain.

Employers also have health and safety responsibilities for their workforce that work on remote or hybrid contracts. They must ensure that workers have a safe home working environment and follow good DSE practices.

The *RICS. Surveying Safety: health and safety principles for property professionals* provides further guidance on the management of health and safety within the surveying sector. It specifics that the organisation should have a recognised corporate structure to identify, manage and monitor health and safety with effective policies and procedures in place to reflect the work undertaken (RICS 2018). The organisation should also provide appropriate and regular health and safety training and promote a positive health and safety culture.

Failure of employers to comply with health and safety requirements could lead to both corporate and personal prosecution. Under the Health and Safety (Enforcing Authority) Regulations 1998, it is the local authority which is responsible for enforcing the legislation in private offices. This is by no means a comprehensive review of health and safety-related matters the reader is advised to use the Health and Safety Executive website to assist on health and safety matters.

Fire safety

Employers are responsible for complying with fire safety law. The Regulatory Reform (Fire Safety) Order 2005 (England and Wales), enforced by local fire and rescue services, sets out fire arrangements in offices. This requires that fire precautions are in place, including risk assessments, fire detection and firefighting systems, planned evacuation procedures, means of escape and an employee training programme (Crown 2015.)

Data protection

The Data Protection Act 2018 updated the previous act from 1998 to reflect the increasing amounts of data being processed in the digital age. The Act is the UK's implementation of the General Data protection Regulation (GDPR). These regulations control how personal data is used. The key data protection principles that every quantity surveyor needs to comply with is that the information used is:

- Fair, lawful and transparent.
- Used for specified and explicit purposes
- Used in a way that is adequate, relevant, and limited to only what is necessary

- Accurate
- Stored only for as long as for required
- Secure. (Gov.Uk 2022)

An additional key principle for organisations is 'accountability', firms need to have effective data protection policies and processes in place to demonstrate continuous compliance.

Employer's liability insurance

An employer is liable to pay compensation for injury to any employees arising out of and in the course of employment, caused by the employer's negligence or that of another member of staff. In order to be able to meet these obligations the employer is required by law to take out a specific Employer's liability insurance under approved policies with authorised insurers. Premiums are based upon the type of employee and the salaries that are paid. Whilst not required by law, many employers also provide life insurance that can be paid to the estate of any employee who dies while in their employment.

Public liability insurance

An owner or lessee of premises may be liable for personal injury or damage to the property of third parties caused by their negligence or that of a member of staff. The insurance cover should be sufficient to cover the different status of individuals, and actions by both principals and employees not only on the premises but anywhere while on business, and in countries abroad if this is appropriate. Increasingly, clients require this and seek confirmation prior to approving the practice.

Professional indemnity insurance

A conventional public liability policy does not normally cover any liability where professional negligence is involved. All RICS regulated firms should have professional indemnity insurance that complies with the Rules of Conduct and the RICS Professional Indemnity requirements (RICS 2021a).

Marketing

In the current climate of changing professional roles and status within the construction industry, the marketing of quantity surveying and other construction-related professions has taken on added importance. The marketing of professional services is a difficult task, however, as it involves selling a service rather than a product. Marketing is about understanding the needs and characteristics of the client base and matching or developing services to meet those needs.

As quantity surveyors seek to diversify the services they offer, it is necessary for them to recognise their specific strengths and weaknesses and those of their competitors, and to identify potential clients (in both existing and new market sectors) together with their needs and requirements. The services provided can then be tailored to meet these requirements and a marketing strategy aligned with the company's business plan can be targeted accordingly.

It is essential to have a clear marketing plan that develops the organisation's digital marketing effectiveness and the presentational skills of key individuals within the organisation to support potential business growth. Public relations and marketing are also seen as vitally important in order to recruit able talent into the profession, including women and ethnic minorities who, within the construction sector generally, are greatly under-represented.

Management systems

Environmental management systems

Quantity surveying companies need to consider the environmental impact of the organisation's activities. A company's environmental policy will state the organisation's intentions to support and enhance its environmental performance. It will include commitments to protect the environment, fulfil the organisation's compliance obligations and seek continuous enhancement of environmental performance (British Standards Institution 2015). Companies will have a structured management approach delivered by an Environmental Management System (EMS). Benefits of EMS will include cost savings through effective management of waste, energy, resources and water. Formal accreditation to BS EN ISO 14001 will demonstrate the practice's commitment to the environment.

Quality management systems

Considerable attention is given nowadays to the aspects of quality of the service provided by consultants and contractors to their clients. Griffiths (2018) identified that effective quality management is key to the business success of any construction organisation by providing services, which directly meet clients and regulatory requirements. Organisations can demonstrate that they operate an effective quality management system by gaining certification that they are compliant with BS EN ISO 9001:2015. Despite widespread adoption of the standard, quality issues still exist within the industry due to its complexity and project nature. Hence, a sector-specific supplementary document BS 99001:2022 has been published, which shifts the focus on time and costs towards good quality management driving both safety and sustainability, for society, economy, and the environment (British Standards Institution 2022).

Document management systems

Quantity surveying organisations will need to give due consideration to the storage, access, auditing of and security of all documentation used to support both project work and its own business function. Electronic document management systems (EDMS) are widely used and there are numerous propriety products on the market. EDMSs offers greater control, access and process efficiency over manual methods and advantages in terms of information retrieval, security, governance, compliance and lower cost of operations (IBM 2022).

The RICS note that a business with an EDMS could achieve BS EN ISO 9000 and BS EN ISO 14000 certification as previously described much more easily (RICS 2022a).

Inhouse data sources

A characteristic of any quantity surveying activity is the need to refer to cost and technical literature. Ensuring that the organisation has subscriptions to published sources of data and access to an inhouse library repository of key literature is essential so that staff are fully aware of current developments in all aspects of the construction industry.

Core material that needs to be readily available includes:

- British Standards Institution Publications
- Construction price e-books
- Building Cost Information Service (BCIS) cost information
- Current editions of all forms of contract
- Wage agreements
- Plant hire rates
- EU directives/publications
- Professional and general construction industry journals
- Technical literature on products, materials and components
- Specifications
- RICS isurv

Building costs records

A certain amount of record-keeping is essential in any office, though there is a tendency to delay preparing it because it is not directly productive. Records have a habit of never being prepared when information is fresh. The surveyor's own cost records will be of much greater value than any amount of published cost data. They relate to live projects and can be stored in a manner that easily facilitates retrieval. Some care needs to be exercised when using them because there could have been special circumstances that might, or might not, need to be reflected in any re-use of the information. It is, however, more likely to reflect the pricing in the local area, and therefore, in this context it is more reliable.

It is good practice to produce a cost analysis of every tender received, and these analyses will form a bank of useful cost records. If prepared in accordance with the principles of analysis drawn up by the BCIS, then some measure of comparability with their cost analyses can be achieved. Different clients may require different forms of cost analysis to suit their particular purpose. Supplementary information on market conditions and specification notes should also be included. These will be necessary in any future comparison of schemes. One of the outcomes of changing procurement trends, in particular the increased use of design and build in lieu of the traditional method, is the reduction in quality of information for analysis purposes.

It is important when compiling cost records to maintain a constant basis. The use of the BCIS principles will greatly assist in this. For instance, it may be desirable or necessary to omit external works, site clearance or preparatory alteration work from the price. The analysed tender sum must, however, make this omission clear.

Refurbishment projects do not lend themselves so well to records of this kind, and it is therefore preferable to provide an analysis in a rather different way, perhaps by identifying the cost centres for this type of project. Although these types of scheme are commonplace today in the United Kingdom, a standard form for their cost analysis has yet to be devised. Their unique nature, in terms of existing design, condition and constraints, is such that estimating by means of previous cost analysis will in any event require considerable adjustment and frequent reversion to resource-based techniques.

When the building scheme is complete, a final cost can be recorded, and this may also be calculated as a cost per functional unit or superficial floor area. No attempt should be made to recalculate the cost analysis, as the project costs by now will be largely historic and the allocation of the final accounts to elements is very time consuming.

Time and cost management

In the increasingly competitive fee and tender market the management of time and cost is of significant importance to a business's profitability. There is a need for staff to be managed in as efficient a way as possible. To assist in this, detailed records of time expended and thereby costs incurred maybe monitored on a regular basis.

Resource allocation

Resources need to be allocated to specific tasks on one or more projects depending upon their size and complexity. Teams may be assembled for document preparation or post-contract administration on large projects. The overall resource allocation needs to be regularly monitored. Certain work will be indeterminate or require immediate attention, which can cause problems in overall planning. Agreement of design and documentation production programmes with members of the design team can assist in preplanning the timing of resource requirements.

Individual time management

Each member of staff needs to manage their time as efficiently as possible. This is particularly important when they are involved on a number of projects at any one time. Effective time management, although achievable in theory, becomes more of a problem in practice. E-mail systems and apps offer time management systems to-do lists and individual activity schedules to be created.

Individual tasks should be prioritised, with the most urgent (not the most straightforward) being addressed first. Where deadlines are critical, a timescale should be applied to each task to provide a target for completion.

Staff time records and diaries

It is also necessary to plan and coordinate the individual staff members' time. This information can be presented on a bar chart for each member of staff. It needs to take staff holidays into account, be revised and reviewed at regular intervals.

Whether or not an organisation has a full job costing system, it is necessary that detailed records are kept of the time worked by each member of staff. This may be required as a basis on which to build up an account for fees for services that will be charged for on a time basis, or to establish the cost for each particular project. Such costs will be used primarily to establish whether a particular project is making a profit or a loss but may also be used to estimate a fee to be quoted for similar future work.

Each member of staff should ensure that their diary is both accessible to others within the organisation and up to date. This will support effective scheduling of meetings and collaborative working practices.

Organisational overview and benchmarking

It is commonplace to have databases or management systems that support the effective management of the overall workload of the organisation and the derived fee income. Such information is useful in helping to decide whether to look for additional staff, or whether there is an under-utilisation of staff for which work in the long term will need to be forthcoming. Such information can also support decision making in the relation to establishing fee levels and profitability of service lines and inter office comparisons.

Benchmarking can be applied across the business sector and throughout the entire hierarchy of a company from strategic level to operational. The aim of benchmarking is to improve the performance of an organisation by:

- Identifying best-known practices relevant to the fulfilment of a company's mission
- Utilising the information obtained from an analysis of best practice to design and expedite a programme of changes to improve company performance. (It is important to stress that benchmarking is not about blindly copying the processes of another company, since in many situations this would fail. It requires an analysis of why performance is better elsewhere, and the translation of the resultant information into an action suited to the company under consideration).

Benchmarking activities will be used to compare inhouse profit margins. In relation to staff the fee earned per fee earner ratio is often used to establish individual's bonuses and the overall composition and balance of junior to senior staff within an organisation.

Developing staff and skills

It is important to consider the training and development of staff to meet the ongoing and future skills and knowledge requirements of the organisation. Education and training of staff on university, full, day-release, undergraduate and postgraduate courses is discussed in Chapter 1. However, it is also important that qualified staff engage in regular continued professional development (CPD) activities to ensure that their knowledge and skills are current. Indeed the RICS recognises the importance of CPD and there is an obligation within the *Rules of Conduct* for companies to ensure that all staff are properly trained to maintain and develop their knowledge and skills, and also to check that they are complying with CPD requirements set by RICS (RICS 2021b).

Finance and accounts

Accounting principles and procedures is a mandatory competency for the quantity surveyor. Accounting can be defined as 'the process of identifying, measuring, classifying, summarising and communicating financial and economic transactions and events to enable users to make informed decisions' (Ebisike 2010). In order to provide monetary advice to clients, it is essential to have a basis understanding of the principle accounting statements that show the company's financial position and performance: the profit and loss accounts, balance sheets and cashflow statements. The UK GAAP (generally accepted accounting practice) is an accounting standard for reporting set by the Financial Reporting Council. Such standards enable users of accounts to receive high-quality understandable financial reporting (FRC 2022a).

The accounts

The primary purpose of financial accounting is to provide a record of all the financial transactions of the business and to establish whether or not the business is making a profit. The accounts will also be used to:

- Determine the distributions to be made to equity holders
- Determine the partners' or company's tax liabilities
- Support an application to a bank for funding
- Determine the value of the business in the event of a sale
- Prove financial stability to clients and suppliers.

All companies are required under the Companies Act 2006 and the Small Business, Enterprise and Employment Act 2015 to produce accounts and to file them annually with the Register of Companies in order that they are available for inspection by any interested party.

Profit and loss account

The profit and loss account records the results of the business' trading income and expenditure over a determined period of time, its trading activity. For a surveying practice, income will represent fees receivable for the supply of surveying services; expenditure is likely to include such items as salaries, travel expenses, rent and insurance. After adjustments have been made for accruals (revenue earned or expenses incurred that have not been paid or received), and prepayments (advanced payments for goods or services not yet provided), an excess of income over expenditure indicates that a profit has been made. The reverse would indicate a loss.

The preparation of the profit and loss account will enable the business to:

- Compare actual performance against budget
- Analyse the performance of different sections within the business
- Assist in forecasting future performance
- Compare performance against other businesses
- Calculate the amount of tax due.

An example of a simple profit and loss account is shown in Fig. 3.3.

Profit and loss account for the year 18 December 2021 to 18 December 2022		
	£	£
Income		
Fees received		300, 000
Expenditure		
Salaries	150, 000	
Rent	50, 000	
Others	30, 000	
Depreciation	20, 000	250, 000
Profit for the period		£50, 000

Fig. 3.3 Example profit and loss account for QS practice.

Balance sheet

The balance sheet (Fig. 3.4) gives a statement of a business' assets and liabilities as at a particular date. It is a statement of the company's net worth at a specific point in time – a 'snapshot'– unlike the profit and loss account, which is for a set period of time. The balance sheet will include all or most of the following:

- *Fixed assets*: those assets held for long-term use by the business, including intangible assets
- *Current assets*: those assets held as part of the business' working capital
- *Liabilities*: amounts owed by the business to suppliers and banks

Balance sheet at 18 December 2022		
	£	£
Fixed assets		
Fixtures and fittings		200, 000
Less: depreciation		20, 000
		180, 000
Current assets		
Fees receivable	60, 000	
Cash at bank	60, 000	
	120, 000	
Current liabilities	50, 000	
Net current assets (working capital)		70, 000
Net assets		£250, 000
Capital		200, 000
Retained profits		50, 000
		£250, 000

Fig. 3.4 Example balance sheet for QS practice.

- *Owner's capital*: shareholder's funds (issued share capital plus reserves) in a limited company, or the partners' capital accounts in partnership.

The various types of asset and liability accounts are considered in more detail below.

Assets

The term 'assets' covers the following:

- *Intangible assets*, which include goodwill, trademarks and licensing agreements, usually at original cost less any subsequent write-offs
- *Fixed assets*, which include land and buildings, fixtures and fittings, equipment and motor vehicles, shown at cost or valuation less accumulated depreciation
- *Current assets*, which are held at the lower cost or new realisable value. Current assets include cash, stock, work in progress, debtors and accruals in respect of payments made in advance.

Depreciation, referred to above, records the loss of value in an asset resulting from usage or age. Depreciation is charged as an expense to the profit and loss account, but is disallowed and therefore added back for tax purposes. Depreciation is recorded as a credit in the balance sheet, reducing the carrying value of the firm or company's fixed assets. For tax purposes, depreciation will be considered under the provisions of the Capital Allowances Act 2001 and The Capital Allowance (Structure and Building Allowance) regulations 2019.

Liabilities

Liabilities include amounts owing for goods and services supplied to the business and amounts due in respect of loans received. Strictly, they also include amounts owed to the business' owners: the business' partners or shareholders. Note that contingent liabilities do not form part of the total liabilities, but appear in the form of a note on the balance sheet as supplementary information.

Capital

Sources of capital may include proprietors' or partners' capital or, for a limited company, proceeds from shares issued. Capital is required to fund the start-up and subsequent operation of the business for the period prior to that period in which sufficient funds are received as payment for work undertaken by the business. The example in Fig. 3.4. identifies an initial capital investment of £200,000; however, this investment is soon represented not by the cash invested but by the business.

To illustrate the movement of cash in terms of receipts and payments, a simple example of a summarised bank account statement is included in Fig. 3.5. This statement is produced periodically, usually monthly, and is the source of the cash postings to the other books of account.

Receipts	£	Payments	£
Capital introduced	200, 000	Salaries	150, 000
Fees received	240, 000	Rent	50, 000
		Other expenses	30, 000
		Fixtures and fittings	150, 000
		Balance carried forward	60, 000
	£440, 000		£440, 000

Fig. 3.5 Bank account summary statement.

Finance

There are several ways in which a business can supplement its finances. The most common way is by borrowing from a bank on an overdraft facility. The lender will be interested in securing both the repayment of the capital lent and the interest accruing on the loan. The lender will therefore require copies of the business' profit and loss account, balance sheet and details of its projected cash flow.

Cashflow statement

It is necessary for a business to predict how well it is likely to perform in financial terms in the future. Budgets are therefore prepared, usually on an annual basis, based on projected income and expenditure. Once a business is established, future projections can be based to a certain extent on the previous year's results. Previous year's cashflow statement supports the setting of budgets as they record the movement of cash and cash equivalents entering and leaving the company. The cashflow statement therefore provides insight into the company's cash management over a determined period. The statement will include financial information on:

- *Operating Cashflow*: this is the net income generated from the provision of the quantity surveying services. A positive cashflow will indicate successful financial performance in the provision of services.
- *Investing Cashflow*: this is the net amount of monies derived from monies used for capital expenditure to improve the company's assets such as property and equipment and also the sale of assets and securities.
- *Financing Cashflow*: this is the net amount derived from monies derived from the issuing of shares etc. and the payment of shareholder dividends.

Managerial accounting

Whilst the primary accounting statements described above are used in financial accounting, cashflow budgets can be used in managerial accounting, this is when the operating cashflow is used to review the income generation and expenditure on an ongoing basis.

This will enable the company to be both agile and proactive in taking financial decisions to inform its business objectives. It is likely that the accounts will also be considered at department or service line level to aid budget considerations, such as staff resourcing.

Annual reports and auditing

At the end of a business' financial year a set of 'end of year' annual report is prepared bringing together all the previous year's financial statements. Such reports, as well as including a financial review, will often include key business activities undertaken by the company during that year. In most circumstances, an independent accountant will audit accounts; indeed, this is a requirement for all larger limited companies under the Companies Act 2006. 'The auditor's objectives are to obtain reasonable assurance about whether the financial statements as a whole are free from material misstatement' (FRC 2022b). The auditor will also consider all other information contained within the annual report to ensure that there are no misstatements or material inconsistencies with the financial statements. Audited accounts will carry more authority with the Inspector of Taxes and are also useful to prove to third parties, including prospective clients, that the financial status of the business has been independently scrutinised.

References

British Standards Institution. *BS EN ISO 14001:2015 Environmental Management Systems – Requirements with Guidance for Use*. BSI Standards Limited. 2015.

British Standards Institution. *Quality Management Systems – Specification for the Application of BS EN ISO 9001:2015 in the Built Environment Sector*. BSI Standards Limited. 2022.

CIPD. *Building Inclusive Workplaces: Assessing the Evidence*. CIPD. 2019. Available at https://www.cipd.co.uk/Images/building-inclusive-workplaces-report-sept-2019_tcm18-64154.pdf (accessed 17/09/2022).

Clack, A. and Gabler, J. *Managing Diversity and Inclusion in the Real Estate Sector*. London and New York. Routledge. 2019.

Crown. *Fire Safety: Risk Assessment: Offices and Shops*. The Stationery Office. 2015.

Ebisike, O. *Real Estate Accounting Made Easy*. Wiley-Blackwell. 2010.

FRC. *Forward to Accounting Standards*. Financial Reporting Council. 2022a. Available at https://www.frc.org.uk/getattachment/b1f55f32-d7a0-492a-9175-6fa9f5613a44/Foreword-to-Accounting-Standards-(January-2022)(1).pdf (accessed 15/05/2022).

FRC. *Description of the Auditor's Responsibility for the Audit of Financial Statements*. 2022b. Available at https://www.frc.org.uk/auditors/audit-assurance/auditor-s-responsibilities-for-the-audit-of-the-fi/description-of-the-auditor%E2%80%99s-responsibilities-for (accessed 26/11/2022).

Gov.uk. *Data Protection*. 2022. Available at https://www.gov.uk/data-protection (accessed 24/09/2022).

Griffiths, A. *Integrated Management Systems for Construction: Quality, Environment and Safety*. London and New York. Routledge. 2018.

IBM. *What is Document Management?* 2022. Available at https://www.ibm.com/uk-en/topics/document-management (accessed 18/09/2022).

Joint Institutes. *Action Plan: Creating a More Diverse, Equitable and Inclusive Built Environment Sector*. 2022. Available at https://www.rics.org/contentassets/66a390389ce942a3aecc8ea447e9e49b/edi-mou_professional-institutes-1.pdf (accessed 18/09/2022).

RICS. *Surveying Safety: Health and Safety Principles for Property Professionals*. 2nd Edition. 2018.

RICS. *Risk, Liability, and Insurance*. 1st Edition. RICS. 2021a.

RICS. *Rules of Conduct*. RICS. 2021b.

RICS. *Do I need an EDMS?* 2022a. Available at https://www.isurv.com/info/53/document_management/242/do_i_need_an_edms (accessed 18/09/2022).

RICS. *Rules for Registration of Firms: Version 7 with Effect from 2 February 2022*. RICS. 2022b.

RICS. *Rules for the Use of the RICS Logo and Designation by Firms: Version 6 with Effect from 2 February 2022*. RICS. 2022c.

United Nations. *Guide to Corporate Sustainability*. 2015. Available at https://www.unglobalcompact.org/library/1151 (accessed 25/09/2022).

4

The Quantity Surveyor and the Law

KEY CONCEPTS

- Appointment of the quantity surveyor (QS)
- Payment of fees
- Professional indemnity insurance
- Professional conduct
- RICS disciplinary procedures
- Legislation applicable to quantity surveyors
  - Employment Act 2008
  - Equality Act 2010
  - Bribery Act 2010
  - Modern Slavery Act 2015
  - Health and Safety at Work Act 1974

LEARNING OUTCOMES

After reading this chapter, you should be able to:

- Understand the importance of the QS's terms of appointment.
- Appreciate that case law supports the need to clearly define the services provided and identify the mechanism to do so.
- Understand the consequences of poor professional conduct.
- Outline the impact legislation has on how quantity surveyors undertake their role.

COMPETENCIES

Competencies covered in this chapter:

- Client care
- Contract practice
- Diversity, inclusion and teamworking
- Ethics, Rules of Conduct and professionalism
- Health and safety

Willis's Practice and Procedure for the Quantity Surveyor, Fourteenth Edition.
Allan Ashworth and Catherine Higgs.

Introduction

The purpose of this chapter is to describe in general terms how the law affects the quantity surveyor in practice, concentrating on the form of agreement between the quantity surveyor and the client, the impact of the demand for collateral warranties and performance bonds, the requirement to maintain professional indemnity insurance cover and the Employment Acts as they relate to the employment of staff. The wider issue of construction law is a complex and ever-changing area, the detail of which falls outside the scope of this book.

The quantity surveyor and the client

The quantity surveyor in private practice provides professional services for a client. The legal relationship existing between them is therefore a contract for services. The nature of this contract controls the respective rights and obligations of the parties. So far as the quantity surveyor is concerned, it determines the duties to be performed, powers and remuneration for the particular work undertaken.

The QS appointment

There is no legal requirement that the agreement or contract between the surveyor and the client should consist of a formal document nor even, indeed, that it be in writing. However, 'RICS members are obliged to record the terms of their appointments in writing' (RICS 2022a). A written record provides clarity of the true intentions of both parties and will reduce the likelihood of dispute. Furthermore, Rule 3 of the RICS *Rules of Conduct* (2021) requires members in ensuring the provision of a quality and diligent service to both agree with the client the scope of service (rule 3.2) and also inform them promptly and seek agreement if changes to the terms of engagement occur (rule 3.4). Professionals can sometimes find themselves in difficulty if they undertake work relying only upon an incomplete agreement, in which important items remain to be settled. However, the law does not recognise the validity of a contract to make a contract and, where any essential element is left for later negotiation, the existing arrangements are unlikely to be recognised as a binding agreement.

The existence and nature of the agreement can be established by the use of a standard form of appointment, such as the RICS *Standard Form of Consultant's Appointment: England and Wales* (RICS 2022b) and *Quantity Surveying Services: England and Wales* (RICS 2022c). It cannot be overstated that whatever practice is adopted, there is a need to ensure that a valid, comprehensive and adequately evidenced contract exists between the respective parties.

Case law relating to QS appointment and payment

The following cases highlight the importance of defining the nature of the services provided by the quantity surveyor and the agreed reimbursement.

In *Courtney & Fairbairn Ltd* v. *Tolaini Bros (Hotels) Ltd* (1975) 1 AER 716; 2 BLR 100, the defendants, wishing to develop a site, agreed with the plaintiffs, a firm of contractors, that if a satisfactory source of finance could be found, they would award the contract for the work to the plaintiffs. No price was fixed for the work but the defendants agreed that they would instruct their quantity surveyor to negotiate fair and reasonable contract sums for the work. Suitable finance was introduced, and the quantity surveyor was instructed as agreed. The quantity surveyor was unable, in the event, to negotiate acceptable prices, resulting in the contract being awarded elsewhere. The plaintiffs sued for damages, claiming that an enforceable contract had been made. It was held that the price was a fundamental element in a construction contract and the absence of agreement in that regard rendered the agreement too uncertain to enforce. This case was, of course, not directly concerned with the provision of professional services but the legal principle illustrated is of general application.

A persuasive incentive to take care arises from the fact that the ability to recover payment for work done will usually depend on the existence of an appropriate contract. Performance alone does not automatically confer a right to remuneration, although where a benefit is conferred, the court will normally require the beneficiary to make some recompense, possibly by way of a *quantum meruit* payment, meaning, as much as is deserved.

In *William Lacey Ltd* v. *Davies* (1957) 2 AER 712, the plaintiff performed certain preliminary work for the defendant connected with the proposed rebuilding of war-damaged premises, in the expectation of being awarded the contract for the work. The defendant subsequently decided to place the contract elsewhere, and eventually sold the site without rebuilding. The plaintiff sued for payment for work already done. In this case, it was held that in respect of the work done, no contract had ever come into existence but, nevertheless, as payment for the work had always been in the contemplation of the parties, an entitlement to some payment on a *quantum meruit* arose.

In this context it is reassuring, from the quantity surveyor's point of view, to note that, where professional services are provided, there is a general presumption that payment was intended. In *H.M. Key & Partners* v. *M.S. Gourgey and Others* (1984) I CLD-02-26, it was said: 'The ordinary presumption is that a professional man does not expect to go unpaid for his services. Before it can be held that he is not to be remunerated there must be an unequivocal and legally enforceable agreement that he will not make a charge'. However, while some recovery of fees may be possible without the formation of a binding contract, the lack of such an agreement enhances the possibility of disputes and litigation. Moreover, if the basis of enforced payment is to be *quantum meruit*, there is no guarantee that the court's evaluation of the services provided will correspond with the practitioner's expectations.

Given that all relevant terms are settled and agreed, and incorporated in a formal contract, the intentions of the parties could still be frustrated by a failure to express the terms clearly.

In *Bushwall Properties Ltd* v. *Vortex Properties Ltd* (1976) 2 AER 283, a contract for the transfer of a substantial site provided for staged payments and corresponding partial legal completions. At each such completion 'a proportionate portion of the land' was to be transferred to the buyer. A dispute arose as to the meaning of this phrase. It was held that, in the circumstances: no certain meaning could be attributed to the phrase; that this represented a substantial element in the contract; hence the entire agreement was too vague to enforce.

However, if a valid contract exists in unambiguous terms, the court will enforce it. It is therefore vital to ensure that the terms are not merely clear but do in fact represent the true understanding of the parties, both at the outset and as the work progresses. When the actual work is in hand with all attendant pressures, it is all too easy to overlook the fact that the obligations undertaken and remuneration involved are controlled by the contract terms. Departure from or misunderstanding of the original intentions, unless covered by suitable amendments of those terms, could have very undesirable consequences as the following case illustrates.

In *Gilbert & Partners* v. *R. Knight* (1968) 2 AER 248; 4 BLR 9, the plaintiffs, a firm of quantity surveyors, agreed for a fee of £30 to arrange tenders, obtain consents for, settle accounts and supervise certain alterations to a dwelling house on behalf of the defendant. Initially work to the value of some £600 was envisaged, but in the course of the alterations the defendant changed her mind and ordered additional work. In the end, work valued at almost four times the amount originally intended was carried out; the plaintiffs continued to supervise throughout and then submitted a bill for £135. This was met with a claim that a fee of £30 only had been agreed. It was held that the original agreement was for an all-in fee covering all work to be done; the plaintiff was entitled to only £30.

The moral is clear: avoidance of difficulty and financial loss is best ensured by accepting an engagement only on precise, mutually agreed and recorded terms, setting out unequivocally what the quantity surveyor is expected to do and what the payment is to be for so doing. If, as often happens, circumstances dictate development and expansion of the initial obligations, the changes must be covered by fresh, legally enforceable agreements. Oral transactions relating to either the original agreement or later amendment of it should always be recorded and confirmed in writing. This is more than an elementary precaution, for it should be borne in mind that what is known as the parol evidence rule will normally preclude any variation of an apparently complete and enforceable existing written contract, by evidence of contrary or additional oral agreement.

Death of the quantity surveyor

Whether the liability to carry out a contract passes to the representatives of a deceased person depends on whether the contract is a personal one. That is, one in which the other party relied on the 'individual skill, competency or other personal qualifications' of the deceased. This is a matter to be decided in each particular case. That said, '*firms with a sole principal must make appropriate arrangements for their professional work to continue in the event of their incapacity, death, absence from or inability to work*' (RICS 2021).

The fact that a contract between a quantity surveyor and the client is a personal contract, if that is the case, does not mean that the quantity surveyor must personally carry out all the work under the contract; this is unless it is obvious from the nature of the contract, for example a contract to act as arbitrator, that the quantity surveyor must act personally in all matters. In other cases, such as the preparation of a bill of quantities and general duties, the quantity surveyor may make use of the skill and labour of others, but takes ultimate responsibility for the accuracy of the work.

Death of the client

The rule referred to in the previous paragraph as to a contract being personal applies equally in the case of the death of the client. Here, the contract is unlikely to be a personal one, and the executors of the client must discharge the client's liabilities under the building contract and for the fees of the professional people employed.

Responsibility for appointment

What has been written so far assumes that the quantity surveyor's appointment arose from direct contact with the client. Additional problems may occur where the appointment arises indirectly from the retained architect or project manager. In such cases, the power to appoint on behalf of the client may subsequently be called into question. There is no general solution to this problem. The actual position will depend on the express and implied terms of the other consultant's contract with the client. If they have express power to appoint, then, of course, no problem arises and the appointment is as valid as if made by the client in person. However, reliance on their possessing, an implied power to appoint would be very unwise. It is clear that the courts do not recognise any general power of appointment or delegation as inherent in an architect's or other consultant's contract with a client.

In *Moresk Cleaners Ltd* v. *T.H. Hicks* (1966) 4 BLR 50, the Official Referee stated bluntly that 'The architect has no power whatever to delegate his duty to anyone else'. That case concerned the delegation of design work but it would seem equally applicable to other unauthorised appointments.

Potential difficulties in the matter can be easily avoided by the simple expedient of ensuring that, where the employment of the quantity surveyor is negotiated by another consultant, the terms of the appointment are conveyed in writing to the client and the client's acceptance thereof is similarly secured. Ratification by the client will then have overcome any deficiencies in the consultant's authority.

Agreement for appointment

The agreement for appointment of a quantity surveyor, whether a standard or non-standard document, will encompass certain general provisions including the following:

- Form of agreement/particulars of appointment
- Consultant's obligations, such as the scope of services to be provided
- Statutory duties, such as complying with CDM Regulations
- Payment and payment procedures
- Limitation of liability
- Collateral warranties
- Professional indemnity insurance requirements
- Assignment
- Copyright (see Chapter 11) and confidentiality
- Dispute procedures.

Other provisions, such as consultant's personal, client's obligations and termination procedures might be included.

It is usually for the scope of services provision within the appointment to refer to a more detailed breakdown of services, such as provided by the RICS *Quantity Surveying Services*. This document not only provides a detailed breakdown of services but also indicates the proposed level of attendance of meetings, both in terms of frequency and the level of seniority of the attendee. If a client requires the quantity surveyor to provide services on their own standard terms and conditions, it must be remembered that these terms are open to negotiation.

Any agreement can be executed either as a simple contract or as a deed. There are important differences, two of which are the most significant as far as the quantity surveyor is concerned: the need for consideration and the limitation period.

In a simple contract there must be consideration. This is a benefit accruing to one party or detriment to the other, most commonly payment of money, provision of goods or performance of work. The liability period, in this case, is limited to six years after Practical Completion. In a speciality contract, a contract executed as a deed; however, there is no need for consideration and the liability period is 12 years. The significance of the latter is that a quantity surveyor who enters into an agreement as a deed doubles the period of exposure to actions for breach of contract.

Responsibility for payment of fees

Where an effective contract exists between the quantity surveyor and the client, provision will be contained relating to the payment of the professional fees involved. The position where no valid agreement exists has already been mentioned, and it was suggested that even in such unfortunate circumstances some remuneration, probably by way of a *quantum meruit*, will usually be forthcoming.

If there are any reservations regarding the financial standing of a potential client, the surveyor must make enquiries, perhaps by taking up bank references, and then trust to commercial judgment. In this connection, it is vital to ensure that the documentation accurately reflects the true identity of the client. This may seem too obvious to mention but misunderstandings can and do occur, particularly in dealings with smaller companies controlled by a sole individual. It is easy to confuse the individual acting on behalf of a company and acting in a personal capacity. The unhappy result may be dependence for payment on a company of doubtful solvency, having imagined that one was acting for an individual of undoubted substance. Finally, it is chastening to reflect that monies owed in respect of professional fees are in no way preferred in the event of insolvency.

Amount and method of payment

Fees are a matter for negotiation, since recommended scales do not exist. Entitlement depends on the terms of agreement under which the services are provided, and any negotiations are constrained by practical rather than legal considerations.

The fee for services can be paid as a percentage of the final building cost of the project (this amount is clearly defined in the standard form of appointment) or as a lump sum. The amount will be influenced by factors such as, the services provided, the level of seniority required, procurement method, contract type and nature of the works, i.e. new or refurbishment. The level of fee will also be influenced by market conditions.

However, where the work involves advising on matters connected with litigation or arbitration, for example in respect of claims, it is not permissible to link the fee to the amount recovered. In *J. Pickering* v. *Sogex Services Ltd* (1982) 20 BLR 66, arrangements of that nature were said to savour of champerty – that is trafficking in litigation – and as such to be unenforceable as contrary to public policy.

Where possible, it is prudent to make provision for the payment of fees by instalments at appropriate intervals. These can be specified within the *Standard Form* and can be linked to project milestones or calendar intervals. The payment provision reflects the requirements of the Housing Grants, Construction and Regeneration Act 1996. Any expenses that will be reimbursed, in addition to the fee and the hourly rate of key personal for any additional services, will also be included. This ensures transparency of the fees and services to be provided (rule 1.6).

Limitation of liability

The maximum aggregate liability is stated within the standard form of appointment. The quantity surveyor will normally negotiate a financial cap on liability equal to their professional liability insurance.

Collateral warranties

A collateral warranty, or duty of care agreement, is a contract that operates alongside another contract and is subsidiary to it. In its simplest form, it provides a contractual undertaking to exercise due skill and care in the performance of certain duties that are the subject of a separate contract. An example of this would be a warranty given by a quantity surveyor to a funding institution to exercise due skill and care when performing professional services under a separate agreement between the quantity surveyor and the client. The purpose of a collateral warranty is to enable the beneficiary to take legal action, with reference to the Contract (Rights of Third Parties) Act 1999, against the party giving the warranty, for breach of contract if the warrantor fails to exercise the requisite level of skill and care in the performance of the duties. The details of any warranty will be stated within the form of appointment. However, the default position is that one is not required (RICS 2019).

Professional indemnity insurance (PII)

It has always been prudent for quantity surveyors to protect themselves against possible claims from their clients for negligence. Such mistakes may not necessarily be those of a principal's own making but those of an employee. The RICS *Rules of Conduct* (RICS 2021) make it mandatory for firms '*to ensure that all previous and current professional work is covered by adequate and appropriate professional indemnity cover*'. The amount of indemnity is specified and will reflect the type of work that is undertaken. Quantity surveyors will negotiate with the client if the level of PII is not appropriate to the services offered and the size of the project.

The PII policy covers claims that are made during the period when the policy is effective, regardless of when the alleged negligence took place. Claims that occur once the policy has expired, even though the alleged event took place sometime previously, will not be covered.

As discussed early, the period of liability is either 6 or 12 years after practical completion and dependent on whether the appointment is or is not made as a deed. A sole practitioner is therefore well advised to maintain such a policy for some time after retirement. The RICS requires its members to maintain run-off cover for a minimum of six years after they retire to cover just such eventualities. Recent court cases suggest that a professional person may be held legally liable for actions for a much longer period than the normal statutory limitation period would otherwise suggest. Professional indemnity insurance, like other insurances, is secured on an annual basis, the cost reflecting the conditions of the insurance market and the policyholder.

Performance bonds

There is an increasing demand for consultants to provide performance bonds, particularly on major projects. Although it is a concern that clients consider it necessary to require such bonds in the pursuit of work, the quantity surveyor might not be in a position to object.

The conditions of the bond are likely to be similar to those required from a contractor (Chapter 14). The value is calculated as a percentage of the total fee, and the conditions under which it can be called upon are stated. In certain instances, 'on-demand' bonds are being requested, whereby payment by the surety can be demanded without the need to prove breach of contract or damages incurred as a consequence. It is therefore important to check the conditions in detail and to ascertain the cost of providing the bond prior to agreeing fee levels and terms of appointment.

The common sources of protection are bank guarantees and surety bonds. The terms 'bonds' and 'guarantees' have similar meanings and are used synonymously within the construction industry. Guarantees are documents that indemnify a beneficiary should a default occur. They are usually provided by banks. Performance bonds are usually issued by insurance companies. A performance bond assures the beneficiary of the performance of the work involved up to the amount stated. Performance bonds are three-party agreements between the bondsman, beneficiary and the principal debtor.

Confidentiality

In addition to the common law duty of confidentiality, under the standard appointment the quantity surveyor cannot disclose any project information to a third party without written consent of the client. This is particularly relevant as many quantity surveyors share project progress photographs on social media. The quantity surveyor must also comply with the Data Protection Act 1998 and the General Data Protection Regulations (GDPR) in relation to any personal data collected that maybe used during the project.

Complaints and dispute procedures

All firms have a professional obligation to publish their complaints-handling procedure. Managing client relationships and reaching agreement when dissatisfaction arises is a core competency of the quantity surveyor. If disputes arise, the standard appointment requires the use of mediation. However, this does not remove the right of the client or the quantity surveyor to refer the dispute to adjudication under the *Housing Grants, Construction and Regeneration Act 1996*. Within the form, parties will also specify whether arbitration or litigation will be used if disputes remain unresolved. Chapter 15 provides further information on litigation and alternative dispute resolution procedures.

Negligence

Where the law is concerned, negligence usually consists either of a careless course of conduct or such conduct, coupled with further circumstances, sufficient to transform it into the tort of negligence itself. As stated earlier, the extent and nature of the duties owed to the client by the quantity surveyor, as well as the powers and authority granted to the client, will be determined by the contract for services between them.

It has always been implied into a professional engagement that the professional person will perform duties with due skill and care. This requirement is reiterated by provisions in the Supply of Goods and Services Act 1982. Lack of care in discharging contractual duties is, and always has been, an actionable breach of contract.

As late as the mid-1960s, when it was so held in *Bagot* v. *Stevens Scanlan & Co* (1964) 3 AER 577, the existence of a contractual link between the parties was believed to confine liability to that existing in contract and to exclude any additional liability in tort. Since then, the position has gradually changed and the courts appeared to recognise virtually concurrent liability in both contract and tort. Thus an aggrieved contracting party was able to sue the other contracting party or parties both in contract and tort. This was illustrated by the decisions in *Midland Bank Trust Co Ltd* v. *Hett, Stubbs & Kemp* (1978) 2 AER 571 and, more immediately relevant to the construction industry, in *Batty* v. *Metropolitan Property Realisations Ltd* (1978) 7 BLR 1.

Liability was also considered to exist independently, where there is no contractual link between the parties, enabling a third party to sue in the tort of negligence. A plaintiff suing in negligence must show that:

- The defendant had a duty of care to the plaintiff, and
- The defendant was in breach of that duty, and
- As a result of the breach the plaintiff suffered damage of the kind that is recoverable.

In the first place, the plaintiff would try to show that the defendant owed a duty of care. From the principles established in *Donoghue* v. *Stevenson* (1932) AER I (the celebrated 'snail in the bottle' case), the courts tended to find the presence of a duty of care in an ever-increasing number of circumstances.

The tentacles of the tort of negligence even extended well beyond normal commercial relationships, at least so far as the professional person was concerned. Since the well-known case of *Hedley Byrne & Co Ltd* v. *Heller & Partners Ltd* (1963) 2 AER 575, any negligent statement or advice, even if given gratuitously, seemed in certain circumstances to afford grounds for action.

In more recent times, in the case of *Junior Books Ltd* v. *The Veitchi Co. Ltd* (1982) 21 BLR 66, the House of Lords held that a specialist flooring subcontractor was liable in negligence for defective flooring to the employer with whom the subcontractor had no contractual relationship. Almost immediately, however, the courts began to retreat from the position by means of a long string of cases which culminated in *Murphy* v. *Brentwood District Council* (1990) 50 BLR 1 which, amongst other things, overturned the 12-year-old decision in *Anns* v. *London Borough of Merton* (1978) 5 BLR 1.

The tortious liability for negligence is therefore reduced, which in itself has led to the growth in the use of collateral warranties.

In the case of *Pantelli Associates Ltd* v. *Corporate City Developments Number Two Ltd* (2010) EWHC 3189 (TCC), the judge, Mr Justice Coulston, struck out the allegations of professional negligence against the quantity surveying firm, Pantelli. He ruled that claims of negligence must be capable of response, i.e. details of negligence should be detailed, it is inadequate to insert 'failing to' as a prefix against the contractual obligations. Also that allegations against any professional must be supported in writing given by a professional with the necessary expertise. However, in terms of the use of expert evidence in professional negligence cases, it was determined in the case of *Wattret v Thomas Sands Consulting* (2015) 3445 (TCC) that the specialist nature of the Technical and Construction Court is considered, resulting in expert evidence being constrained to matters relating to quantity surveying practice rather than that of the law. The role of the quantity surveyor as expert witness is discussed in Chapter 15.

There is little other case law concerning successful claims of the negligence of a quantity surveyor. However, it is always advisable to limit the risk involved. The best safeguard is discretion and a reluctance to express opinions or proffer advice on professional matters, unless one is reasonably acquainted with the relevant facts and has had the opportunity to give them proper consideration. Indeed, 'Members … must provide good-quality and diligent service' is one of the five rules of the RICS's Global Rules of Conduct. It recommends the following behaviour:

- Agree with the client the scope of the services (rule 3.2)
- Work is carried out with due care, skill and diligence and in accordance with RICS technical standards (rule 3.5)
- Act to prevent others being misled about their professional opinion (rule 1.7) (RICS 2021)

The RICS also provides a series of mandatory Professional Statements, Codes of practice, guidance notes and information papers intended to embody 'best practice'. If an allegation of professional negligence is made against a surveyor, the court would likely take into account whether the surveyor acted within these guidelines when deciding whether they had acted with reasonable competence.

Unbefitting professional behaviour

The public is increasingly sceptical of self-regulated professions. There are too many examples of where professional bodies have failed to thoroughly investigate a member who, on the surface at least, has failed to properly safeguard either a client or public interest. The RICS, in common with other professional bodies, has a set of bye-laws, regulations and rules of conduct to which every chartered surveyor must adhere. The RICS describes itself as a global professional body that represents, regulates and promotes chartered surveyors.

Of course, the vast majority of chartered surveyors go through their professional careers upholding the highest standards of professional conduct. Under the Royal Charter and bye-laws, all RICS members are expected to comply with regulations governing their conduct.

The RICS bye-law 5.2 states that every member shall conduct themselves in a manner befitting membership of the institution and comply with regulations and rules laid down to govern the manner in which their profession is conducted (RICS 2020a). However, in common with many of the other professional bodies, and despite various attempts to do so, there is no precise definition of what this means in practice. The lack of such definition allows a professional body to move with the times as practices and procedures change. For example, acceptable practices in the past might be unacceptable to the profession today where standards have changed. In the case of discrimination, this might not have been much of a consideration fifty years ago, but today this is high on the agenda.

Disciplinary procedures

On receipt of a complaint, allegation or information about an alleged contravention, regardless of the originator or source, the RICS is empowered under the disciplinary rules to investigate as appropriate. The rules do not require a complaint to be made to initiate an investigation. It is a fundamental aspect of self-regulation that the RICS itself is able to investigate matters based on information from a variety of sources. Where they are satisfied that there is evidence of a breach of the bye-laws, rules or other regulations, and that the misconduct is sufficiently serious and in the public interest to do so, disciplinary action will be taken in accordance with the Regulatory Tribunal Rules (RICS 2020b). These rules stipulate the following courses of action:

- A fixed penalty is imposed
- A Regulatory Compliance Order, which may consist a caution, reprimand or require the surveyor or firm to take certain actions; this may also involve the administration of costs and a fine.
- The consideration of the case by a Single Member of the Regulatory Tribunal; this will be for cases such as noncompliance with CPD requirements, cases where members have a custodial sentence and those that are not of public interest.
- or an RICS Disciplinary Panel

Disciplinary panels take place for more serious breaches. The panel for the hearing will normally be held in public and panel members are chosen from the Regulatory Tribunal members consisting of three members; of which two are non-RICS members. The panel can decide to expel the surveyor from membership of the RICS. Members do have a right to appeal. Other than in exceptional circumstances, the results of all cases are publicised on the RICS website.

Other legislation requirements

The following section briefly highlights other legislation and regulations that the quantity surveyor needs to be aware of:

Contracts of employment

There are certain legal requirements relating to the employment of staff. The basic relationship between the employer and the individual employee is defined by the contract of employment. This is a starting point for determining the rights and liabilities of the parties. Although these rights originated from different statutes, they are now consolidated in the Employment Protection (Consolidation) Act 1978, though further amendments were introduced in the Employment Act of 1990 and the Employment Act 2008. An Employment Relations Act was introduced in 1999 and amended in 2004; this act recognises trade unions and the taking of industrial action. An important feature of these rights is that they are not normally enforced in the courts but in employment tribunals.

An employment contract is made as soon as a job offer is accepted. A written statement of employment particulars must be given to each employee within two months of commencing employment. It should cover matters regarding the conditions of employment, including hours of work, salary, holiday entitlement, sick leave, termination, and the procedures to be followed in the event of any grievance arising.

Another Act worthy of note is the Equality Act 2010, whereby a person cannot be discriminated against because of his or her sexual orientation, marital status or age. The Act also covers contractual terms *and* conditions of employment in addition to pay, making it unlawful for an employer to treat someone differently because of their sex, colour, ethnic or national origins or nationality. This Act covers not only recruitment but also promotion and other non-contractual aspects of employment. From October 2010, the Equality Act also replaced most of the Disability Discrimination Act 2005 (DDA). The Disability Equality Duty in the DDA continues to apply. The Act aims to protect disabled people and prevent disability discrimination, which has consequences for the QS at both a personal and professional level. This Act, in relation to disability, is addressed separately below.

An employer must give the employee the amount of notice to which he or she is entitled under the contract of employment. This will relate to the employee's length of service up to a maximum of 12 weeks, although the contract may specify a longer period. Employees may be dismissed for acts of misconduct, but the employment tribunal must be satisfied that the employer acted reasonably should a complaint be brought to them. Some of the following might be considered as misconduct:

- Absenteeism
- Abusive language
- Disloyalty
- Disobedience
- Drinking
- Using drugs and smoking
- Attitude
- Personal appearance
- Theft or dishonesty
- Violence or fighting.

In order for an employee to bring a case for unfair dismissal, an employee must be able to show at least one of the following:

- Employer ends employment without notice
- A fixed-term contract ends without being renewed

- An employer forces an employee to resign; this is known as constructive dismissal
- An employer refuses to take back a woman returning to work after pregnancy
- An employer gives a choice of resignation or being dismissed.

An Employment Tribunal will investigate whether the employer was acting reasonably after examining all of the facts of the case. It will want to establish that:

- Warnings were given
- Adequate notice was provided for a disciplinary hearing
- The employee had the opportunity to comment on the evidence
- A decision was not made by someone who had not heard the employee's view
- An appeal was decided by someone who was not already involved.

Sometimes a job comes to an end because the firm has no more work or because the kind of work undertaken by the employee has ceased or diminished. In these circumstances, the employee will normally be entitled to redundancy payments

Equality Act 2010

Seeking to protect and enhance the rights and opportunities of disabled persons, this Act has consequences for the quantity surveyor both as an employer/employee and as a practising professional.

In the field of employment, employers must make facilities available for those of their staff who are registered disabled, as true for the quantity surveyor as in any other field. Whilst *occupational exceptions* may be appropriate for certain situations on site, in the office setting an employer will be expected to construct or adapt their premises accordingly.

In their work with and for clients, quantity surveyors should be aware of the latest legislation, particularly that regarding buildings to which the public have access. There will surely be cost consequences arising out of the current legal requirements concerning access, lighting, signage and the like. As the UK moves steadily towards a growing proportion of elderly people, this will present further challenges in the future in the design and costing of construction works, challenges for which the quantity surveyor must be prepared.

Bribery Act 2010

The quantity surveyor is required to comply with all relevant requirements under the Bribery Act 2010 (rule 1.12). Whilst it is acceptable to participate in bona fide hospitality events, it is important that any receipt of both financial or nonfinancial gifts from parties, such as contractors or developers cannot be construed as a bribe. Bribery can be defined as 'giving someone financial or other advantage to encourage that person to perform their function or activities improperly' (Ministry of Justice n.d.). The RICS Professional Statement *Countering bribery and corruption, money laundering and terrorist financing* (2019) requires all members to declare items such as gifts, hospitality and entertainment to their employers. Also included within this professional statement is the mandatory requirement for quantity surveyors to not facilitate or be complicit in money laundering or terrorist financing activities (RICS 2019).

Modern Slavery Act 2015

The construction industry is an environment where modern slavery, human trafficking and forced labour can exist. The high reliance on migrant labour, use of agency labour, fragmentation of supply chains and a price-driven culture increase the risk of exploitation within the sector (CIOB 2018). Companies with an annual global turnover more than £36 m are required to prepare a slavery and human trafficking statement. Such a statement will provide guidance on how to identify modern slavery and where to go to seek advice. Alternatively the modern slavery and exploitation helpline (https://www.modernslaveryhelpline.org/) provides advice and support.

Health and Safety at Work Act 1974

The quantity surveyor has a 'direct responsibility to ensure corporate health and safety policies are practised effectively and competently … and that they assume individual behavioural responsibility for their own work, their colleagues and other's health and safety during the course of their work' (RICS 2018). Under the Construction (Design and Management) Regulations 2015, the quantity surveyor when visiting site, as they are under the control of the contractor, must engage in any site safety induction and follow site rules and procedures.

References

CIOB. *Construction and the Modern Slavery Act. The Chartered Institute of Building*. CIOB. 2018.

Ministry of Justice. *The Bribery Act 2010; Quick Start Guide*. n.d. Available at https://www.justice.gov.uk/downloads/legislation/bribery-act-2010-quick-start-guide.pdf (accessed 14/05/2022).

RICS. *Surveying Safety: Health and Safety Principles for Property Professionals*. 2nd Edition. 2018.

RICS. *Countering Bribery and Corruption, Money Laundering and Terrorist Financing*. 1st Edition. RICS. 2019.

RICS. *RICS Regulatory Tribunal Rules: Version 1 with Effect from 2 March 2020*. RICS. 2020a.

RICS. *Royal Institution of Chartered Surveyors Bye-Laws*. RICS. 2020b.

RICS. *Rules of Conduct*. RICS. 2021.

RICS. *Standard Form of Consultant's Appointment Explanatory Notes: England and Wales*. RICS. 2022a.

RICS. *Standard Form of Consultants' Appointment: England and Wales*. RICS. 2022b.

RICS. *Quantity Surveying Services: England and Wales*. RICS. 2022c.

5

Cost Control

KEY CONCEPTS

- Project cost control
- Cost advice and reporting
- Pre-contract estimating methods
- Cost planning
- Accuracy of cost estimates
- Lean construction and cost control.

LEARNING OUTCOMES

After reading this chapter, you should be able to:

- Understand the purpose and importance of cost control.
- Understand the principles and practice of estimating techniques.
- Appreciate factors to be considered in providing cost estimates and their level of accuracy.
- Appreciate the benefits of the use of BIM in cost planning.

COMPETENCIES

Competencies covered in this chapter:

- Capital allowances
- Data management
- Design economics and cost planning
- Quantification and costing (of construction works)
- Risk management.

Introduction

Anyone proposing to construct a building or engineering structure will need to know in advance the probable costs involved in the works. These costs include the cost of the works carried out on site by the contractor, professional fees, and any taxes that may be due to the government.

Willis's Practice and Procedure for the Quantity Surveyor, Fourteenth Edition.
Allan Ashworth and Catherine Higgs.

In addition to these sums the client, or promoter on civil engineering projects, will also need to make provision for the costs of site acquisition and other development costs, together with the fitting out and furnishings that are required in the completed project. Many of these costs are often excluded from the normal process of construction project cost control. It is one of the duties of the client's quantity surveyor to ensure that the building to be constructed is carefully controlled in terms of costs arising throughout the entire design and construction process. The focus of cost control must be balanced with the importance of value in terms of what is being provided for the client. This chapter is concerned only with project cost control, i.e. those associated with capital construction costs; other aspects of cost are considered in Chapter 6.

Project cost control

Cost control in the construction industry is a term used for the 'process of planning, predicting, and controlling the costs of building(s) and takes place throughout the duration of the construction project' (RICS 2021). Hence, the process starts at project inception, when indicative costs will be required, through the stage when an early price estimate is prepared and the tender process undertaken by the contractors through to the final completion and agreement of the final account for the project. Fig. 5.1 relates the cost estimates and cost plans prepared by the quantity surveyor to both the RIBA Plan of Works and the Government's Gate Review process (known within the sector as 'OGC Gateways'). The RIBA Plan of Work is an industry-wide recognised framework that organises the process of the management, design and administration of building projects into seven sequential steps. The alternative to the RIBA Plan of Work is the Gateway Process; this is used for government projects, such as defence, education and health developments. In addition, this function is related to the RICS Information Stages within the *RICS Cost Prediction Global Professional Standard* (RICS 2020). If the initial estimate is within the budget and a realistic cost limit has been set and approved by the client, the project then moves forward to the design stage. During the various stages of design, the architect or engineer will want to consider alternative solutions that meet the client's overall aims and objectives.

The quantity surveyor will offer cost advice for the comparative design solutions of the alternative materials to be used or the form of construction to be adopted. In doing so, the whole life cost in relation to the initial capital costs may be considered (Chapter 6). The quantity surveyor will also provide advice on the cost implications of the design morphology and procurement. This stage, known as cost planning, has been developed in further detail and could be described as a system of relating the design of buildings and other structures to their costs. This considers quality, utility and appearance. The cost is planned within a combination of the budget provided by the client, and the design and construction considerations determined by the design team.

The good practices of cost planning further require the quantity surveyor to allocate the estimated costs into subdivisions, defined as elements, of the building. These element costs

RIBA Plan of Work 2020 Stages	RICS Formal cost estimating and cost planning stages	RICS information stages	OGC Gateways
0 Strategic Definition	Rough order of cost estimate	Level 1 Estimate	1 Business justification
1 Preparation and Briefing	Order of cost estimate (s)	Level 2 Estimate	2 Delivery strategies
	Elemental cost estimate		3A Design brief and concept approval
2 Concept Design	Formal cost plan 1	Level 3 Estimate	
3 Spatial Coordination	Formal cost plan 2	Level 4 Estimate	3B Detailed design approval
4 Technical Design	Formal cost plan 3	Level 5 Estimate	3C Investment decision
	Pre-tender estimate Pricing documents (for obtaining tender prices) Post tender estimate	Level 5 Estimate(s)	
5 Manufacturing and Construction			
6 Handover	Formal cost plan (renew/maintain) (measured in accordance with NRM3)	Level 6 Estimate	4 Readiness for Service
7 Use			5 Operation review and benefits realisation

Fig. 5.1 RICS Formal cost estimating and cost planning stages (*Source:* Adapted from RICS 2021, NRM1, Table 1.1).

can be compared against the element costs of other similar projects from the quantity surveyor's cost library records. The contract documents may also be prepared on this basis to facilitate easier preparation of the cost analysis.

Cost control does not stop at tender stage but continues up to the agreement of the final account and the issue of the final certificate for the works. Post-contract cost control and procedures are described in more detail in Chapters 11–13.

Cost advice

Quantity surveyors throughout the design and construction process are required to advise the client on any cost implications that could arise. Such advice will be necessary irrespective of the procurement method used for contractor selection or tendering purposes. However, the advice will be especially crucial during the project's inception. During this time, major decisions are taken that affect the size of the project and the quality of the works, if only in outline form. The cost advice given must therefore be as reliable as possible, so that clients can proceed with the greatest amount of confidence. Where the advice is inaccurate, it could cause a client to proceed with a project that cannot be subsequently afforded or, because of a too high forecast of probable costs, result in a project being prematurely aborted.

Quantity surveyors are the recognised professionals within the construction industry as cost and value consultants. Their skills in measurement and valuation are without equal. Quantity surveyors recognise the importance of providing sound, reliable and realistic cost information that will contribute to the overall success of the project. In this context it is important to have an awareness of both design maturity and construction methodology processes. Quantity surveyors are the industry's experts on building costs and must perform their duties in accordance with the *RICS Cost Prediction Professional Standard*. Failure to carry out these duties properly, in the context of an expert, may result in disciplinary proceedings and could provide grounds for liability for negligence (Chapter 4).

Cost reporting

When communicating cost advice to clients, it is important to consider the client's experience with construction projects and the project context in terms of design maturity, assumptions and the level of risks around uncertainties. Clarity of communication of the information is key, and the level of explanation and support provided to the client to interpret the costs reported needs to be adapted for the client's experience and need. The *RICS Cost Prediction Professional Standard* identifies the following principles that must be adhered to:

- Based upon the briefing and information provided, produce a reliable prediction of costs appropriate for the needs and best interests of the client, the size and complexity of the project and the project stage.
- Consider and agree with the client the appropriate method for reporting costs, recommending the use of ICMS where that would be in the best interests of the client.
- Provide the sources of the data on which the cost prediction is based and a commentary on the dependability of the cost data, unless the data source is subject to confidentiality provisions.
- Identify the key assumptions, including any exclusions and how they can be managed, as well as timing and methodology constraints, made in deriving the predicted cost and the grounds for making them.
- State the change in predicted cost since the last report and the reason(s) for the change.

- Commensurate with the size and complexity of the project, provide an estimate of the accuracy or level of uncertainty of the cost prediction and how this can be improved through management action (RICS 2021).

New Rules of Measurement (NRM1)

The introduction of the New Rules of Measurement suite of documents (NRM1, NRM2 and NRM3) in May 2009 provided a consistent industry-wide standard set of measurement rules that are understandable by all those involved in construction projects. Each document provides measurement rules for building works: NRM1 for cost estimates and plans, NRM2 for preparation of bills of quantities (Chapter 11) and NRM3 for whole life costing (Chapter 6). A standardised structured methodology enables the quantity surveyor to provide effective and accurate advice to the project team. A codified framework approach provides versatility, so that costs reported in the cost plans can be realigned, reallocated and reconciled with pricing documents used throughout the design, construction and maintenance of building projects. This format also enables integration of cost data produced early in the project life cycle with building information models used later in the design process.

The *International Cost Management Standard (ICMS): Global Consistency in Presenting Construction Life Cycle and Carbon Emissions* is a principle-based standard that provides a high-level classification framework to enable global consistency in classifying, defining, measuring, recording, analysing, presenting and comparing entire life cycle costs and carbon emissions of construction projects (ICMS Coalition 2021). Both the NRM1 and NRM3 elemental breakdown structures map to the ICMS Level 4 (cost sub-group level).

The *NRM Order of Cost Estimating and Cost Planning for Capital Works* (NRM1), the focus of this chapter, provides guidance on the quantification of building works for preparing cost estimates and cost plans (RICS 2021). The rules also provide guidance relating to those costs that are not reflected in the measurable building work items, such as consultant fees and risk allowances.

Pre-contract estimating methods

The Order of Cost Estimate will support the evolution of the design of the project, ensuring that proposals are financially viable. They will be used to establish the cost limit of the project. Whilst these are primarily concerned with the initial building costs, the quantity surveyor should consider the impact of whole life costs (see Chapter 6) when considering value for money in the longer term. The architect must then ensure that the design can be constructed within this cost limit. These are common procedures to be adopted in connection with public buildings such as houses, schools and hospitals. They are also used in the private sector where a developer may need to place a limit on the building cost, based upon the other costs involved, such as the land cost and the selling price of a house or the commercial value of the project.

The estimating methods used for the preparation of the building works estimate section of an order of costs estimate are the functional unit method, floor area method and the elemental method. Although they are often referred to as approximate estimating methods, this needs to be read in the context of the way in which they are calculated rather than in terms of the intended accuracy of price alone. The degree of accuracy will depend upon the type of information provided, the quality of relevant available pricing information and the skills and experience of the quantity surveyor who prepares the estimate of cost. Familiarity with the type of project and the location of the site are important factors to consider. At this stage in the design process, typically less than 30% of project information is available (RICS 2021).

Functional unit method

The unit method of approximate estimating is only generally used when the quantity surveyor is unable to establish the floor area of a project, and it consists of choosing a standard unit of accommodation and multiplying it by an appropriate cost per unit. The technique is based upon the fact that there is usually some close relationship between the cost of a construction project and the number of functional units, based on the user requirements, that it accommodates. The standard units may, for example, represent the cost per theatre seat, hospital bed space or car park space. Appendix A in the NRM1 provides a list of commonly used functional units and the units of measurements. Retail space can be more complex, so if using this method for shopping centres, for example, reference should be made to the *International Property Measurement Standards: Retail Buildings*. Such estimates can only be very approximate, so cost sources such as *Spon's Architects' and Builders' Price Book* will provide a unit price range. Functional unit cost data is also available from the Building Cost Information Service (BCIS). As with the floor area method, the external work is measured separately, based on the net site area (site area less the footprint of the building).

Floor area (superficial) method

This type of estimate, sometimes referred to as the superficial method, is fairly straightforward and quick to calculate and costs are expressed in a way that can be readily understood by a client. The area of each of the floors is measured and then multiplied by a cost per square metre. In order to provide some measure of comparability between various schemes, the floor areas are calculated from the internal dimensions of the building; that is, within the enclosing walls. Table 2.0 of the *Code for Measuring Practice* provides further guidance on areas included and excluded from the calculation (RICS 2015).

The floor area rate is based on historic data of similar projects, and the quantity surveyor will use judgement to adjust the rate to align with the current project. The best records for use in any form of approximate estimating are those that have been derived from the quantity surveyor's own previous projects, inhouse data and cost models. However, extensive use is also made of industry cost databases to provide, if nothing else, a benchmark on the estimated costs. These are also especially useful where a quantity surveyor has no previous records on which to base an estimate. In these circumstances, the *Building Cost Information Service* (BCIS) database may need to be consulted. This type of information must, however, be treated with considerable caution since it can easily produce misleading results. Also, it can rarely be used without some form of adjustment. Buildings within the same category, such as schools or offices, have an obvious basic similarity, which should enable costs within each category to be more comparable than for buildings from different categories.

Within the estimate, it is important to consider varying site conditions, both the topography and any site restrictions and access. A steeply sloping site must make the cost of a building greater than for the same building on a flat site. The nature of the ground conditions, and whether they necessitate expensive foundations or difficult methods of working, must also be considered. The construction design, details and function of the space to be used will also have an important influence. A single-storey garage of normal height will merit a different rate per square metre than one that is constructed for maintenance of double-decker buses. Again, a requirement for, say, a 20 m clear span is a different matter than allowing stanchions at 5 m or 7 m intervals. An overall price per square metre will be affected by the number of storeys. A two-storey building of the same plan area has the same roof and probably much the same foundations and drains as a single storey of that area but has double the floor area. However, the prices for high buildings are increased by the extra time involved in hoisting materials to the upper floors, the danger and reduced outputs from working at heights and the use of expensive plant such as tower cranes. Increases in building services costs as the number of storeys increase must also be taken into account.

The shape of a building on plan also has an important bearing on cost. A little experimental comparison of the length of enclosing walls for different shapes of the same floor area will show that a square plan shape is more economical than a long and narrow rectangle, and that such a rectangle is less expensive than an L-shaped plan. There are, however, exceptions to these general rules.

The overall specification, including sustainability requirements, will also have to be taken into consideration (see Chapter 18).

For projects offering different standards of accommodation it would be preferable to price these independently. A variety of rates could therefore be required, depending upon the different functions or uses of the parts of the building. There might also be the possibility of having to include items of work that do not relate to the floor area, and these will have to be priced separately.

Elemental method

The elemental method provides for a more detailed approximate estimate than any of the aforementioned methods. This method considers the major elements of the building. The *Elemental Standard Form of Cost Analysis (SFCA)* defines each element and provides rules for allocating costs to their functional element (BCIS 2012). Fig. 5.2 identifies the group elements in accordance with NRM1 Appendix B, and each group element can be further broken down into a series of elements. For example, Internal Finishes can be separated as walls, floor and ceiling finishes. The choice and the number of elements will be dependent on the design and specification information the quantity surveyor has at the time of the estimate. An elemental unit rate for this work can be readily built up, making the appropriate project adjustments on a basis obtained from inhouse and BCIS data.

This method does provide a more reliable means of approximate estimating, but it also involves more time and effort than the alternative methods. NRM1 provides measurement

Group Element
0. Facilitating works
1. Substructure
2. Superstructure
3. Internal finishes
4. Fittings, furnishings, and equipment
5. Services
6. Prefabricated buildings and building units
7. Work to existing buildings
8. External works

Fig. 5.2 NRM1 group elements for elemental cost planning (*Source:* Adapted from Appendix B, NRM1, 2021).

rules and guidance which reflect the level of design information available. For order of cost estimates, at the early stages in the RIBA Plan of Works (0 and 1) or the OGC Gateways (1 and 2) (Fig. 5.1) Table 2.2 of NRM1 sets out the unit of measurement for each element. As drawings supplied at this stage are usually the floor and roof plans, elevations and sections, the majority of elements are measured by area. Fig. 5.3 provides an example of a builder's work estimate, which is a constituent part of the order of cost estimate. Addition of an estimate for the main contractor's preliminaries, and overheads and profits will provide the Works Cost Estimate.

To establish the cost limit of the project (excluding inflation), the following costs need to be included:

- Project and design team fees
- Risk allowances
  - Design development risk
  - Construction risks
  - Employer change risks
  - Employers' other risks.

These will be discussed in more detail later in the chapter.

General considerations

The selection of appropriate rates for pre-tender estimating for the Building Works Estimate depends upon a large variety of factors. Some of these can be considered objectively, but in other circumstances only experience or expert judgement will be sufficient. Some of these considerations are as follows.

Market conditions

In order to get the price as near as possible to the tender sum, the quantity surveyor must be able to interpret trends in prices based upon past data and current circumstances. This demands a great deal of skill and some measure of luck! Allowances may also need to be

Job No. 123		Date 14 August 2023	
New Church of England Primary School Blankchester Diocesan Board of Finance BUILDING WORKS ESTIMATE to support ORDER OF COST ESTIMATE (note elements listed in accordance with NRM1) (Fixed price 65 weeks)			
Summary	**Floor area cost per m²**	**1200 m² Elemental cost**	
	£	£	£
1. SUPERSTRUCTURE	132.00	158400	158400
2. SUPERSTRUCTURE			903600
2.1 Frame	100.00	120000	
2.3 Roof	268.00	321600	
2.5 External walls	138.00	165600	
2.6 Windows and external doors	127.00	152400	
2.7 Internal walls and partitions	80.00	96000	
2.8 Internal doors	40.00	48000	
	885.00		
3. INTERNAL FINISHES			126000
3.1 Wall finishes	25.00	30000	
3.2 Floor finishes	52.00	62400	
3.3 Ceiling finishes	28.00	33600	
	105.00		
4. FITTINGS & FURNISHINGS	150.00	180000	180000
5. SERVICES INSTALLATIONS			1680000
5.1 Sanitary appliances 5.3 Disposal installations 5.4 Water installations 5.5 Heating source 5.8 Electrical installation 5.9 Gas installation 5.12 Communication installations	1100	1320000	
5.14 Builders work in connection	300.00	36000	
	1400.00		
8. EXTERNAL WORKS[1]			300000
8.1 Site works 8.2 Drainage 8.3 External services		300000	
BUILDING WORKS ESTIMATE			**3348000**

Fig. 5.3 Building works estimate example.

1 External works have been calculated separately and incorporated into estimate.

made for changes in contractual conditions, type of client, labour availability, workloads, and the general state of the industry.

Design economics

When using previous costs or cost analyses, changes in these costs for design variables such as shape, height and building size will need to be considered.

Quality factors

Cost information is deemed to be based upon defined or assumed standards of quality. Where the quality in the proposed project is different from this, then changes in the proposed estimate rates will be required. The surveyor should always provide quality indications with any cost advice.

Engineering services

These have become an ever-increasing proportion of building project costs. Their cost importance is such that they now need to be considered in detail. On large schemes it is now usual to employ specialist engineering services quantity surveyors.

External works

There are considerable differences between building sites and hence few established cost relationships for this building element. The size of the site and the nature of the work to be carried out will be important factors to consider.

Cost planning

Cost planning is not simply a method of pre-tender estimating but seeks also to offer a controlling mechanism during the design stage. Its objectives in providing cost advice are to control expenditure and to offer to the client better value for money. It attempts to keep the designer fully informed of all the cost implications of the design iterations as the work proceeds. The cost implications associated with procurement, programming and construction decisions will also be accounted for. Full cost planning will incorporate the attributes of whole life costing, value, and risk management.

The cost planning process commences with the preparation of the Formal Cost Plan 1 (see Fig. 5.1) which sets indicative cost targets for each element as described above. As the design evolves, through RIBA Stage 2 and 3, these cost targets are checked against the developing design to ensure alignment with the overall budget. The quantity surveyor will assist design decision-making by evaluating the comparative costs for alternative construction and specification proposals. Areas of uncertainty, such as risks, provisional sums and the like, will also be identified and appropriate cost allowances made. During this iterative process the quantity surveyor will update the formal cost plan (Formal Cost Plan 2) with a reconciliation against the previous cost report and publish to the client following the mandatory guidelines. The prudent quantity surveyor will also always be looking for ways of simplifying the details without altering the design, in an attempt to reduce the expected costs. Not only will the building construction be considered, but also the ease or otherwise with which the design can actually be built. This process should also result in fewer abortive designs, and the seeking of value for money should not cease at tender stage but should continue throughout the post-contract cost-control procedure.

The iterative process of refining the cost plan continues as the design, specification and construction information becomes more complete (RIBA Stage 4). The increase in the level of project information detail enables the quantity surveyor to refine the level of accuracy of the building works estimate within the cost plan (Formal Cost Plan 3). Focus will be on the cost significant elements and sub-elements. These will be measured using approximate quantities based on the quantity of the element rather than the g.i.f.a. in accordance with the tabulated rules in Part 4 of the NRM1. Inhouse data is most likely to be used to calculate the composite rates used within the cost plan. This cost plan can then be used to appraise tender submissions.

Additions to the building works estimate

Appendix D and E of the NRM1 provides templates for the Formal Cost Plans. In addition to the building works, the quantity surveyor will make additions for:

- Main contractors preliminaries, overheads and profits
- Projected design team and professional fees
- Other project costs
- Risk allowances
- Inflation

As with the building works costs, ascertaining these additions will be an iterative process; the allowances being refined as more information is available on matters such as the procurement strategy, and construction methods.

Main contractor's costs

In the early stages of the design, the main contractor's preliminaries will be based on a percentage addition to the cost of the building works estimate; this information will be obtained by analysis of previous projects. NRM1 Part 4 Group Element 9 provides rules for the measurement of the preliminaries when more project information is available. Likewise, the estimated cost for main contractor's overheads and profits. These can be calculated as either a single or separate cost centre and will be a percentage addition based on the sum of the building works estimate and preliminary estimate. This subtotal is referred to as the Works Cost Estimate.

Design and professional fees

Design and professional fees will reflect the procurement method, services provided over the duration of the project and allocation of design liability. When actual fee levels are not known, a percentage allocation will be attributed to this cost centre based on the Works Cost Estimate. The percentage applied can be derived from knowledge of the level of services provided and the fees on similar projects.

Other project costs

Other development and project costs also need to be included, such as insurances, planning fees, contributions associated with Section 106 and 278 Agreements and marketing of the finished asset. The latter cost item, for example, will be established by consultation with the client. Group Element 12 of the NRM1 provides measurement rules for these costs.

Risk allowances

Traditionally, quantity surveyors had presented their clients with single-price estimates, even though it was apparent that on virtually every project differences would occur between this sum and the final account.

Risk is measurable and can therefore be accounted for within an elemental cost plan. Uncertainty is more difficult to assess, as it represents unknown events that cannot be even assessed or costed. Risk is ascertained by undertaking formal risk analysis to support the cost plan. Different techniques can be applied, such as Monte Carlo simulation, to assess the risk involved. The risk will of course not be eliminated but at least it can be managed rather than ignored. Risk is considered in more detail in Chapter 8.

During the design and construction process, aspects of the design will still be evolving, and a contingency sum will need to be provided for possible additional costs that were not envisaged. Traditionally, a contingency sum would have been included to cover both price and design risk. NRM1 recommends a separate allowance for risks associated with uncertainty related to the procurement, design and construction of the projects. Separate allowances are permitted for design development risks, construction risks, employer change risks and employer other risks (NRM1 Group Element 13). These design risks are reassessed regularly as the project information becomes more comprehensive. The sums allowed are therefore likely to be larger at the early stages of the design than at the tender stage.

Inflation

As the project is priced at current prices, some addition is required to allow for possible increased costs. This is normally added as two separate amounts – the first up to the date of tender return, and a second to allow for increased costs during construction, from the date of tender return to the midpoint of the construction period. The cost of both tender and construction inflation will be applied as a percentage, ascertained by the use of inhouse data or BCIS indices. The cost limit is normally stated as firstly including the tender inflation and the addition of the construction inflation to the resultant cost limit. To forecast these sums in periods of high inflation is difficult.

The price risk factor within the cost plan is related to market conditions. A market that is more volatile, such as in periods of high inflation or where world events impact on resource costs will result in a larger percentage being allowed in the cost plan.

Exclusions associated with specialist advice

Cost plans normally exclude Value Added Tax (VAT) even on those projects where VAT will be charged. VAT is charged on the supply of goods and services in the UK and on the import of certain goods and services into the UK. It applies where the supplies are made in the course of business by a taxable person. VAT Notice 708 covers Buildings and Construction. The application of standard, reduced or zero-rated charges is complex, and the client is therefore recommended to seek specialist professional advice.

It is also important that consideration is given at the strategic stage of a project to the eligibility of financial assistance and its positive impact on project funding. Such assistance might be in relation to capital allowances, land remediation and government grants. As with VAT, it is recommended that the client seek specialist advice in these areas.

Capital allowances are the main form of tax relief on capital cost expenditure, enabling capital sums to be offset against taxable income. The amount of relief on capital allowances is contained in the Capital Allowances Act 2001 and The Capital Allowance (Structure and Building Allowance) regulations 2019. The relief is available for a number of years and this is accelerated where the location of the business has been designated as an Enterprise Zone. Capital allowances are also available on capital expenditure on plant and machinery, such as parts of a building considered 'integral' (Gov.uk 2022). Capital allowances are not automated but need to be substantiated and claimed for with the HM Revenue and Customs. It is advisable to consult with a capital allowance specialist to maximise tax relief benefits.

Land remediation relief is a corporation tax incentive for property expenditure and can be claimed for money spent on decontaminating land that may occur on brownfield sites. Expert advice should be sought as such relief will improve project viability.

In building development some consideration must also be given towards the financial assistance that may be available to encourage socially desirable development to be undertaken. Financial assistance may arise for one of the following reasons:

- Urban renewal programmes
- Regeneration of industrial areas
- Investing in jobs to benefit areas of high unemployment
- Land reclamation schemes
- Property improvement, such as housing improvements
- Slum clearances and derelict land clearance.

The aim is to encourage private companies or public managing agents to develop areas either as a means of improving the standards and amenities or through investing in projects that will help in wealth creation. At the same time, it could help to reduce unemployment in a region. Financial assistance is therefore targeted to areas in which it may otherwise be difficult to encourage companies to invest. Financial assistance may be a combination of taxation allowances, reduced rental and business rates, amenity grants and Enterprise Zone benefits.

Accuracy of approximate estimates

Whether or not the process of cost planning at the design stage is used, an analysis of tenders will be valuable for future approximate estimating purposes. Quantity surveyors must attempt to anticipate matters that will affect the level of future tender sums. They cannot, however, forecast exceptional swings in existing tendencies, just as they cannot forecast the future. They must, however, be aware of all current published trends. These publications will include those specifically for the construction industry as well as publications of a more general nature.

An understanding of the accuracy or reliability of the estimate is of importance to ensure appropriate caveats and clarifications are given when reporting costs. Accuracy is dependent on the comprehensiveness of the information on which it is based. Hence as the design process progresses, improvements in accuracy will be gained. Hence, estimates should be shown as a range of values rather than as a single lump sum. Alternatively, confidence limits could be offered as a measure of an estimate's reliability. The following factors are said to have some influence upon the accuracy of estimating:

- Quality of the design information
- Amount, type, quality and accessibility of cost data
- Type of project, as some schemes are easier to estimate than others
- Project size, as accuracy increases marginally with size
- Stability of market conditions
- Familiarity with a particular type of project or client.

As early estimates are a forecast of a contractor's tender sum, accuracy is also influenced by the inaccuracy in cost-estimating practices in contracting organisations. Causes of inaccuracy include the estimator's lack of practical knowledge of the construction process, insufficient time to prepare cost estimates, poor tender documentation and the wide variability of subcontractors' prices (Akintoye and Fitzgerald 2000). Furthermore, Ashworth and Perera conclude that unless human interaction is removed, improvements in the level of forecasting is unlikely (Ashworth and Perera 2015).

Cost data

The production of a realistic cost plan appropriate for the particular stage of the design relies on the reliability of the cost data used. It is important to understand the quality of the source of the cost data, its robustness and maturity. This is where the quantity surveyor uses judgement and experience of both costs and it's influences on previous projects and their application within the context of the proposed project. The same skill is required for adjusting for factors such as inflation, location and other drivers. This process is supported by BCIS datasets. As reflective practitioners, quantity surveyors should also be conscious of bias: whether there is a tendency to be over optimistic, rely too heavily on one piece of information or be complacement with the familiar. These biases may lead to misperception of risk and uncertainty of the project and poor decision-making (RICS 2021).

Cost models

Cost models are computer-based systems that support the development of cost estimates and cost control for the proposed construction project. Ashworth and Perera (2015) details the development of the cost model, its purpose and classification as either traditional, statistical, risk, knowledge based, resource based or whole life cost models. These models vary from a simple build-up of costs to simulations (see Chapter 7). Many organisations have developed cost planning applications built around spreadsheets supported by inhouse cost data. Such models vary in analytical sophistication and can use key design and cost parameters to effectively produce early stage cost estimates.

BIM and cost estimating

As quality of design information has been identified as a major influence on cost estimating, the use of building information modelling should have a positive impact on the accuracy of estimates. This is because information is always consistent with design, for example when a design change is made, quantification data is also changed. Speeding up quantification will also allow those responsible for estimating to focus on the higher value project specific factors, such as impact on construction assemblies and risks (Autodesk 2007). Fig. 5.4 aligns the level of information detail in the BIM model with cost prediction levels showing the potential of BIM-enabled cost estimating.

As cost planning is an iterative process, the use of BIM means that the cost impacts of changes can be more effectively and efficiently ascertained. This can also better inform the decision-making when alternative design options are considered. However, it is important

Cost prediction level	BIM Level of Detail	Level of Detail Content
Level 1 estimate	1	Block model with performance requirements; site constraints
Level 2 estimate	2	Concept of mass model including basic areas and volumes, orientation and cost.
Level 3 estimate	3	Generalised systems with approximate quantities, size, shape, location and orientation.
Level 4 estimate	4	Technical design model. Accurate and coordinated modelled elements that can be used to estimate costs and check regulatory compliance.
Level 5 estimate	5	Model suitable for fabrication and assembly, with accurate requirements and specific components.
Level 6 estimate	6	Details of how the project has been constructed, for use in operations and maintenance
	7	Asset information model for operations, maintenance and ongoing monitoring

Fig. 5.4 Cost predictions and BIM levels (*Source:* Adapted from Table 5, RICS 2021).

to consider the level of detail that would be contained within the model and its impact on the cost planning methodology; Smith et al. (2016) identify three approaches for various levels of BIM-integrated cost estimating:

- *Estimating based on model objects* where cost data is added to the object's parametric information and therefore is fully integrated.
- *Estimating based on model data export* involving exporting quantities from the module into BIM-compliant software applications or inhouse spreadsheets. Further definition or manipulation of quantities is normally required.
- *Estimating based on simulation of construction processes* enabling the quantity surveyor to analyse the relationships amongst resources, cost and cashflow mechanisms.

Babatunde et al. (2019) identified the top-ranked drivers for adoption of BIM-based detailed cost estimate by consulting firms using BIM as automation of quantities, enhanced qualities of decisions, time saving in preparation of detailed estimates, accurate construction sequencing and clash detection and cost saving in preparation of detailed estimates. The automation of quantities will minimise inaccuracies of the cost plan due to human error. BIM has the potential to not just support costing the building work but can be used to analyse risk scenarios and support value engineering processes implemented during the project design.

Client's cash flow

In addition to the client's prime concern with the total project costs, the timing of cash flow is also important since this will affect borrowing requirements. The client's quantity surveyor will prepare an expenditure cash flow profile based on the contractor's programme of activities, and any subsequent changes or revisions to this programme. On large and complex projects, and in periods of high inflation, the timing of payments, based upon different constructional techniques and methods, might result in a different contract sum representing a better economic choice for the project as a whole.

Contractor's cost control

The contractor, having priced the project successfully enough to win the contract through tendering, will ensure that the work can be completed for the estimated costs. One of the duties of the contractor's quantity surveyor is to monitor the expenditure and advise management of action that should be taken. This process also includes the cost of subcontractors, as these are likely to form a significant part of the main contractor's total expenditure. The contractor's quantity surveyor will also comment on the profitability of different site operations. Wherever a site instruction suggests a different construction process from that originally envisaged, then details of the costs of the site operations are recorded. The contractor's quantity surveyor will also advise on the cost implications of the alternative construction methods that might be employed.

Recognising that there may be accuracy errors when estimating even with common work items for the reasons discussed above, contractors need to satisfy themselves if wide variation between costs and prices arise. This will be done for two reasons: first, in an attempt to recoup, where possible, some of the loss; and second, to remedy such estimating or procedural errors in any future work. There are various additional reasons why such discrepancies arise:

- Character of the work is different from that envisaged at the time of tender
- Conditions for executing the work have changed
- Adverse weather conditions severely disrupted the work
- Inefficient use of resources
- Excessive wastage of materials
- Plant standing idle for long periods
- Plant being incorrectly selected
- Delays due to a lack of accurate design information.

Often when the project is disrupted by the client or designer this can have a knock-on effect on the overall efficiency and output of the contractor's resources. Contractors may sometimes suggest that they always work to a high level of efficiency. This is not always the case, and losses occur due to their own inefficiency. Costing systems that indicate that a project or site operation has lost money are of limited use if a contractor is unable to remedy the situation. A contractor needs to be able to ascertain which part of the job is inefficient and to know as soon as it begins to lose money. The objectives, therefore, of a contractor's cost control system are to:

- Carry out the works so that the planned profits are achieved
- Provide feedback for use in future estimating
- Cost each stage or building operation, with information being available in sufficient time so that possible corrective action can be taken
- Achieve the benefits suggested within a reasonable level of administration charges.

Contractor's cash flow

Contractors are not, as is sometimes supposed, singularly concerned with profit or turnover. Other factors also need to be considered in assessing the worth of a company or the viability of a new project. Shareholders, for example, are primarily concerned with the rate of their return on the capital invested. Contractors have become more acutely aware of the need to maintain a flow of cash through the company. Cash is important for day-to-day existence, and some contractors have suffered liquidation or bankruptcy not because their work was unprofitable but because of cash flow problems in the short term. In periods of high inflation, poor cash flows can result in reduced profits, which in their turn reduce shareholders' return. A correct balance between the objectives of cash flow, profit, return and turnover is required. In addition, inflation and interest charges will also have an impact on these items.

The role of lean construction in cost control

Cost control or controlling construction costs are methods that are used to increase added value in construction projects. In some cases, this may be the development and application of proper and effective systems. In other situations, it might mean reviewing some of the actual practices of organisation and management that are employed in both design and construction. For example, design and build is often championed as a least cost solution. This may be correct under certain conditions and especially where the design content is minimal, and the process used for construction is relatively routine in nature.

Lean construction is a high-level process that establishes strategic long-term collaborative alliances and working practices that reduce capital costs (Pasquire 2022). Lean construction methods are, in many ways, like cost planning and value management. Each of these techniques or practices aim to reduce the unnecessary costs from construction projects. In today's world, they will seek to do this within a context of whole life costing. Cost planning was introduced in the 1950s, and whilst it sought to reduce expenditure it did this only against a principle of setting cost limits. Both value management and lean construction set no cost limits and thus claim to offer more added value to the client.

Lean construction is a derivative of the lean manufacturing process. This concept has been popularised since the early 1980s in the manufacturing sector. The original thinking was developed from Japan (Womack and Jones 2003), although its principles have since been adopted worldwide. It is concerned with the elimination of waste activities and processes that create no added value. It is about doing more for less. It fits neatly within the philosophy of John Ruskin (1819–1900): *It is not the cheaper things in life that we wish to possess, but expensive things that cost less.*

Lean production is the generic version of the Toyota Production System. Automobile manufacturing has seen spectacular advances in productivity, quality and cost reduction (Howell 1999). The construction industry, by comparison, has not yet achieved these advances. Whilst it is possible to learn and adapt successful methodologies from other industries, it should be recognised that construction is a different activity.

The concepts involved in lean construction are identified in Fig. 5.5. This suggests that lean construction is a combination of current management thinking combined with proven developments in both the UK and worldwide. It recognises that it is a process of continuous improvement, of doing more for less and that this will result in a competitive advantage for construction firms who adopt this process. The practices involved include:

- Elimination of waste, whether this is in design or construction practices
- Prefabrication of construction components off-site, where these can be manufactured under factory conditions to a higher standard than site working is able to achieve
- Standardisation of building components to allow cost efficiencies to be made by using mass production techniques
- Partnering with a smaller number of subcontractors and suppliers and keeping successful teams together
- Benchmarking of practices against leading edge companies to maintain the most up-to-date practices
- Total quality management that will result in getting it right the first time and to a quality that cannot be matched
- Just-in-time management practices to keep investment and storage costs to a minimum
- Supply chain management practices where all firms involved are working to the same requirements

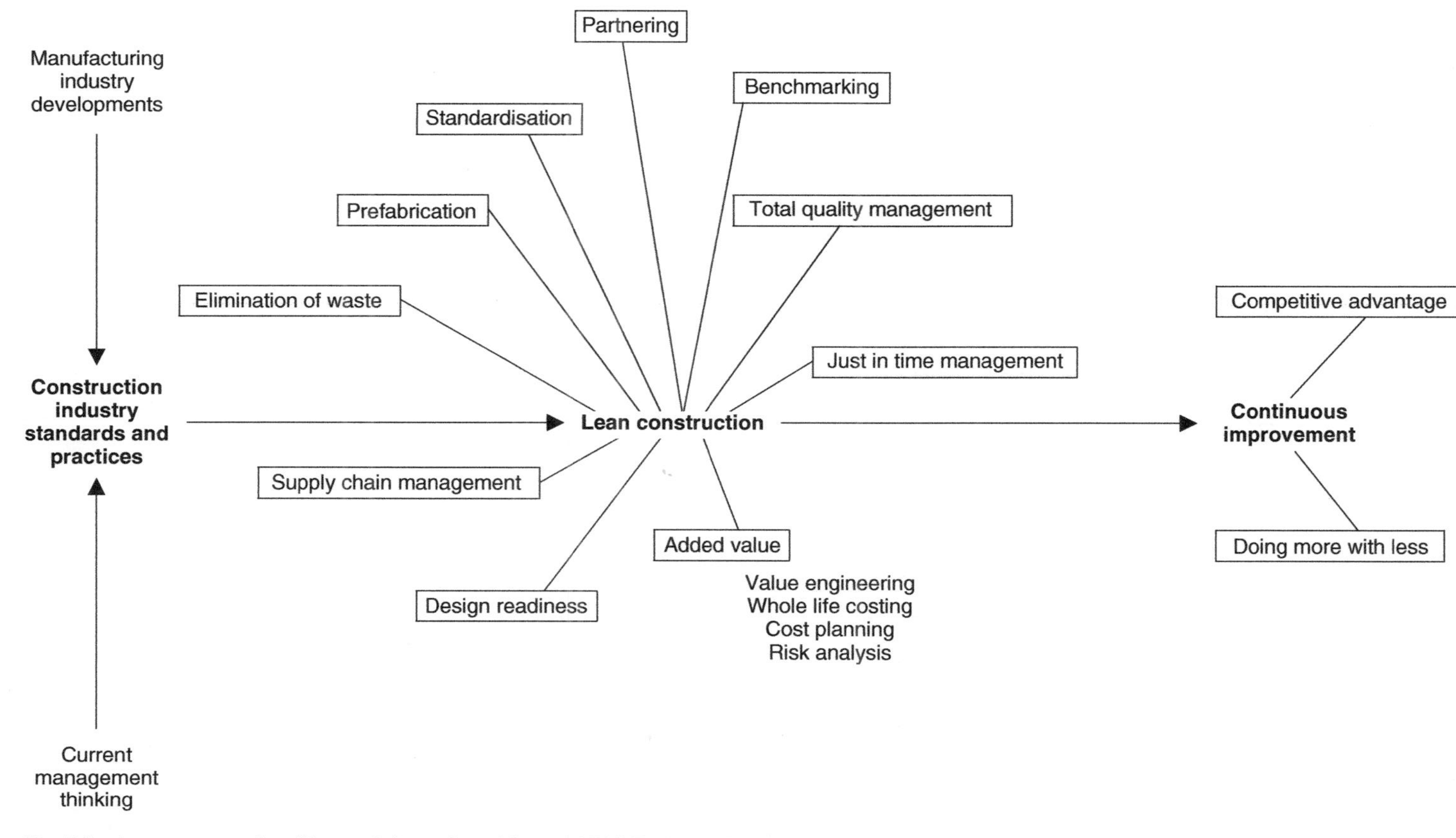

Fig. 5.5 Lean construction (*Source:* Ashworth and Perera 2015/Taylor & Francis).

- Design readiness to allow the constructor the best opportunity of organising and carrying out the work efficiently and effectively. This also has the potential for reducing the amount of time contractors spend on site
- Adding value through the use of a range of techniques that focus on a holistic approach to the project.

Lean construction can be pursued through a number of different approaches. The lean principles, first identified by Womack (1991) as follows:

- The elimination of all kinds of waste. This includes not just the waste of materials on site, but all aspects, functions or activities that do not add value to the project
- Precisely specify **value** from the perspective of the ultimate customer
- Clearly identify the process that delivers what the customer values. This is sometimes referred to as the **value stream**
- Eliminate all non-added value steps or stages in the process
- Make the remaining added value steps **flow** without interruption through managing the interfaces between the different steps
- Let the customer **pull**; do not make anything until it is needed, then make it quickly. Adopt the philosophy of *just in time management* to reduce stockpiles and storage costs
- Pursue **perfection** through continuous improvement (Construction Excellence 2004).

With reference to the above, Corfe (2013) further refines these principles – five principles of Lean, which provide a route map to embedding lean practice as value, value stream, flow, pull and perfection.

In addition, Male (2002) has recognised that lean thinking has more obvious applications for large, regular procuring clients who can provide some continuity of work and are able to integrate the supply chain.

The use of lean production techniques must be placed within the context of the construction industry. The difference between mass production in factories with generally bespoke buildings on a construction site must not be minimised. Furthermore, lean manufacturing appears to achieve the greatest improvements in efficiency and quality when design and manufacture occur in close proximity. With traditional procurement arrangements in the construction industry this is frequently not the case.

The vision of lean construction stretches across many traditional boundaries. It challenges our current practices. It has required changes in work practices and in understanding new roles and different responsibilities. It has required a change in attitudes and culture. It recognises that technology has an important role to play today. It expects that those involved in a construction project will share a common purpose. It is setting new standards that can be measured to indicate improvement.

Building Information Modelling as a process enables collaboration and, in doing so, brings improvement to the process of project delivery. Dave et al. (2013) identified the conceptual connections between BIM and Lean in its ability to support reduction of waste and adding value through 3D clash detection, the 4D capability to facilitate collaborative planning and the use of integrated software to analyse and evaluate whole life costs and carbon (Dave et al. 2013). Bhattacharaya and Mathur (2019) further identify the contributions towards Lean through enhanced decision-making, predictive simulation and risk analysis. BIM compared to traditional practices enables lean construction – is a vision of the future.

References

Akintoye, A. and Fitzgerald, E. A survey of current cost estimating practices. *Construction Management and Economics*, 18(2), 161–172. 2000.

Ashworth, A. and Perera, S. *Cost Studies of Buildings*. 6th Edition. Routledge. 2015.

Autodesk. *BIM and Cost Estimating*. Autodesk. 2007.

Babatunde, S, Pereda, S., Ekundayo, D. and Adeleye, T. An investigation into BIM-based cost estimating and drivers to the adoption of BIM in quantity surveying practices. *Journal of Financial Management of Property and Construction*, 25(1). 2019. Available at https://www.emerald.com/insight/content/doi/10.1108/JFMPC-05-2019-0042/full/html (accessed 24/04/2022).

BCIS. *Elemental Standard form of Cost Analysis: Principles, Instructions, Elements and Definitions*. 4th (NRM) Edition. BCIS. 2012.

Bhattacharya, S. and Mathur, A. Synergising lean objectives through BIM to enhance productivity and performance. *International Journal of Productivity and Performance Management*, 2022. https://doi.org/10.1108/IJPPM-04-2021-0199.

Construction Excellence. *Lean Construction*. Construction Excellence. 2004.

Corfe, C. *Implementing Lean in Construction: Lean and the Sustainability Agenda*. CIRIA. 2013.

Dave, B., Koskela, L., Kiviniemi, A., Owen, R. and Tzortzopoulos, P. *Implementing Lean in Construction: Lean Construction and BIM*. CIRIA. 2013.

GOV.UK. *Claim Capital Allowances*. Crown. Available at https://www.gov.uk/capital-allowances/what-you-can-claim-on (accessed on 23/04/2022).

Howell, G.A. What is lean construction? In: *Proceedings of the Seventh Annual Conference of the International Group for Lean Construction, IGLC-7*. University of California, Berkeley. 1999.

ICMS Coalition. *International Cost Management Standard (ICMS): Global Consistency in Presenting Construction Life Cycle and Carbon Emissions*. 3rd Edition. ICMS Coalition. 2021.

Male S. Building the business value case. In: *Best Value in Construction* (edited by J. Kelly, R. Morledge and S. Wilkinson). Blackwell Publishing. 2002.

Pasquire, C. *Lean Construction*. 2022. Available at https://www.isurv.com/site/scripts/documents.php?categoryID=56 (accessed 30/06/2022).

RICS. *Code for Measuring Practice*. Royal Institution of Chartered Surveyors. 2015. reinstated 2018.

RICS. *Cost Prediction*. Royal Institution of Chartered Surveyors. 2020.

RICS. *Order of Cost Estimating and Cost Planning for Capital Building Works: NRM1*. 3rd Edition. RICS. 2021.

Smith, J., Jaggar, D. Love, P. and Olatunje, O. *Building Cost Planning for the Design Team*. 3rd Edition. Routledge. 2016.

Womack, J. *The Machine that Changed the World*. Simon and Schuster. 1991.

Womack, J. and Jones, D. *Lean Thinking*. 2nd Edition. Simon and Schuster. 2003.

6

Whole Life Costing

KEY CONCEPTS
• Whole life costs (WLC) • Whole life costs and the environment • Life cycle costs (LCC) • WLC and LCC factors • LCC calculations

LEARNING OUTCOMES
After reading this chapter, you should be able to: • Understand the importance of whole life costs, their influence on environmental impact and their use through all phases of a building life. • Understand LCC in terms of the 'CROME' acronym and the use of NRM3 in supporting cost advice. • Appreciate the levels of LCC studies and factors to be considered in LCC calculations.

COMPETENCIES
Competencies covered in this chapter: • Data management • Design economics and cost planning • Sustainability

Introduction

It has long been recognised that to evaluate the costs of buildings and engineering structures on the basis of their initial capital costs alone are unsatisfactory. Some consideration must also be given to the costs-in-use, which will accrue throughout the life of the building or structure. The use of whole life costing for this purpose is an obvious idea, in

Willis's Practice and Procedure for the Quantity Surveyor, Fourteenth Edition.
Allan Ashworth and Catherine Higgs.

that all costs arising from an investment are relevant to that decision. The image of the whole life of a building or structure is one of progression through a number of phases, with the pursuit of an analysis of the economic whole life cost as the central theme of the evaluation. The proper consideration of the whole life costs is likely to result in a project that offers the client both the best value solution for money and its environmental impact. The earlier that whole life costing is considered, the greater will be the potential benefits.

BS 8544 : 2013 defines the whole life cost (WLC) as 'all significant and relevant initial and future costs and benefits of an asset, throughout its life cycle, while fulfilling the performance requirements' (BSI 2013).

Figure 6.1 details the initial and future costs to be considered and its relationship with life cycle costing. Whilst life cycle costs are associated with initial and future costs of the building, WLC are much broader, including additional client costs and revenue streams associated with the building asset. These will be discussed in more detail later in the chapter. The quantity surveyor role is therefore to ascertain the life cycle costs that can then be used by the client's team to support whole life cost evaluation. As stated, these can be significant; research by Gleeds states that over a 20-year property life cycle for every unit of capital spent on construction, three units will be spent on maintenance and 30 on operational costs (Gleeds 2022).

Each building has a useful life. Planning costs in terms of that usage is as important as establishing costs at the inception. Companies recognise that their operation is not just measured in financial metrics but in a wider sustainability context resulting from an increased awareness by clients of the benefits of whole life costings.

There are a variety of factors that influence the whole life costs of building. These include the:

- Identification of costs incurred during a building or engineering structure's life and the inter-relationship with its use and maintenance

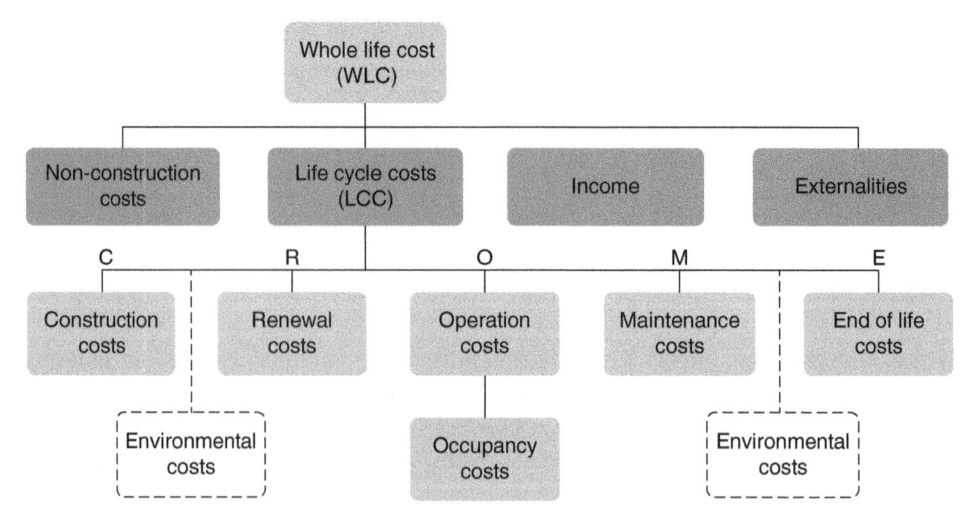

Fig. 6.1 Cost categories of WLC and LCC (*Source:* RICS 2021/Royal Institution of Chartered Surveyors RICS).

- Appreciation of available forecasting techniques and their use as a planning tool
- Consideration of the effects of time on the accuracy of cost advice with particular regard to technological advancement and government policy.
- Understanding of the importance of the application of risk analysis techniques in the validation of cost advice
- Importance of long life, loose fit and low energy in managing the flexibility of designs during the life of buildings.

Government policy

Construction 2025 set the target of a 33% reduction in both the initial cost of construction and the whole life of assets (HM Government 2013). However pre-dating this the UK government had taken a decision to make all construction procurement choices on the basis of whole life costs with the focus of PFI and PPP on long-term risk management and long-term maintenance and operation making comprehensive whole life costing a necessity. Government procurement outside of PFI and PPP also has an emphasis on whole life costs at all levels, including local authority and housing association procurement.

Government procurement practices have had a huge influence on the construction industry as a whole, with major client organisations adopting similar policies. This has been reinforced by the move towards partnering that engenders longer-term strategies and risk management processes. These too are naturally supported by whole life costing practices. Within the *Construction Playbook*, part of the Governments Construction Sector Deal, 2019, the delivery model assessment for public works includes an assessment of whole life costs requiring benchmarking and development of a Should Cost Model (SCM). A SCM provides a forecast of what the project 'should' cost over its whole life including both the build and the expected design life (Crown 2020). This model should also be linked to the whole life carbon assessment, supporting the government's commitment for better, faster and greener building solutions.

Whole life costing and environmental impact

Whole life value (WLV) is a term that has been developed by the Building Research Establishment (BRE) to describe the various aspects of sustainability in the design, construction, use, demolition and, where appropriate, the reuse of a built asset. The WLV of an asset represents the optimum balance of stakeholder's aspirations, needs and requirements, and the costs over the life of the asset (Bourke et al. 2005). It involves achieving compromise and synergy between the following three sets of values:

- *Economic value.* This is focused on economically sound sustainability: growth-oriented and seeks to safeguard the opportunities of future generations.
- *Social value.* This is concerned with aspects of a societal nature and covers a wider scope of social, cultural, ethical and juridical impacts.
- *Environmental value.* This focuses on environmental aspects of development such as pollution, waste and CO_2 emissions. These issues involve the initial manufacture of

construction materials, the construction of the project, its use and eventual replacement. In this context, value is maximised when environmental pressures are minimised to the level of the carrying capacity of ecological systems while using natural resources effectively and safeguarding natural capital and its productivity.

Whole life costing is an integral tool in assessing the economic aspect of whole life value. The importance of the Life Cycle Cost component of WLC (see Fig. 6.1) is further recognised within the *BREEAM UK New Construction* objectives that challenge the market to provide innovative, cost effective solutions that minimise the environmental impact of buildings (BRE 2018). Indeed under MAN 02, 4 credits of the 21 credits available for the Management section can be achieved for evidence of life cycle costing and service life planning and an enhanced understanding of capital costs. The Government Soft Landings Golden Thread also sets environmental, social and economic targets and measures. This focuses the design of buildings to ensure all assets perform to the optimum level; in doing so key project success measures are set, one of which will include operational costs.

Whole life costing applications

The following are some of the advantages of using whole life costing:

- It gives an emphasis to a whole or total cost approach undertaken during the acquisition of a capital cost project or asset, rather than merely concentrating on the initial capital costs alone
- It considers the initial capital costs, repairs, running and replacement costs and expresses these in comparable terms. It can allow for different solutions of the different variables involved and sets up hypotheses to test the confidence of the results achieved
- It is an asset management tool that allows the operating costs of premises to be evaluated at frequent intervals
- It allows, for example, changes in working practices, such as hours of operation, introduction of new plant or machinery, use of maintenance analysis, to be properly evaluated as tools of facilities management.

Whilst there has been an emphasis upon the use of whole life costing during the pre-contract period, its use can be extended throughout every phase of a building's life, as follows.

At inception. Whole life costing can be used as a component part of an investment appraisal. The technique is used to balance the associated costs of construction and maintenance with income or rental values.

During the design stage (Stage 2). It is used to evaluate the different design options in order to assess their economic impact throughout the project's life. It is frequently used alongside value management and other similar techniques. The technique focuses on those areas where economic benefits can be achieved.

At procurement. Sustainability requirements and evaluation criteria should be included within the tender bid in relation to whole life costing practices and contractors required to demonstrate their experience and competence to fulfil sustainability practices throughout their supply chain.

During the construction stage. During this phase there are many different areas that can be considered for its application. It can be applied to the contractor's construction methods, which can have an influence upon the timing of cash flows and hence the time value of such payments. The contractor can apply the principles to the purchase, lease or hire of the construction plant and equipment. Construction managers and contractors' surveyors are able to offer an input to the scrutiny of the design, if involved sufficiently early in the project's life to be able to identify whole life cost implications of the design, manufacture and construction process.

During the project's use and occupation. It is a physical asset management tool. Costs-in-use do not remain uniform or static throughout a project's life, and therefore need to be reviewed at frequent intervals to assess their implications. Taxation rates and allowances will change and have an influence upon the facilities management policies being used.

In energy conservation. Whole life costing is an appropriate technique to be used in the energy audit of premises. The energy audit requires a detailed study and investigation of the premises, recording of outputs and other data, tariff documentation and appropriate monitoring systems.

Whole life costs

Figure 6.1 identified the four cost categories for WLCs, as Non Construction Costs, Life Cycle Costs, Income and Externalities. Table 3.1 of *PD156865 The Standardised Method of Life Cycle Costing for Construction Procurement (BSI* 2008*)* outlines these as:

Nonconstruction Costs: These are costs outside those associated with the construction costs of the building.

- Acquisition of land and enabling works
- Finance costs, such as interest charges
- Rental costs
- Client support costs, such as those associated with the client's strategic estate management activities
- Taxes on non -construction costs

Life Cycle Costs: These are the costs of the asset throughout its life cycle and will be the focus of the quantity surveyor's WLC expertise; the key aspects being defined by the RICS 'CROME' acronym:

- C – Construction costs
- R – Renewal costs
- O – Operation and occupancy costs
- M – Maintenance costs, and
- E – End-of-life costs (RICS 2021).

Income: such as from

- The sale of the constructed asset and land at the end of life.
- Third-party income generated during the occupancy.

Externalities: the inclusion of this cost category recognises that the costs and benefits of any project are wider than just financial consideration. These impacts maybe on the environment, culture, health, social care, justice and security, summarised as the social or public value (HM Treasury 2022).

NRM3

The *NRM3: Order of cost estimating and cost planning for building maintenance work* provides a structured approach to measure the renewal and maintenance aspects of the built asset and also the costs associated with any environmental improvements that occur during the life of the asset. Primarily it focuses on the R and M of CROME (as detailed above). Any adaptations, retrofitting or conservation work carried out subsequent to the initial build will also be included (C). NRM3 provides a common data structure linking the construction elements to the relevant renewal and maintenance items. The element cost structure aligns with both published asset expectancy life data and industry standard preventative asset maintenance task schedules. NRM3's codified framework is the same as NRM1, allowing cost advice for renewal costs and maintenance work (NRM3) to be aligned with cost advice for the capital costs (NRM1).

Together, NRM1 and NRM3 provide the basis for life cycle cost management, enabling more effective and accurate cost advice to be given to clients and other project team members, as well as facilitating better control of the combined construction renewal and maintenance costs. For example, it provides information for:

- Input into life cycle cost plans in a structured way, facilitating meaningful comparison and more robust data analysis
- Advising clients on appropriate cashflow requirements for annual budgeting and informing the forecasting of the forward maintenance and renewal plans
- Informing the implementation of the maintenance strategy and procurement (Green 2012).

Part 3 of NRM3 provides guidance on measurement rules for Order of Cost Estimates for the building maintenance work, whilst Part 6 provides tabulated rules for elemental cost plans. As detailed in Fig. 5.1 (Chapter 5) at Stage 6 'Handover' and Stage 7 'Use the Formal Cost Plan 4' will be prepared by the quantity surveyor based on the project's as-built information. NRM3 data structure is aligned with COBieII data classification, ensuring interoperability for BIM modelling.

Life cycle costing

Life cycle costing (LCC) is 'the methodology for the systematic economic evaluation of life cycle costs over a period of analysis' (BSI 2013); in the UK, LCC is carried out in accordance with *The Standardised Method of Life Cycle Costing for Construction Procurement* (BSI 2008).

In addition to initial capital construction costs, other costs accrue as changes occur throughout the project's life (Fig. 6.2).

Terminology description	
Maintenance	Regular ongoing work to ensure that the fabric and engineering services are retained to minimum standards. It is frequently of a minor nature.
Repairs	Associated with the rectification of building components that have failed or become damaged through use and misuse.
Renewal	The upgrading of a building to meet modem standards.
Adaptation	Frequently includes building work associated with conversions, such as a change in function and typically includes alterations and extensions.
Renovation	Repairing and rebuilding.
Retrofitting	The replacement of building components with new components that were not available at the time of the original construction.
Conservation	Building works carried out to retain the original features or restore a building to its original concept.

Fig. 6.2 Building renewal.

In addition to the items shown in Fig. 6.2, there are many other costs-in-use that must also be considered in connection with the life cycle cost of a project. The UK LCC data structure defines these as:

- *Maintenance costs*: both preventative and corrective maintenance, including services maintenance and redecoration (see Chapter 17)
- *Operating costs*: such as fuel for heating, lighting and communications, business rates and insurances, facilities management and cleaning
- *Occupancy costs*: such as reception and IT services.
- *End of life costs*: such as dilapidation inspections and demolition.

These costs-in-use are significant when considering LCC. For example, a study of a primary school (g.i.f.a. 1500 m^2) shows that the renewal, maintenance and operational costs contribute to 46% of the LCC of the asset (RICS 2020). Cleaning costs are also a major cost in the running of a building and will be influenced by factors such as the building's use, hours of operation, frequency and level of cleaning required and the time allowed for the cleaning staff. The cleaning cost as a percentage of maintenance costs (which includes renewal and maintenance of building fabric and services and redecoration) is estimated at 117% for hospitals and 94% for hotels compared to 70% for offices (BCIS 2021).

These LCC costs can be associated with either:

One off activity costs. These are costs associated with end-of-life replacement of items, decommissioning costs. They may also include residual value at demolition.

Recurring activity costs. These are annual or regular costs such as energy, staff maintenance contracts and rent payments.

The BCIS *Building Running Costs Online* is an estimating expenditure tool that includes component costs, maintenance costs and wage rates to support the quantity surveyor in providing life cycle cost advice.

Levels of life cycle studies

LCC, like WLCs, can be undertaken at a number of levels of study:

- Whole building level
- Multiple element level (e.g., all windows and doors)
- Element level (e.g., a roof or external walls)
- System level (e.g., Heating system)
- Component or sub-component level (e.g. a boiler). (BSI 2008)

The level of study will be influenced by the level of design information available; an elemental LCC plan will be used during the concept stage of design (RIBA Stage 2) to influence design decisions. Whilst at the technical stage (RIBA Stage 4), component level studies are carried out as part of an option appraisal. Indeed, BREEAM awards credit for LCC plans at each of these stages to 'influence building and system design and specification to maximise life cycle costing and maximise critical value' (BRE 2018). Of critical value are those elements that have the greatest impact on minimising the overall costs during the life of the building. Such elements will be services and wall and floor finishes, where there will be a greater LCC relative to the initial capital costs.

Elements such as substructure, frame, upper floors and stairs are normally designed to last the life of the asset with minimum or no replacement, so LCC plans at component level are generally not carried out. The maintenance descriptor within NRM3 Part 6 tabulated rules of measurement is a useful indicator of sub-element items where renewal and maintenance works are not applicable.

Periods of analysis

The client should determine the period of analysis for the LCC as it will be governed by the reasons for development and how the built asset fits into the client's business portfolio and strategic objectives. It is now generally accepted that the life time horizon should be more related to current use expectations associated with the buildings or structure's own cycle or related to the cyclical effect of population movements associated with the project. Different buildings have different life expectancies, for example a fast food outlet building will have a shorter renewal time than that of a school. Clients may specify a 20-, 30-, 50- or 60-year period. BREEAM recommends a default design life of 60 years when the life expectation has not been formally established. For the reasons to be discussed in the next section, the RICS guidance states that analysis periods more than 30 years should be treated with caution (RICS 2016).

Main factors to consider in WLC and LCC

Building life

Over time buildings decay and become obsolete and require maintenance, repair, adaptation and modernisation. There also lies a varied pattern of existence, where buildings are subject to periods of occupancy, vacancy, modification and extension. The useful life of any

Condition	Definition	Examples
Physical Deterioration	Deterioration beyond normal repair	Structural decay of building components
Forms of Obsolescence:		
Technological	Advances in sciences and engineering result in outdated building	Office buildings, in 2000's unable to accommodate modern communications technology
Functional	Original designed use of the building is no longer required	Cotton mills converted into shopping units, chapels converted into warehouses
Economic	Cost objectives can be achieved in a better way	Site value is worth more than the value of the activities current on the site
Social	Changes in the needs of society result in the lack of use for certain types of buildings	Multi-storey flats unsuitable for family accommodation in Britain
Legal	Legislation resulting in the prohibitive use of buildings unless major changes are introduced	Asbestos materials, Fire Regulations
Aesthetic	Style of architecture is no longer fashionable	Office building designs of the 1960s
Environmental	Noncompliance with current environmental targets	Assets with high greenhouse gas emissions, waste and pollution products

Fig. 6.3 Building obsolescence (*Source:* Adapted from RICS 2016 and Pourebrahimi et al. 2020).

building is governed by a number of factors, such as the methods of construction envisaged at the initial design and the way that the building is cared for whilst in use, e.g. the amount of wear and tear and the levels of maintenance that are applied. Figure 6.3 identifies types of building obsolescence that influences the life span of a building.

Deterioration and obsolescence in buildings

A distinction needs to be made between obsolescence and the deterioration of buildings (Fig. 6.4). The physical deterioration of buildings is largely a function of time and use. Whilst it can be controlled to some extent by selecting the appropriate materials and components at the design stage and through correct maintenance whilst in use, deterioration is inevitable as an ageing process. Obsolescence is much more difficult to control since it is concerned with uncertain events such as the prediction of changes in fashion, technological development and innovation in the design and use of buildings. Deterioration eventually results in an absolute loss of use of a facility, whereas with buildings that become obsolete it is accepted that better facilities are available elsewhere. Whilst deterioration in

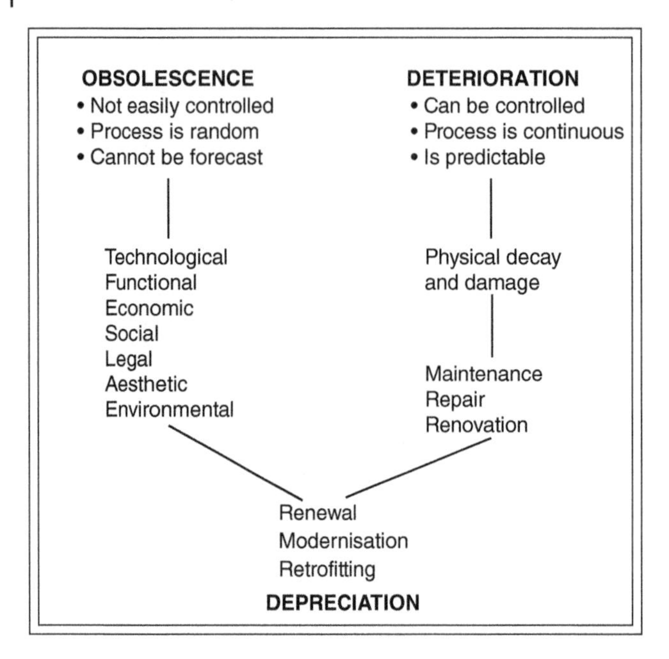

Fig. 6.4 Obsolescence, deterioration and depreciation (*Source:* Adapted from Flanagan et al. 1989).

buildings can be remedied at a price, obsolescence is much less easy to resolve. Obsolescence can be defined as value decline that is not caused directly by use or passage of time (Ashworth and Perera 2015).

Due to the large investment that is required in a building, demolition due to obsolescence is a last resort and will only take place where the building is either not capable of renewal or is in a wrong or decaying environment. In some cases, the site on which the building stands may be required for other purposes.

Long life, loose fit and low energy

Physical deterioration occurs more slowly than the various forms of functional and other types of obsolescence. The blame for a great deal of obsolescence is due to inflexible planning and designing buildings that were unsuitable for adaptation should their original function cease. However, most building clients commissioning buildings require bespoke design solutions to meet their individual needs. Some of these solutions, perhaps decades later when the property is no longer required, are difficult to adapt to changing circumstances. Obsolescence is also to some extent coupled with population movement, which may make even the most adaptable structure obsolete if it is in an area of declining desirability or usefulness. Over five decades ago, the architect Alex Gordon (1977) coined the phrase 'Long life, loose fit and low energy', describing these as the '3 Ls principle'. He recognised that obsolescence rather than physical decay was the major reason for buildings falling into disuse. He emphasised the importance of designing flexible solutions to clients' problems so that these could easily be adapted to suit the changing needs of work, leisure and living spaces. Since 1977, changes throughout society have been accelerating, emphasising the responsibility to ensure that the scarce resources are used with as much longevity in mind as possible.

Component life

The life span of the individual materials and components has a contributory effect upon the life span of the building. The choice of component must also recognise the difference between those parts of the building with long, stable life and those parts where constant change, wide variation in aesthetic character and short life are the principal characteristics. There seems to be little merit in including building components with long life in situations where rapid change and modernisation are to be expected. Life cycle costing is concerned not only with how long a component will last but for how long a component will be retained. The particular circumstances of each case will have a significant influence upon component longevity. These will include:

- Correct choice of component specification
- Use of appropriate design details
- Installation in accordance with the manufacturer's directions, relevant Codes of Practice and British Standards
- Compliance with the conditions of any relevant third-party assurance certificates
- Appropriate use by owners, users and third parties
- Frequency and standards of maintenance.

The management policies used by owners or occupiers are perhaps the most crucial factors in determining the length of component life.

Table 9 of RICS *Whole Life Carbon Assessment for the Built Environment* provides information on indicative component life spans. For example, roof coverings – 30 years; lighting fittings – 20 years and painting and walls and ceilings – 10 years (RICS 2017). A module within the BCIS Building Running Cost Online service provides the quantity surveyor with access to survey data to support life cycle costs, which includes statistical information for the life expectancy of 300 building components.

Discount rate

The selection of an appropriate discount rate to be used in whole life and life cycle costing calculations depends upon a wide range of different factors. The discount rate to be chosen will depend to some extent upon the financial status of the client. For example, public sector clients are generally able to obtain preferential rates of interest. Consideration should also be given to whether the finance is borrowed or obtained from retained profits. The choice of a discount rate is frequently inferred to mean the opportunity cost of capital. The choice of a discount rate is one of the more critical variables in the analysis.

Choosing an appropriate discount rate is a difficult task. The Government's *Green Book* recommends the 'social time preference' rate as the discount rate. This is a declining long term discount rate that incorporates the effects of general inflation. It recommends a 3.5% discount rate for years 0–30, 3.0% between years 31 and 75 and if the building life expectancy exceeds this 2.50% (HM Treasury 2022).

Taxation

Since whole life costing is an attempt to aggregate the initial and recurring costs of building, taxation must also be considered. The effects of taxation could result in building projects becoming more economically viable. The application of taxation to initial and future costs is not uniform and is subject to frequent change. This must be considered when calculating the whole life costs of buildings.

Calculations

One of the apparent difficulties of using both whole life costing in practice is the mathematics associated with the evaluation. An understanding of the principles involved in discounting the value of future receipts and payments is an essential feature of such an evaluation. Although the arithmetic associated with discounting may appear complicated, the concept is simple. This is that the capital in the hand today is worth more than the capital at some time in the future. Even ignoring inflation, it would be more beneficial to choose to receive an amount of money today than the same amount next year.

If the current rate of interest is 5% per annum, then £100 invested today will yield £105 in twelve months' time. Conversely, £100 to be received next year has what is known as a present value (PV) of:

$$\pounds 100 \times \frac{100}{105} = \pounds 95.24 \text{ today}$$

This example is the principle associated with discounting or discounted cash flow calculations. Whilst the amounts can be calculated using valuation formulae or equations, it is still common to use standard valuation tables to allow comparisons between money spent or received at different times. It is necessary to convert these sums to a common timescale and valuation tables are a means of making this conversion.

Amounts are generally expressed in the following ways:

- Actual values, e.g., capital costs
- Present values of a lump sum to be paid at some time in the future. The amount is calculated using the 'Present value of £1 table'. (See replacement cost below)
- Present value of a regular annual amount. The amount is calculated using the 'Present value of £1 per annum table'. (See cleaning costs below).

Each of the above can also be represented as the 'annual equivalent' amount by converting present values to annual sums by using the 'Present value of £1 per annum table'. The technique of discounting is best explained by the following example.

A restaurant proprietor is seeking advice on floor finishes for a new restaurant that is being constructed. The data on the floor finishes are shown below. The situation implies a time horizon of 30 years (or multiple of 30 years for comparison purposes). A discount rate of 5% has been selected (see Fig. 6.5).

Floor finish	Initial cost	Annual maintenance	Annual cleaning	Expected life
A	£ 20,000	£ 750	£ 1000	15 years
B	£ 30,000	£ 500	£ 500	30 years

		Discounted amount	Notes
Floor finish A			
Initial cost		20 000	
Maintenance	750 × 15.3725	11 529	Use PV of £1 per annum table
Cleaning	1000 × 15.3725	15 373	Use PV of £1 per annum table
Replacement (year 15)			
	20 000 × 0.481017	9 620	Use PV of £1 table
Net Present Value (NPV)		56 522	
Floor finish B			
Initial cost		30 000	
Maintenance	500 × 15.3725	7 686	Use PV of £1 per annum table
Cleaning	500 × 15.3725	7 686	Use PV of £1 per annum table
Replacement life = 30 years		0	
Net Present Value (NPV)		45 372	

Fig. 6.5 Life cycle costs of floor finishes.

The calculation in Fig. 6.5 suggests that floor finish B is the preferred choice based upon the data provided. However, this is not done on the comparison of the cost calculation alone. Other subjective factors must also be considered. In this example these might include the following:

- As well as being more durable, expensive construction is generally more pleasant in appearance
- Future replacement or repairs may be inconvenient
- Replacement or repairs may be difficult and therefore expensive (costs and revenue)
- The saving of money on a specific item could involve costs out of all proportion to the possible savings
- Obsolescence may not be a factor to be considered
- Prestige associated with more expensive construction
- A client may be more concerned with higher initial costs than the reduction of future costs in use.

Life cycle worked examples

The calculations to support life cost calculations as discussed above can be modelled effectively using a spreadsheet. The BSRIA guide -*Life Cycle Costing*- includes a worked example of a comparison of life cycle costs for three alternative solutions for upgrading heating and ventilation systems for an office (Churcher and Tse 2016). The worked solution provides guidance on specific spreadsheet functions such as net present values and look up tables. Worked examples of LCC can also be found in Appendix B of the RICS *Life Cycle Costing* and Annex D *of The Standardised Method of Life Cycle Costing for Construction Procurement*.

Sensitivity analysis

The importance of attempting to account for future costs-in-use in an economic appraisal of any construction project has already been established in theory. The problem of its application in practice is twofold: the known or predictable and the unknown or uncertain. The application of statistical analysis to life expectancies, discount factors and other data to be used in the calculations should not be underestimated. This is the assessment of risk, where sensitivity analysis can be used to interpret the results and provide confidence limits to such assessments. The sensitivity analysis will test the impact of changes of input variables, e.g. the timing of equipment replacement or an increase in the annual maintenance costs to inform decision-making in the viable choice of alternative design solutions and component choice.

Reporting to the client

In accordance with Rule 3.6 of the *Rules of Conduct*, it is mandatory that quantity surveyors in respect to LCC communicate to the client the material information on which the WLC or LCC calculation and sensitivity analysis advice is based. The RICS *Life Cycle Costing* Professional Guidance states that in all situations the report should include:

- The source of information and data
- All assumptions on which decisions are based
- The methods used for the calculations.

It is best practice to include graphical representation within the report.

BIM and LCC

Researchers have identified the benefits of BIM in supporting life cycle costing. A BIM environment has the potential of including life cycle data to be added to object attributes, enabling the life cycle assessment process to be carried out collaboratively and dynamically throughout the design process (Pittard and Sell 2016). Such life cycle data would include expected service life information, installation, replacement, and service cost information and possibly recyclability of component information. Lu et al. (2018) also identified the possibility of embedding a LCC calculation model structure within existing BIM technology. However, there are still barriers relating to data management that need to be overcome to fully integrate life cycle calculations and BIM (Saridaki et al. 2019). Therefore, in practice, it is most likely that quantity surveyors will import cost data and analysis LCC within a spreadsheet environment.

References

Ashworth, A and Perera, S. *Cost Studies of Buildings*. 6th Edition. Routledge. 2015.
BCIS. *Cleaning a Major Contributor to FM Costs*. BCIS News. 2021. Available at https://service.bcis.co.uk/BCISOnline/News/DmsContent/3387?returnUrl=%2FBCISOnline%2FSearch%2FSearch%3FsearchTerm%3Dcleaning%2520a%2520major%2520contributor%2520to%2520FM%2520costs&returnText=Go%20back%20to%20search%20results (accessed 27/06/2022).
BRE. *BREEAM UK New Construction Non-domestic Building (England)*. BRE Global Ltd. 2018.
BSI. *PD156865 The Standardised Method of Life Cycle Costing for Construction Procurement*. British Standards Institution. 2008.
BSI. *8544:2013 Guide for Life Cycle Costing of Maintenance During the in Use Phase of Buildings*. British Standards Institution. 2013.
Bourke, K., Ramdas, V., Singh, S., Green, A., Crudgington, A. and Mooranah, D. *Achieving Whole Life Value in Infrastructure and Building*. BRE. 2005.
Crown. *The Construction Playbook*. Cabinet Office. 2020.
Churcher, D and Tse, P. *A BSRIA Guide: Life Cycle Costing*. BSRIA Ltd. 2016.
Flanagan, R., Norman, G., Meadows, J. and Robinson, G. *Life Cycle Costing: Theory and Practice*. Blackwell Science. 1989.
Gleeds. *Operations and Facilities Management*. 2022. Available at https://gb.gleeds.com/services/asset-management/operations-management/ (accessed 27/06/2022).
Gordon, A. *The Three 'Ls' Principle: Long Life, Loose Fit, Low Energy. Chartered Surveyor Building and Quantity Surveying*, Quarterly. 1977.
Green, A. A measure of success. *Construction Journal*. RICS. June–July 2012.
HM Government. *Construction 2025: Industry Strategy: Government and Industry in Partnership*. HM Government. 2013.
HM Treasury. *The Green Book: Central Government Guidance on Appraisal and Evaluation*. Crown. 2022. Available at https://assets.publishing.service.gov.uk/government/uploads/system/uploads/attachment_data/file/1063330/Green_Book_2022.pdf (accessed 28/05/2022).
Lu, W., Lai, CC. and Tse, T. *BIM and Big Data for Construction Cost Management*. London and New York. Routledge. 2018.
Pittard, S and Sell, P. *BIM and Quantity Surveying*. London and New York. Routledge. 2016.
Pourebrahimi, M., Eghbali, S.R. and Pereira Roders, A. Identifying building obsolescence: towards increasing buildings' service life. *International Journal of Building Pathology and Adaptation*, 38(5), 635–652. 2020. https://doi.org/10.1108/IJBPA-08-2019-0068.
RICS. *Life Cycle Costing*. RICS. 2016.
RICS. *Whole Life Carbon Assessment for the Built Environment*. 1st Edition. RICS. 2017.
RICS. *ICMS 2 and BCIS Life Cycle Cost Data*. 2020. Available at https://www.rics.org/uk/news-insight/latest-news/news-opinion/icms-2-and-bcis-life-cycle-cost-data/ (accessed 28/05/2022).
RICS. *NRM3- Order of Cost Estimating and Cost Planning for Building Maintenance Works*. 2nd Edition. RICS Books. 2021.
Saridaki, M., Psarra, M. and Haugbølle, K. Implementing life-cycle costing: data integration between design models and cost calculations. *Journal of Information Technology in Construction (ITcon)*, 24. 14–32. 2019. http://www.itcon.org/2019/2.

7

Value Management

KEY CONCEPTS

- 'Value'
- Value management
- Value engineering
- Value management decision points
- The eight phases of the job plan
- Functional analysis

LEARNING OUTCOMES

After reading this chapter, you should be able to:

- Understand 'value' and differentiate between value management and value engineering.
- Understand how value management is performed and how value is enhanced by its application.
- Explain the role the quantity surveyor has at each stage of the job plan.
- Understand the application of functional analysis and functional analysis diagramming techniques.

COMPETENCIES

Competencies covered in this chapter:

- Design economics and cost planning
- Project feasibility analysis

Introduction

The opportunity that value management (VM) affords the practising surveyor to improve value to the client has now been well demonstrated in the UK construction sector. Value management has been recognised as an important component that is essential to the success of projects in providing the foundation for improving value for money in

Willis's Practice and Procedure for the Quantity Surveyor, Fourteenth Edition.
Allan Ashworth and Catherine Higgs.

construction. Although this may be regarded as a specialist area, many quantity surveying practices provide it as one of the growing areas of professional service. In this chapter, value methodology is outlined and supported with examples that will aid the reader in understanding how the service may be performed and how value can be enhanced by its application. The execution of value management requires an understanding of the processes involved and an understanding of how to determine an appropriate VM approach. Whilst this knowledge is essential to the practice of value management, the principles and philosophy provide surveyors with additional tools and techniques, and possibly new ways of thinking, all of which can be used in other areas of professional activity. Knowledge of value methodology is important to the surveyor, either from the perspective of actual service provision, or from an appreciation of its benefits and application when advising clients.

Background

Value management developed from the demands of the manufacturing industry in the USA during World War II. Lawrence Miles (1972), an electrical engineer with the General Electric Company, adopted a functional approach to the purchasing requirements of his company and developed the value analysis concept. This involved the function analysis of a component in terms of what it did and invited a search for an alternative solution to the provision of that functional requirement at a lower cost. The use of the concept further developed during the 1940s and 1950s and grew within the USA, becoming a procedure that could be used during the design or engineering stages. The term value engineering was initiated in 1954 by the US military, an organisation that has a long history of involvement with value techniques. Value engineering spread to the UK manufacturing industry during the 1960s and, at around the same time, was introduced to the US construction industry.

The value management concept was first used within the UK construction industry in the 1980s. Although manufacturers around the world today use value techniques, their use within the construction sector is largely restricted to the USA, UK and Australia. However, interest from other parts of the world is growing, for example Far East countries such as Hong Kong and Malaysia. The status of value management is now recognised by legislation in the USA and New South Wales, Australia. In the UK, one of the challenges facing value management implementation is government and the National Audit office do not currently mandate the use of BS EN 12973:2020 Value Management.

Terminology

The terminology used in value management may be confusing to those being introduced to the subject for the first time. For example, the terms value management, value engineering and value analysis are frequently used synonymously, particularly value management and value engineering, to mean the entire concept. Although the semantics are considered unimportant, it is vital that the meaning of each term is understood in the context in which it is read or discussed. Therefore, the following definitions are provided as those used throughout this text:

Value management. This is the overarching term used to describe the total philosophy and extent of the practice and techniques.

BS EN 12973:2020 defines value management as 'an underlying concept applied within existing management systems and approaches based on **value** and function-oriented thinking, behaviour and methods, particularly dedicated to motivating people, developing skills and promoting synergies and innovation, with the aim of maximizing the overall performance of the organization'. More simply, value management can be defined as 'a strategy for identifying the project that provides the best value for money through the best use of the limited resources that are available' (Ashworth and Perera 2015).

$$\text{Value is} \quad \frac{\text{Satisfaction of Needs}}{\text{Consumption of Resources}} \quad \text{(BSI 2020)}$$

Each project will have its own unique value profile (CIH 2020). This can be established using *The Value Toolkit*, which will be discussed later in this chapter.

Within construction, Higham et al. propose the following definitions as most aligned to current practice:

Value management is the process of making explicit the functional benefits of projects and appraising those benefits against value systems defined by clients, and

Value engineering as an organised approach to providing the necessary functions of buildings, elements or components at lowest costs whilst maintaining the specified level of quality. (Higham et al. 2017)

Benefits of value management

The most apparent benefit coming from VM application is added value through an improvement in decision-making and, importantly, because VM clients can see what is actually happening. The OGC's publication on *Value management in construction: Case studies* (Office of Government Commerce 2009) identifies the principle benefits of value management for seven projects as:

- Improved definition of project and articulation of value
- Clearer brief and improved decision making
- Enhanced value and benefits for end users
- Reduced costs, improved affordability and value for money
- Improved productivity, efficiency, collaboration and trust reduced waste and deficit.
- Earlier management involvement
- Benefits realised where previous methods failed.

The Institute of Value Management (IVM) states the use of best practice VM on eight case studies with an aggregate construction budget of £2.9 bn achieved total cost savings of £546 m (19%) in addition to nonfinancial benefits, such as social and environmental (IVM 2022a).

Value management and the RIBA plan of work

Central to the value management process is the structured workshop, or series of structured workshops, at which group decisions are made. As the definitions above suggest, value management is used early in the project life cycle to ascertain the client's strategic ambitions for the project, for example the functional, performance, financial or environmental requirements. As the design and construction process of the asset becomes more defined, value engineering (VE) will be used to test that these requirements are still being met. With reference to Fig. 7.1, value engineering will be a continuous process through VM2 to VM5. Depending on the complexity of the project, VM0 and VM1 maybe combined at stage 1, and VM3 and VM4 at stage 4. VM should not be viewed as stand-alone processes but rather as part of a larger project management process. Their consideration should be included in the project's periodic assessment, where the question of value can be evaluated with other aspects of the project's development and progress (Oke et al. 2022).

Figure 7.2 illustrates the likely opportunity to revise or make changes to the design (Ashworth and Perera 2015). In terms of value management, the potential to maximise cost savings is therefore early on in the project life cycle.

An important aspect of the value management process is the job plan. The structured job plan approach was developed by Lawrence Miles (1972) and, whilst academics and practitioners have refined and contextualised a variety of differing methods, it is normally adhered to in all approaches to value management in some form. Whilst this suggests a rigid approach, it should be regarded as an outline; situations and projects will differ and make their own demands on workshop approaches and activity. For example, a workshop is perhaps more efficient when limited to the information, creative and evaluation phases (described below) where time is restricted or resources are not available, with provision for development and presentation in a follow-up meeting. Likewise, the

Stage	RIBA Plan of Work 2020	Value management Key Decision Points
0	Strategic definition	Value Management 0 (VM0) *Defining the business need*
1	Preparation and brief	Value Management 1 (VM1) *Understanding, challenging and validating the business need*
2	Concept design	Value Management 2 (VM2) *Deciding the best business solution*
3	Spatial coordination	Value Management 3 (VM3) *Re-validating the project objectives and design development*
4	Technical design	Value Management 4 (VM4) *Signing off the design*
5	Manufacturing and construction	Value Management 5 (VM5) *Keeping the project on track*

Fig. 7.1 Value management decision points (*Source:* Adapted from Owen 2021).

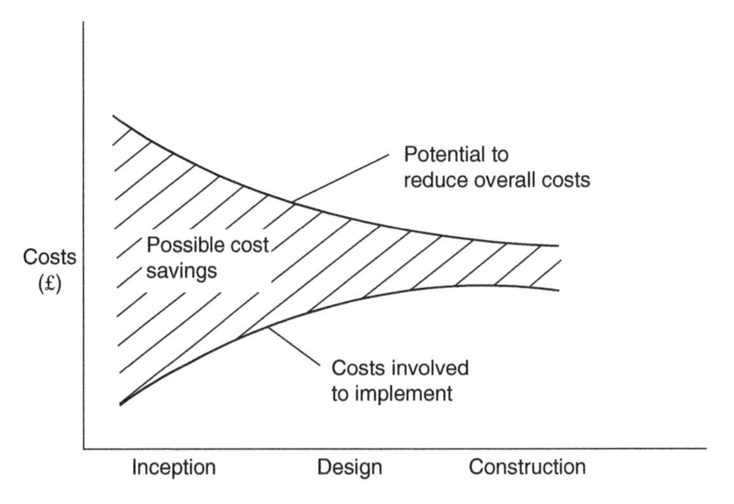

Fig. 7.2 Opportunity to change a design (*Source:* Ashworth and Perera 2015/Taylor & Francis).

job plan structure indicates a logical and sequential path, which in practice may vary and necessitate iterative action. The eight stages of the job plan (SAVE 2022) discussed later in the chapter are:

Pre-workshop phase

Workshop phase consisting of:

- Information phase
- Function analysis phase
- Creative phase
- Evaluation phase
- Development phase
- Presentation phase

Post-workshop phase

Key aspects of value management

Although value management does not provide a strict discipline or science to which every practitioner must rigidly adhere, in addition to the use of a structured job plan, it is considered to have some essential components, as follows.

- *The involvement of a multidisciplinary team.* Value management is a *multidiscipline* exercise in that it ideally requires the participation of consultants from all relevant design disciplines and client representatives who share a common interest (and thus are *stakeholders*) in the success of a project. To be effective, and it is regarded as critical to the success of value management, the team should have an appropriate mix of experience, knowledge and skills and, dependent upon workshop objectives, a range of stakeholder perspectives.

- *Maintenance of the basic project function.* A comment often made by quantity surveying practitioners is that they carry out value management as part of their traditional duties. This belief is held due to the contribution the quantity surveyor frequently makes relating to the establishment of cost reductions. This reveals a fundamental misunderstanding of value management. The task of reducing costs is usually carried out by the quantity surveyor, irrespective of function. Value management attends to the elimination of unnecessary costs, costs that relate to elements of a design that provide no function. The difference between cutting costs and cutting unnecessary costs is fundamental to the value management approach (Ashworth and Hogg 2000).
- *The management of a competent facilitator.* The facilitator is central to the success of the value management process. The role of the facilitator includes advising upon the selection of the VM team, coordinating pre-workshop activities, deciding upon the most appropriate timing and duration of workshops, workshop management and preparing reports. Workshop management can be a difficult challenge requiring a variety of skills, including:
  - The ability to determine and adhere to an appropriate agenda
  - Identifying and gainfully using the characteristics of team members
  - Promoting the positive interaction of participants
  - Motivating and directing workshop activity
  - Overseeing the function analysis process
  - Encouraging an atmosphere conducive to creativity whilst at the same time maintaining a disciplined structure.

The job plan

When examining the practice of value management, there is a tendency to focus upon the activities of the workshop and pay too little attention to the preparatory and completion stages. Since the essence of the value management philosophy is held in the workshop activities, this may be understandable. However, it is important to stress the importance of both the pre- and post-workshop activities of a value management study. For example, at the pre-workshop stage there is a need to identify the nature of workshop events, and during the post-workshop phase, action is essential if the client is to benefit from the achievements of the study.

Pre-workshop stage

The pre-workshop stage is necessary to:

- Establish why the client wishes to undertake a value management study, and the expected outcomes. Learning about the problem before action is taken is an important step in the value management process. Although a client may have identified a problem at the onset, a preliminary investigation might reveal a different problem to that outlined. For example, the reduction in building costs may be seen as the solution to a budget shortfall. However, the real problem might lie in the need to enhance project revenue and thus

establish a higher project budget. Although this example is fundamental, the suggestion is also quite radical in terms of our traditional approach.
- Decide upon the nature of the workshop including duration, timing, core activities and location. Although the benefits of value management may be extended to any problem or decision-making situation, specific project needs and resources have unique demands. For example, the detail relating to a study at the concept stage for a new rail tunnel between Portsmouth and the Isle of Wight will clearly be different from that required in connection with the need to attain savings, one week into an inner city refurbishment project. Factors such as project complexity, size, public profile, available resources and programme could all have a significant bearing.
- Determine the composition of the value management team. As stated, stakeholders' participation is a fundamental requirement of value management since value needs to be improved in terms of their perspectives, not those of consultants or a single client agent. The term 'stakeholder' can be considered to incorporate any person with an interest in the proposed project. The composition of the team will vary with each project, not only in terms of personnel, but also the number and experience of participants. For example, in the above mentioned comparison between the Portsmouth/Isle of Wight Tunnel and inner city refurbishment project, the tunnel project would likely attract a greater number of stakeholders, many with little relevant experience. In such situations, briefing members of the value management team prior to the workshop with regard to the project, study objectives and the value management process will be an important factor. In any value management study, briefing the participants of the study objectives will be beneficial.
- Collect information and circulate to the selected workshop participants. This may necessitate substantial research and the support of client representatives and members of the design team. Throughout the value management process, the support of senior management is seen as crucial to a successful outcome.

An example of the type of pre-workshop activity that may occur is shown in the scenario below.

Scenario

A cinema organisation proposes to construct a new multiplex cinema in a major city location. They have been advised to undertake a value management study at feasibility stage as a response to a lower than anticipated project return. A facilitator has been appointed (This term is commonly used in lieu of value manager and has been used throughout this example).

Following appointment, the facilitator meets with the client and ascertains the following information:

- A site has been obtained and outline planning permission has been given
- The client's architect has been appointed to the proposed scheme and has prepared an outline sketch proposal. The quantity surveyors have prepared a feasibility report.

It is agreed that a first value management workshop should be held two weeks hence. Following this initial briefing, the facilitator also recommends a two-day event that will be held in the business suite of a hotel located close to the proposed development.

Study team

In cognisance of the information obtained from the client, the facilitator recommends a value management team comprising:

- Project consultants: architect, quantity surveyor, structural engineer, services engineer. Each of these consultants has been involved in other cinema developments for the client
- Client stakeholders: commercial director, sales and marketing director, regional operations director.

An important aspect of the team composition in this study is the presence of decision takers. Absence of the need to refer back or be concerned about the approval of a senior authority before giving advice or taking decisions will contribute to the successful outcome of the study.

Information gathering and briefing

The facilitator arranges a meeting with members of the design team and the client organisation to discuss details of the project. At this meeting, information required for the workshop is identified and key information is collected for distribution to all study participants. The design team is also briefed with regard to the study and its objectives. Their contribution to the workshop is outlined, namely brief presentations from each member at commencement of the study. The quantity surveyor is advised that cost advice will be required at key points during the workshop and is requested to review, in advance, the budget estimate and to provide an elemental cost comparison (see Chapter 5, Cost Control) with similar developments previously executed by the client. A spatial cost model showing a breakdown of costs for key areas in the cinema is also requested. The zones to be considered include entrance lobby, ticket sales, pre-entertainment areas, refreshment/merchandise areas, theatres, staff and management areas, ancillary space provision and general circulation space. It is agreed that an additional quantity surveying representative will also attend to assist with cost advice needed during the workshop.

In addition to briefing the participants as to the purpose and format of the workshop, the facilitator is able to obtain feedback with regard to particular views and perspectives of each stakeholder. This is important to the success of the study, allowing some forethought as to workshop management and identifying some items of focus.

Workshop phases

It is of some importance to note that the arrangements for the workshop that have been outlined can and should vary depending upon individual situations. In the USA, the VM workshop is five days; however, this format is rarely used in the UK (RICS 2017). It is more likely that in the UK workshops will be one day (Hunter and Kelly 2007). Spellacy et al. suggest that each phase within the workshop would last between 2 and 4 hours (Spellacy et al. 2021). Whilst it is possible that in practice many studies would benefit from more time, value can be gained from workshops of shorter duration. Clearly, the required scope and depth of the study will be major determinants.

The information phase

The workshop begins with an information phase in which particulars of the project or problem (dependant on the stage of the project) are presented to the value management team. If the value management study relates to a proposed building, this can include a contribution from the client's representatives, the architect, structural engineer, quantity surveyor and other members of the design team who will offer details of project background, aims (distinguishing between a client's needs and wants) and constraints (e.g., site, budget, time). Although the primary aim of the information phase is to provide all team members with sufficient detail to allow a good understanding of the project, it also serves as a team-building opportunity that, if well managed, is useful in preparing a good base for the remainder of the workshop.

Function analysis phase

A feature of this phase of the workshop is some form of function analysis that is frequently carried out via the production of a function logic diagram (see the function analysis section later in this chapter). This results in an enhanced project understanding and allows unnecessary costs to be identified in terms of cost worth and forms a focus of further study. An example of a FAST diagram relating to the cinema scenario outlined above is also considered in the function analysis section.

Creative phase

Once the VM team has a good insight into the project or problem, including an understanding of its functional needs, the participants are requested, in the creative phase, to engender alternative solutions and ideas. This part of the proceedings is usually performed with the aid of brainstorming and other creative thinking methods to stimulate members to generate ideas that will improve value. With adherence to the key brainstorming rules – that as many ideas are produced as possible and that participants are reserved in their judgement of the suggestions until the creative phase is complete – many suggestions will be generated for future evaluation.

The evaluation phase

A range of methods are used during the evaluation phase to judge the merits of the proposals made during the creative phase. How best to obtain the agreement of workshop participants as to the selection of ideas for further development is a matter that needs to be resolved. Since the work carried out in the development phase is likely to be very detailed and time consuming, only those ideas able to demonstrate good value improvement should be selected. The method used to evaluate the ideas generated in the creative phase will be situation dependent For example, it will be influenced by the stage in the project – VM0-VM1 (Figure 7.1), available time, workshop team, project complexity and may rely on a democratic procedure or the facilitator's ability to get open accord. One technique that is occasionally used is 'championing', which depends upon team members volunteering to 'champion' a particular idea (i.e., accept responsibility for its development). Therefore, ideas without champions are rejected and thus the best ideas retained. The outcome of the process, irrespective of the method used, is to carry forward the most beneficial ideas to the development stage.

VMA VALUE MANAGEMENT ASSOCIATES								
Project:		Sheet:					Date:	
		Impact			Opp.		Comment	
Nr	Description	T	£	Q		Action		
1	Redesign entrance/ticket dispensing area	2	5	3	5	D		
2	Reduce storey height	2	3	1	1	R		
3								
4								
5								
6								
7								
8								
9								
10								
11								
...								

Notes :
Impact (T = time, £ = cost, Q = quality): 1 = low; 5 = high
Opp. (i.e. Opportunity) – ease of implementation: 1 = impossible; 5 = easy
Aclioo: Accept, Develop, Reject

Fig. 7.3 Value proposal: evaluation form.

One approach that may be used to evaluate brainstormed suggestions is to provide qualitative estimates as to the likely impact in terms of time, cost and quality and other value criteria, and ease of implementation for each proposal. This procedure will allow the identification of those items that offer high value gain and are relatively easy to incorporate within the design. An example of a value proposal evaluation form is shown in Fig. 7.3.

In this example, the suggestion 'Redesign entrance/ticket dispensing area' is considered likely to have a very high impact upon cost and could be incorporated within the design easily. Alternatively, the suggestion 'Reduce storey height' is considered to have a much lower cost impact and will be impossible to achieve. Clearly, suggestion one will be worthy of development and suggestion two rejection.

The development phase

The evaluation of ideas generated in the creative phase has probably been based upon no more than an outline perception by this stage of the workshop. The development phase accommodates the further work that is necessary to establish whether an idea should become a firm proposal or, if the workshop is at briefing stage, to consider the incorporation of ideas within a revised brief. This detailed work is time consuming and will probably involve much technical input. It is nevertheless essential to execute this developmental work before the presentation of formal proposals. Because of the time involved during this development phase, it is often advantageous to complete the related work outside the confines of the workshop and present it at a subsequent meeting. The work performed during this phase will include the preparation of alternative designs and cost exercises to justify the merits and feasibility of the new proposals. The services of non-consultant stakeholders are therefore unlikely to be required at this stage.

Whilst details of whole life costing techniques are outlined elsewhere in this book, it is important to further emphasise their importance in the preparation of adequate value management proposals. In the life of a building, occupancy costs will be more significant than initial costs. Therefore, when considering alternative design proposals within the value management process, whole life costs must be considered to assist the decision-making process.

It is usual to submit proposals on pro-forma such as the one shown in Fig. 7.4. The form serves to illustrate the level and type of information that will be prepared in the development stage. In addition to the information contained on the form, additional detail in the form of drawings, calculations, etc. may also be relevant and necessary for the client to give full consideration to the design change proposed.

The presentation phase

The objective of the presentation phase is to present the team's proposals to the client representatives. The presentation of proposals, which will probably include adjustments to original design proposals, is something that may be very sensitive to consultants and possibly the client.

The presentation normally occurs at the end of the value management workshop process and is intended to communicate the proposals to the client representatives. Decisions relating to the proposals will probably be deferred until after workshop closure. Although the proposals may have been prepared in detail and with as much accuracy as possible, some review is likely to be necessary and further investigation and consultation may be required.

Post-workshop stage

The completion and monitoring of an implementation plan to ensure post-study action is taken as an essential component of successful value management. Although the benefits of value management include team building and 'buy in' to a project, project value improvement necessitates action beyond words. Despite this, in some situations, the post-workshop stage may fail to fulfil the findings and undertakings of the value management study. These will be included in a detailed report, but reports do not guarantee action.

The post-workshop phase is improved if action determined in the value management study is fully accepted by participants (rather than imposed) and is carefully monitored. Personnel need to be identified to follow up the action outlined in the report and a post-workshop meeting, at which the outcome of the action phase is reviewed, will promote success.

The activities included in the post-workshop stage include the following.

Report

Following completion of the value management study, it is necessary to prepare and submit a detailed report of the activities of the workshop. This will include a review of the value management process carried out, indicating key aspects such as project background, study objectives, value management team membership, information base, timing and duration, workshop activities, function logic diagrams and other supporting models and details of the workshop proposals.

<table>
<tr><td colspan="6">VMA VALUE MANAGEMENT ASSOCIATES</td></tr>
<tr><td>Project</td><td colspan="3">New City Cinema</td><td>Date: 11/07/23</td><td rowspan="2">Impact considerations</td></tr>
<tr><td>Proposal:</td><td colspan="3">Revise reception/entrance area</td><td>Ref: 21 A</td></tr>
<tr><td colspan="5">Detail:

The original entrance/reception area accommodates a ticket counter and lobby which is designed to handle all customers. With reference to the FAST diagram, the system to receive viewers/dispense tickets/inspect tickets is high cost in terms of both designed accommodation and operation. There is an increasing trend toward auto-dispensers and inspection systems which is changing the space requirements and staffing location and levels.

The new design proposal shows a revised entrance layout which accommodates IT displays/ticket dispensers at 12 customer stations. Anticipated reduction in staffing - at reception, 1 at entrance.</td><td>Time:
No revision to exterior. Interior redesign: project at concept - minimal time implications

Resources:
No significant resource implications to implement</td></tr>
<tr><td colspan="5">Rationale/advantage/disadvantages:

Entrance to the cinema is more rapid and comfortable and requires less space that can be dedicated to ancillary entertainment and sales. This will improve revenue and reduce running costs. As a result of reduced staffing levels, customer care may be seen to reduce and security problems may increase.</td><td>Action:

❑ Reject
❑ Develop
❑ Accept</td></tr>
<tr><td>Cost summary</td><td colspan="2">Drawing refs:</td><td colspan="2">Design status: Outline/working</td><td>Implementation details:</td></tr>
<tr><td></td><td>Initial cost</td><td>NPV running</td><td>NPV maintenance</td><td>Total cost</td><td rowspan="4"></td></tr>
<tr><td>Original design</td><td></td><td></td><td></td><td></td></tr>
<tr><td>This proposal</td><td></td><td></td><td></td><td></td></tr>
<tr><td>Saving/add</td><td></td><td></td><td></td><td></td></tr>
</table>

Notes:

1 The accuracy of costs provided will depend upon the level of information and. time available for the calculations.

2 The net present value of running costs and net present value of maintenance costs must relate to a realistic business life projection.

3 The proposal should be prepared in such a way as to allow various members of the design team and. client organisation to understand. the rationale.

4 In addition to providing details of the proposal, the form acts as a checklist and prompter of action and implementation.

5 When considering the proposal, consideration should be given to the impact of accepting the change in terms of time or resource implications, e.g. time to redesign and associated fees, planning approvals, delivery dates ...

Fig. 7.4 Value management proposal form.

Implementation

Irrespective of the level of success achieved during the workshop stage, the real measure of success in a value management study lies in the extent of implementation of the proposals. This will be dependent upon several factors including the level of client support and commitment, the attitude of members of the design team and the time available.

Implementation will occur via a detailed response to the facilitator's report from the client body. This will indicate those proposals outlined in the report which are accepted, rejected or requiring further development before a decision can be made. Following this, it is good practice to hold a further meeting at which all outstanding matters can be clarified and confirmed.

Role of the quantity surveyor in the value management workshop

During each phase of the VM workshop, the quantity surveyor, as discussed above, has a key role in providing cost advice, in terms of both capital and whole life costs, relating to proposed solutions that could help clients achieve best value. It is an important output of the first workshop, VM0 (see Fig. 7.1) to establish and understand 'value' in the context of the project. The Construction Innovation Hub's *Value Toolkit* provides a suite of tools to support faster value-based decision-making. It recognises that value does not just mean costs but suggests a wider project value profile recognising the wider sustainability impact of the project. This value profile consists of five categories: natural, social, human, manufactures and financial. Converting the value profile into a set of indices helps inform whether choices and options are value creating or value destroying (CIH 2020).

BIM and value management

The BIM environment provides the ideal platform to perform cost and value management exercises (Pittard and Sell 2016). Fig. 7.5 provides an outline of the potential use of BIM at each phase of the Job Plan.

At the pre-workshop and information stage, the commercial content of the Employers Information Requirements (EIR) will detail the expected purposes of the information within the BIM model, such as cost, operational, maintenance and repair data. Defining these model requirements will align with value management value criteria. As BIM includes visualisation, simulations and clash detection, it has the potential to strengthen the function analysis of value engineering (Xiaojuan et al. 2021). The focus of the creative phase is brainstorming of ideas from the VM team and therefore BIM does not directly support this stage of the workshop. Within the evaluation phase, the use of integrated BIM software packages that can support whole life cost analysis, embodied energy analysis, lighting analyse etc. increases the potential of using quantitative data to support the VM's team's qualitative findings in identifying high value gains and the impact of the proposed solution on the design. At the development stage, an integrated approach enables more accurate analysis of cost and function information, quicker decision-making and analysis of design change impacts

Job Plan Phase	Use of BIM
Pre-workshop	Concepts of EIR of BIM increases the understanding of the overall project (Punnyasoma et al. 2021)
Information phase	
Function Analysis phase	BIM simulations and clash detection supports function analysis methods (Xiaojuan, 2021)
Creative phase	Not applicable
Evaluation phase	Integrated BIM software provides quantitative data to identify high value gains
Development phase	BIM enables more accurate and quicker analysis of cost and function information to inform design development (Xiaojuan, 2021)
Presentation	BIM models can effectively communicate the outcomes of the VM process

Fig. 7.5 Use of BIM during the VM workshop.

on costs and other value criteria. It is at this stage where BIM solutions can be maximally utilised for design development (Xiaojuan et al. 2021). Allowing more rapid analysis of alternative design proposals is one way in which BIM may support VM and VE (RICS 2017). At the presentation stage, the BIM model and outputs can effectively communicate to the client the outcome of the value management and value engineering exercises.

Function analysis

Function analysis is a recognised value management method. It is a structured approach for creative thinking to identify functions and classify them. This enables the identification of criteria to facilitate the evaluation of alternative solutions (BSI 2020).

Function analysis is generally recognised as a distinguishing feature of value management, dating back to Larry Miles and value analysis. However, the analysis of function should not be seen purely as an integral part of the value management process but should be recognised as a useful methodology that may be used in the absence of the value management framework. For example, there is nothing to prevent a site team (or a single individual) examining a process or aspect of the design with the use of an internally executed function analysis exercise. This may be relatively quick to perform and be helpful in searching for value improvement or the mitigation of a problem. This is not to suggest that this is a substitute for a formally executed value management study. This informal approach to function analysis will clearly have limitations; an individual or small and unrepresentative team will have a restricted view of function, and in any event, it is only one part of value management.

The main reason for the use of the simple and effective technique of function analysis is clear. When applying the method to a building component or element, the question 'What does it do?' as opposed to 'What is it?' is asked. Therefore, when looking for alternatives, we search for something that will provide the required function rather than try to find a substitute for the previous solution. Establishing the true functions of a product and considering the costs of each function identified can in itself be an illuminating exercise. The potential impact of the approach can be seen when considering the wristwatch. Most people would state that its function is to indicate the time. If this were simply the case, why do people pay a wide range of prices for such a function? If you can obtain this function for £3.50, why spend £2000? The answer lies in the additional functions that some people require; improving image (esteem function), extending life/providing date (additional use functions) may well justify, to some, an additional expenditure of £1996.50.

Figure 7.8 shows the range of application of function analysis, from the strategic level to that of the individual component. The applicability of function analysis within the construction industry is determined by level of operation. Although a one-week value management study examining, in detail, the design of a door closer may provide value gain to a manufacturer (in a production run of 100,000), there can be no place for such examination within the context of a new building. The relative costs involved are insignificant and, in any event, suggested design changes are likely to lead to non-standard components resulting in cost increase rather than decrease. However, the application of function analysis at a higher level of abstraction, for example a high cost element or entire building, is valid. Figure 2 in RICS *Value management and value engineering* provides an additional example of a function analysis of an office development at concept stage.

The matrix shown in Fig. 7.6 (Ashworth and Hogg 2000) further explains some of the principles of function analysis. It shows a hypothetical analysis of the costs of a softwood window relative to functional requirements. Note that the values and function allocation indicated are entirely notional and are there purely to serve the explanation. The logic shown in the table is related to cost/worth and is based upon the convenient assumption that all costs can be allocated to some particular function. Worth is defined as the least cost necessary to provide the function. Thus, if we focus upon the functional costs of the 'casement' we can observe the following:

- The minimum cost of a casement to serve the basic functions of 'control ventilation', 'exclude moisture' 'retain heat' and 'transmit light' is £60 (assumed to be present in all windows). This amount has been allocated in equal amounts to each of the functions.
- An additional cost of £30 is attributed to the increased specification of softwood – the function of which is to 'extend life' (and reduce the maintenance) of the window.
- An additional cost of £35 is attributed to the window style, say of Georgian appearance in small panes incorporating moulded sections, the function of which is to 'enhance appearance'.
- Function is not always as it first may seem. The function of 'permit ventilation' is achieved by forming an opening; with regard to ventilation, the purpose of the casement is to 'control ventilation'.

	Function											
Component	Permit ventilation	Control ventilation	Exclude moisture	Retain heat	Transmit light	Improve security	Reduce sound	Reduce glare	Extend life	Assist cleaning	Enhance appearance	Component cost £
Lintel	15	—	—	—	15	—	—	—	—	—	—	30
Opening	10	—	—	—	10	—	—	—	—	—	—	20
Frame	—	5	5	5	5	—	—	—	10	—	5	35
Casement	—	15	15	15	15	—	—	—	30	—	35	125
Ironmongery	—	10	—	—	—	5	—	—	—	20	5	40
Glass	—	5	5	5	—	5	10	5	—	—	10	45
Paint	—	—	—	—	—	—	—	—	5	—	10	15
Function cost	25	35	25	25	45	10	10	5	45	20	65	310

Fig. 7.6 Functional matrix 'softwood window' (*Source:* Ashworth and Hogg 2000/Taylor & Francis).

- The functions can be divided into basic – those which are essential (shaded) in Fig. 7.6 and secondary – those which are not essential (but possibly unavoidable or necessary to sell the product), often provided in response to the design solution (e.g. reduce glare). The choice of what is basic and secondary may be subjective and dependent upon individual perception (hence the need for stakeholder participation). The separation of basic and secondary functions assists with the understanding of a project and may identify areas to target for value improvement.
- In terms of cost/worth, if we only require the basic functions as shown (as stated above this is a subjective view), the window is only worth £165 (the sum of the shaded function totals). Thus, if we have expended £310 and do not require 'reduce sound', 'reduce glare', 'extend life', 'assist cleaning', and 'enhance appearance' (e.g., we live in a bungalow, in a quiet rural hamlet, surrounded by trees and have simple aesthetic tastes) we have not achieved good value.

This shows the way in which function analysis could be used to identify unnecessary costs or to highlight a disparity of expenditure, possibly leading to substitution with a design alternative. The benefits of function analysis in practice on a component such as a softwood window will probably be minor and therefore not worthwhile; however, the benefits of the technique can be seen (Ashworth and Hogg 2000).

Function analysis diagramming techniques

A common method of performing function analysis is with the aid of function analysis diagramming techniques such as FAST (function analysis systems technique) or a value hierarchy. Please note that these diagrams are a means of carrying out and expanding the usefulness of a function analysis; as shown above, function analysis may be performed without such techniques.

Function analysis diagramming techniques can be used across the varying levels of a project development, for example at a strategic stage to possibly identify the building need, at built solution level to assist in the briefing process, and at an elemental level to identify unnecessary costs. In practice, however, within the construction domain the use of the technique at technical level is doubtful. Examples of these, and how they may inter-relate, are shown in Fig. 7.8.

Cinema scenario

As part of the information phase of the value management study, a FAST diagram is constructed. A simplified example is provided in Fig. 7.7. The following notes describe the main features:

- The diagram is constructed by using the identified functions and applying 'How/Why' intuitive logic to develop a structured model. Functions are identified by the value management team and conveyed in the form of verb–noun descriptions. In practice, the use of Post-it Notes solves the problems associated with continual development and amendment of the diagram.

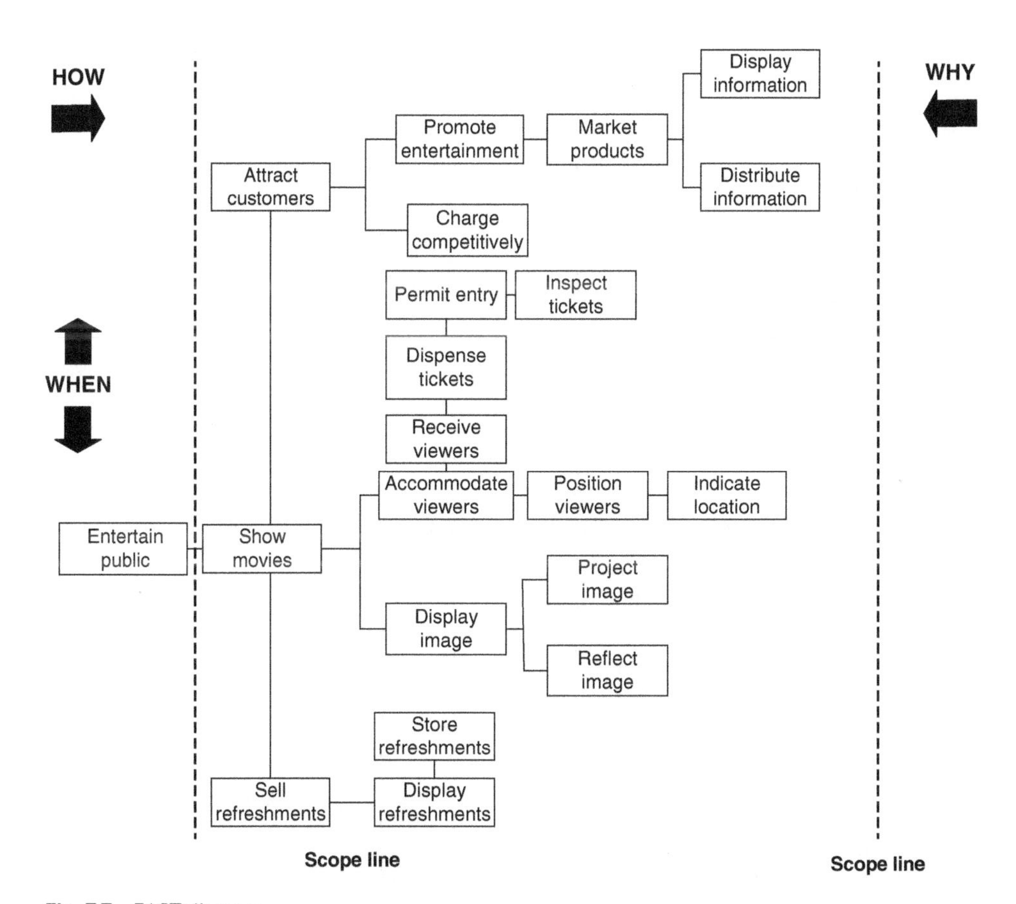

Fig. 7.7 FAST diagram.

- Applying the questions 'How?' and 'Why?' to the relationship between identified functions tests the logic of the diagram. For example, the function 'show movies' is served jointly by 'display image' and 'accommodate viewers'; likewise, the reason for display images and accommodating viewers is to show movies. A further question which may also be used when constructing the diagrams is 'When?'. In the cinema example, 'When' accommodating viewers, there is also the need to receive viewers and dispense tickets. This approach assists in the practical construction of the diagram and in the maintenance of the applied logic.
- The left scope line has been established with 'Show movies' as the basic function. This is effectively the limit of the study. Although a higher level of abstraction would provide the opportunity to analyse other alternatives to the function 'entertain public', or even beyond, e.g. 'improve profit', the core business activity of the client in this study is to 'show movies'.
- The scope line to the right of the diagram provides the limit at which further dilution to the question 'How' would serve no practical purpose. Design solutions generally appear to the right of this line.

To construct FAST diagrams can be a difficult task and the notes above are intended to provide an outline explanation only. In practice, the use of FAST diagramming techniques will require training. Simpler forms of function/logic diagrams may be more easily used and are more common. These follow the principles of How/Why logic in the form of a tree diagram but are generally looser in terms of adherence to convention. The outline examples shown in Fig. 7.8 are indicative examples.

The use of function logic diagrams helps to describe the problem environment and allows participants in a value management workshop to understand the relationship between project objectives and solutions. The application of costs to a function logic diagram will also assist in the identification of high cost functions, which may help in directing further value management activity.

Figure 7.8 shows the inter-relationship between the various levels of use of function logic diagrams.

Supporting the case for value management

The IVM suggests value management delivers a return on investment of 10 : 1 and 100 : 1 (IVM 2022b).

Despite the acclaim given to value management, and although its application within the domain of the large quantity surveying practice is increasing, the extent of its growth and application is inconsistent. This may be due to several reasons including:

- There is insufficient time to carry it out
- Clients are unwilling to pay for the service
- Clients do not request the service
- Belief that the quantity surveyor provides the service already
- Value management skills are unavailable

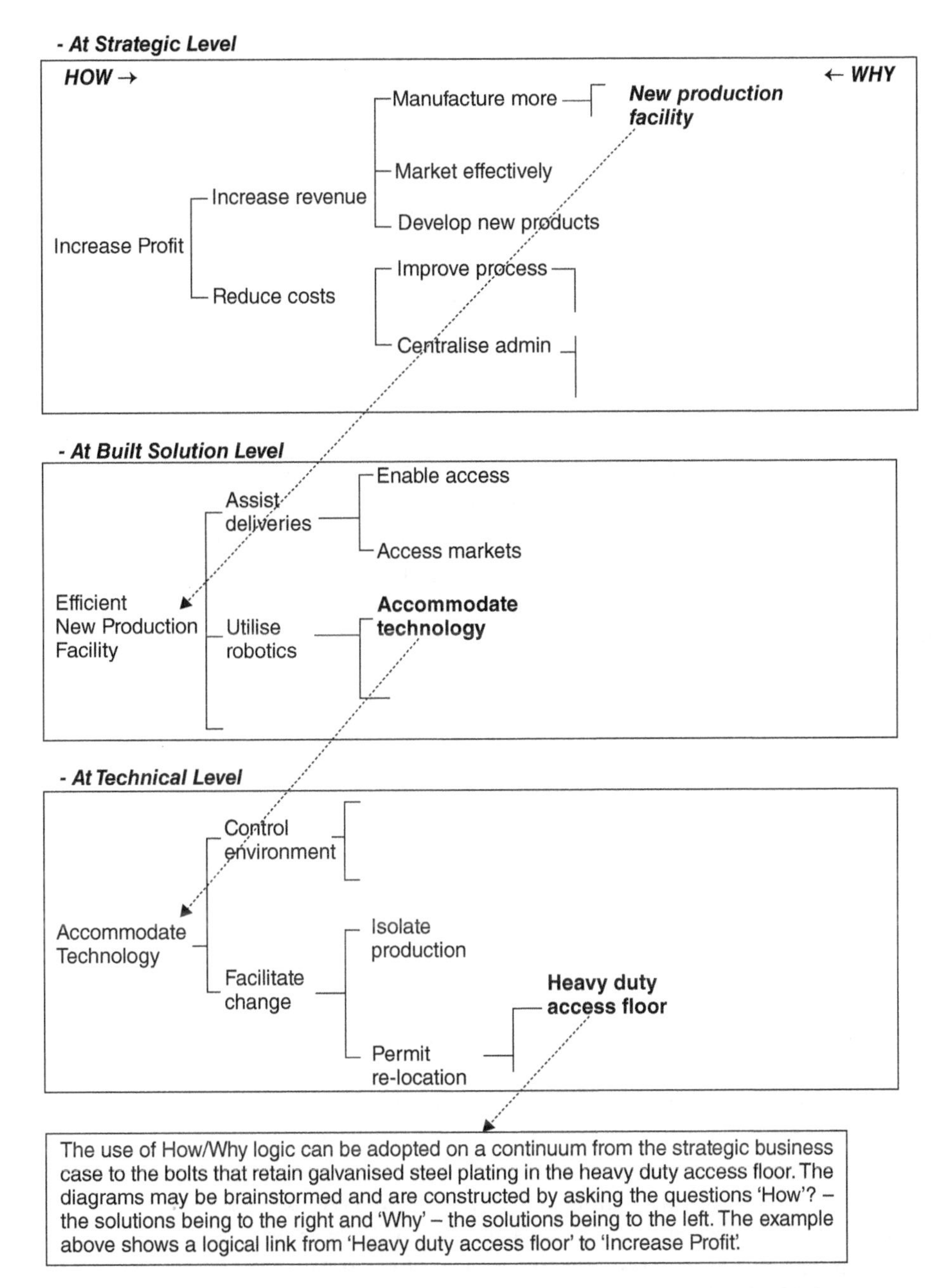

Fig. 7.8 The use of HOW/WHY logic and inter-relationships at varying levels of indenture (*Source:* Ashworth and Hogg 2000).

Within the design team, it is reasonable to believe that the quantity surveyor is in a position of great influence with regard to the implementation of value management and this is reflected within the *Quantity Surveying Scope of Services*, which includes the supplementary service provision of 'Facilitate, set up and manage value engineering services' (RICS 2022).

Value management can add great value to a project and the quantity surveyor should appreciate the full nature of the process and provide opportunity for its application in practice.

Professional development and accreditation

The 'Design economics and cost planning' competency requires a quantity surveyor to apply value engineering processes and the use of value management techniques. In addition, for more experienced quantity surveyors, there are several value management societies in various parts of the world that aim to assist value managers by promoting, developing and controlling standards of practice and qualifications. The Institute of Value Management (IVM) in the UK has introduced a programme of formal training which provides practising and prospective value managers with the opportunity to obtain a recognised education and comprehensive training, resulting in a UK qualification that recognises the competence to lead value studies. The IVM website (www.ivm.org.uk) also provides some useful information on the benefits of VM, an overview of techniques and case study summaries. By far the most established international VM organisation is the Society of American Value Engineers (www.value-eng.org).

References

Ashworth, A. and Perera, S. *Cost Studies of Buildings*.6th Edition. London and New York. Routledge. 2015.

Ashworth, A. and Hogg, K.I. *Added Value in Design and Construction*. Pearson. 2000.

BSI. *BS EN 12973:2020 Value Management*. British Standards Institution. 2020.

CIH. *An Introduction to the Value Toolkit*. Construction Innovation Hub. 2020.

Higham, A., Bridge, C. and Farell, P. *Project Finance For Construction*. London and New York. Routledge. 2017.

Hunter, K. and Kelly, J. *Is One Day Enough? The Argument for Shorter VM/VE Studies*. Society of American Value Engineers. 2007.

Institute of Value Management. *Value Management – Benefits in Action – Case Studies*. 2022a. Available at https://ivm.org.uk/case_studies/value-management-delivers-real-benefits/ (accesses 08/06/2022).

Institute of Value Management. *Value Management – Benefits in Action – Case Studies*. Institute of Value Management. 2022b. Available at https://ivm.org.uk/case_studies/value-management-delivers-real-benefits/ (accessed 08/06/2022).

Miles L.D. *Techniques for Value Analysis and Engineering*. McGraw-Hill. 1972.

Office of Government Commerce. *Value Management in Construction Case Studies*. OGC. 2009.

Oke, A.E., Stephen, S.S. and Aigbavboa, C.O. General summary on value management globally. *In*: *Value Management Implementation in Construction*. Emerald Publishing Limited, Bingley. 211–226. 2022. Available at https://www.emerald.com/insight/content/doi/10.1108/978-1-80262-407-620221018/full/html (accessed 10/06/2022).

Owen, K. Getting value management right. *RICS Construction Journal*, 04. 2021. Available at https://ww3.rics.org/uk/en/journals/construction-journal/getting-value-management-right.html (accessed 30/05/2022).

Pittard, S and Sell, P. *BIM and Quantity Surveying*. London and New York. Routledge. 2016.

Punnyasoma, J., Jayasena, H. and Tennakoon, T. 2019. Use of BIM Solutions to Facilitate Value Management. In: Sandanayake, Y.G., Gunatilake, S. and Waidyasekara, A. (eds). *Proceedings of the 8th World Construction Symposium*, Colombo, Sri Lanka, 8–10 November 2019, pp. 598–607. Available at http://dl.lib.uom.lk/handle/123/15377 (accessed 11/06/2022).

RICS. *Quantity Surveying Services*. RICS.AQ 2022.

RICS. *Value Management and Value Engineering*. 1st Edition. RICS. 2017.

SAVE 2022 *About the Value Methodology*. Available at https://www.value-eng.org/page/AboutVM (accessed 10/06/2022).

Spellacy, J., Edwards, D.J., Roberts, C.J., Hayhow, S. and Shelbourn, M. An investigation into the role of the quantity surveyor in the value management workshop process. *Journal of Engineering, Design and Technology*, 19(2), 423–445. 2021. Available at https://www.emerald.com/insight/content/doi/10.1108/JEDT-07-2020-0289/full/html (accessed 07/06/2022).

Xiaojuan, L., Wang, C. and Alashwal, A. Case study on BIM and value engineering integration for construction cost control. *Advances in Civil Engineering*, 8849303. 2021. Available at https://www.hindawi.com/journals/ace/2021/8849303/#abstract (accessed 11/06/2022).

8

Risk Management

KEY CONCEPTS

- Risk and cost management
- Risk management
  - Risk identification
  - Risk analysis
  - Risk evaluation
  - Risk plan
  - Risk monitoring and reporting
- Risk management workshops
- Risk management techniques

LEARNING OUTCOMES

After reading this chapter, you should be able to:

- Understand how quantity surveyors allow for risks in cost planning.
- Discuss the application of risk management to identify, analysis, evaluate and manage risks during the design and construction process.
- Appreciate the range of risk management techniques, both qualitative and quantitative, and their appropriate use.

COMPETENCIES

Competencies covered in this chapter:

- Design economics and cost planning
- Project feasibility analysis
- Project finance (control and reporting)
- Risk management

Willis's Practice and Procedure for the Quantity Surveyor, Fourteenth Edition.
Allan Ashworth and Catherine Higgs.

Introduction

Risk management is a practice that many of us use instinctively at a personal level on a regular basis; however, the importance of active risk management in construction demands greater attention. The complexity and scale of most building projects is such that good risk management in the construction industry requires more than purely common sense and instinct.

'A risk can be defined as an uncertain event or circumstance, that if it occurs, will affect the outcome of a project' (RICS 2015). In resolving uncertainty a better outcome may be achieved, for example a cost saving; therefore, risks must be considered as having both positive and negative outcomes.

Construction projects are full of risks and include those that relate to external commercial factors, design, construction and operational matters. Risk management provides the opportunity to control the occurrence and impact of risk factors and provides the project team and clients with better information upon which to make decisions.

The risk management principles outlined in this chapter can, in the main, be transferred to any stage of a project's development. It is likely that the project manager will provide the core service of preparing the risk management strategy and maintain the project risk register; however, as part of the team, the quantity surveyor will make a valuable contribution to risk mitigation.

Management of risk and the cost management process

The NRM Suite identifies and values risks to increase the transparency of cost management; this is in lieu of the use of contingency sums. The contingency provision was often seen to be an inadequate device with which to protect clients against risk. The *Order of Cost Estimate* recommends that risk allowances are not a standard percentage but a considered assessment of the risk, taking into account the completeness of the design. Risk allowances are identified as either design development risks, construction risks, employer change risks and employer other risks. The cost plans allow for risks in Group Element 13 (see Chapter 5). In preparing the Bill of Quantities NRM2 suggests that to give the client cost certainty, a Schedule of Construction Risks is included. These risk allowances should be based on the results of a formal risk analysis. Where the project is to include maintenance and renewal works, NRM3 includes risk allowances for design development and associated installation risks, maintenance risk (Chapter 17), client change risks and client other risks.

The benefits that risk analysis may bring to the cost management process are not restricted to the client. Contractors, particularly in the light of current procurement trends, are well used to dealing with more than construction risk, including that relating to design. Risk management should therefore be a key concern to contracting organisations.

To assist in the decision-making process

Generally, when a quantity surveyor gives cost advice to clients, the one thing that they are usually sure of is that the advice is not given with certainty. At the Strategic Definition stage, when a quantity surveyor advises a client that a project will cost £5,000,000, this is given as an approximation. As consultants, estimates are given on the basis of a forecast of a yet unknown party's forecast of costs (the contractors) usually in an uncertain timescale and with uncertain design information. It is therefore common practice to provide range estimates, e.g. £4,800,000–£5,200,000, and to bring uncertainty to the client's attention. Indeed, the RICS *Cost Prediction* mandatory requirements states 'an estimate of the accuracy or the level of uncertainty of the cost prediction' should be reported (RICS 2020). Therefore a most likely cost, together with minimum and maximum costs, is an alternative method. Nevertheless, the decision to progress with a project will be made upon the basis of this information and will require interpretation by the client. Simulation, discussed in detail later in this chapter, is a technique which provides clients with a much more comprehensive view of project risk. A probabilistic cost forecast provides an open view on outturn cost likelihood and reduces the relative vagueness associated with traditional single point estimating. Although this might be regarded as a tool that would normally be restricted to dedicated project management, the quantity surveyor should understand its application and benefits.

To assist in the choice of procurement

Risk management and the choice of procurement are interlinked, so risk allocation needs to be considered as part of the process of selecting an appropriate procurement route since different routes entail differing degrees of risk transfer. The aim is to select a procurement route that achieves the optimum balance of risk, benefit and funding for a particular project. The level of risk transfer ranges from PFI projects where most scope for risk is transferred to the private sector to traditional procurement where it is allocated to the client (Chapter 9).

Whilst JCT contracts do not include formal risk management processes within the conditions, NEC4 includes a commitment to risk analysis and the management of risk within the contract core clauses.

The application of risk management

In principle, risk management is a straightforward process in that it requires the evaluation of risk and the execution of a risk management strategy. The assessment of risk first entails risk identification, followed by the analysis of risks identified. This imparts a level of understanding that is needed to facilitate the adoption of a suitable risk management plan. The successful management of risk requires:

- Focus upon the most significant risks
- Consideration of the various *risk management options*
- An understanding of effective *risk allocation*
- An appreciation of the factors which may have an impact on a party's *willingness to accept risk*
- An appreciation of the *response* of a party if and when a risk happens.

Engagement with Risk (Lewis et al. 2014) identifies an effective communication of project risk to all relevant stakeholders, the promotion of a project culture where there is effective behaviour of teams to identify timely risks and adopt appropriate management solutions, and a greater understanding of the interrelationship of project risks, will positively influence any risk management strategy.

The following British Standards provide guidelines on risk management, its implementation and assessment techniques.

- BS ISO 31000:2018 Risk management – Guidelines
- BS EN IEC 31010:2019 Risk management – Risk assessment techniques
- BS ISO 31100:2021 Risk management – Code of practice for the implementation of BS ISO 31000:2018.

Risk identification

There are a variety of methods that the project team can use to identify risks. A team workshop could be held, which brings together specialists from a variety of relevant disciplines, promoting a wide project viewpoint. At which participants 'brainstorm' risks that they consider could have an impact upon the project. Historical data and the lessons learnt from previous experience may be used. Also, as with value management (see Chapter 7), the use of checklists may assist in providing structure to the thought processes; structures such as SWOT and PESTLE can be used. Checklists enable a common understanding of the project risks by all stakeholders. An indicative example of a project checklist is provided in Fig. 8.1.

A checklist such as the one shown in Fig. 8.1 will be useful in directing attention to predetermined and recognised categories of risk and thus assisting with the identification of those which are project specific. The examples of categories given will incorporate a large range of risks; some categories in particular are wide in their potential scope. Lester (2017) recommends that the long list of risks identified can be split into four main categories:

> *Organisation*, such as management, resourcing and health and safety,
> *Environment*, such as security, weather, and pressure groups,
> *Technical*, such as technology, construction, and commissioning, and
> *Financial*, such as funding, inflation, and the financial stability of parties to the contract.

There is some danger with checklists that their use could limit deliberation to those categories contained in the list and it should be borne in mind that this could result in ruling out some major, and possibly significant, items. It is important to realise that the process of risk identification is not likely to result in the discovery of all possible risks. This is not a practical objective.

When identifying risks, it is important to appreciate exactly what the process is trying to establish. To facilitate the process of risk analysis and risk management, it is necessary to think about the possible sources of the risk, not merely the risk event. For example, if we consider the scenario of a basement excavation, one risk event could be that of the collapse of an adjacent road during the course of excavations. To allow proper risk analysis and risk management, the sources of this risk event should be understood. In this case, these may include inadequate direction of the workforce, accidental damage to an existing retaining wall, inadequate shoring design, inadequate shoring construction, or vandalism – all of

Risk category	Indicative examples
Health and Safety	Collapse of sides of trench excavations, surrounding infrastructure or striking existing services resulting in delays, additional cost and possible injury. Incomplete client information in relation to asbestos within existing building.
Disputes	Disruption to a third party's business due to noise, dust, restricted access or construction traffic resulting in reduced sales, financial loss and possible litigation
Price	Increased inflationary pressures causing a severe increase in building costs and excessive financial loss
Payment	Delay in the payment by the main contractor to subcontractors causing a reduction in works progress and resultant programme delays
Supervision	Delays in the issue of drawings or instructions by the architect resulting in abortive work, delays, additional costs and contractual claims from the contractor
Materials	Non-availability of matching materials required in a refurbishment project resulting in possible redesign, programme delay and additional expense
Labour	Non-availability of labour due to the construction of another nearby major project which causes a regional shortage of specialist subcontractors
Design	Errors in the design due to lack of communications between structural engineers and architect resulting in abortive work, or possibly building failure

Fig. 8.1 Checklist of risk categories.

which may be independently assessed and managed in some detail. Figure 8.2 provides a summary of the existing risks and outlines possible considerations and actions with the assumption that this risk belongs entirely to the contractor.

Risk analysis

To begin with, it is important to be aware that problems in construction do not restrict themselves to cost, although all problems could have a cost effect. In numerous situations, time or schedule risk is of more significance than pure cost and, in some cases, quality may be the most important priority. Therefore, it is essential that risk analysis addresses the needs of a given situation and centres upon applicable areas of concern. *BS ISO 31000 states* 'the purpose of risk analysis is to comprehend the nature of risk and its characteristics including, where appropriate, the level of risk' (BSI 2018).

There is a range of risk analysis tools that may be used to evaluate the identified risks. The choice of the most appropriate approach will depend on project size, type in terms of what is needed by the stakeholders and opportunity, in terms of expertise and software availability. Techniques, can be categorised as qualitative and quantitative. Qualitative methods are those that assess the risk in terms of impact and probability only and normally

Collapse of adjacent road during excavations		
Identified sources	Considerations	Possible actions
Inadequate direction of the workforce/Accidental damage to existing retaining wall/Inadequate shoring construction	All of these items more or less relate to one source, that of direction of the workforce, although each will require separate consideration. Potential high impact (in terms of time, cost and personal injury). Implications of Statutory Health and Safety transgression. Insurance?	Enforce quality control procedures. Allocate a reliable supervisor to the task. If subcontracted out, ensure the reliability and good standing of the subcontractor and that risk adequately transferred. Verify insurance provisions for such occurrences. Verify arrangements with municipal engineers as to condition of existing road, location of drains etc.
Vandalism	Potential high impact as above, also inner-city location suggests above average likelihood.	Strictly enforce health and safety procedures including proper protection of the site and the works. Verify insurance provisions for such occurrences.
Inadequate shoring design	As above. The appointment of experienced engineers should reduce the likelihood.	Transfer the risk relating to the design of the temporary works, by appointing external consultants. Verify insurance provision.

Fig. 8.2 Scenario of a construction project – basement excavations to residential development.

apply only to a single risk event. Quantitative methods use verified data to ascertain the impact in terms of cost or time of an interrelated set of risks. Examples of some of the approaches are as follows.

The risk management workshop – qualitative

The benefit of the risk management workshop are they elicit information and the thoughts of a wider range of experienced stakeholders. It is possibly the most simple and most effective aspect of risk analysis due to the appraisal of risk that becomes possible during the course of structured team discussions. Workshops offer the means whereby risks can be identified, assessed and attended to; great advantage may be obtained without any element of more complex and demanding quantitative assessment.

Probability/impact tables (a 'semi-quantitative' approach)

One uncomplicated method of assessing risk is the use of the probability impact ranking matrix (P/I matrix). This simple process involves the weighting of a qualitative assessment and hence is termed 'semi-quantitative' analysis. An example of a P/I matrix – indicating the consideration of risks relating to a possible schedule delay – is shown in Fig. 8.3.

	Impact				
Probability	Very low - 1	Low - 2	Medium - 3	High - 4	Very high - 5
Very high - 5		5	9		
High - 4			8		
Medium - 3		2,3	6		
Low - 2				7,10	1
Very low - 1				4	

Identified Risk
1. Existing road collapse
2. Cut through gas mains
3. Cut through power supply
4. Labour dispute
5. Excavation equipment breakdown
6. Delay in piling rig delivery
7. Collapse of large sewer
8. Delay due to vandals
9. Delays due to hard rock
10. Planning delays

Fig. 8.3 Probability impact (P/I) matrix showing indicative consideration of schedule risk relating to the basement excavations outlined above.

To make the use of the P/I matrix effective, it is essential that the team has a shared understanding of the descriptors; therefore it is important that they are clearly defined. For example, 'very high probability' equates to more than, say, a 75% chance of occurrence. It is clear that this technique should be designed and applied to meet the circumstances of specific project requirements. The example shown in Fig. 8.3, which examines aspects affecting time, may also be adapted and utilised to assess risks affecting cost and quality. By multiplying the likelihood (probability) of the risk by the impact, a total risk rating can be established. Use of this simple, but subjective, technique permits risks to be positioned in terms of severity (Fig. 8.4) and therefore allows the management team to be more purposeful and focused upon the most significant issues. There is a need to concentrate upon those risks that will have a high impact if they arise and have a high chance of occurrence. It is both inefficient and impossible to spend time on all risks, and matters that are inconsequential or of an extremely low incidence should be put to one side.

Risk identified in Fig 8.3	Total Risk Rating Probability × Impact	Ranking
1	2 × 5 = 10	3
2	3 × 2 = 6	8
3	3 × 2 = 6	8
4	1 × 4 = 4	10
5	5 × 2 = 10	3
6	3 × 3 = 9	5
7	2 × 4 = 8	6
8	4 × 3 = 12	2
9	5 × 3 = 15	1
10	2 × 4 = 8	6

Fig. 8.4 Ranking of identified risks established in the probability impact matrix (Fig. 8.3).

Significance	Commercial	Education	Refurbishment
1	Variation by client	Variation by client	Change in scope of works
2	Change in design	Extremely bad weather	Change in design
3	Change in scope of works	Change in scope of works	Third party delay
4	Unexpected site conditions	Unexpected site conditions	Variation by client

Fig. 8.5 Perceptions of risk impacts on costs (*Source:* Adapted from Odeyinda et al. 2009).

Odeyinda et al. study (2009), assessing the perceived risk impacts on different building types on the variability of costs between tender and final account, determined the most significant risks as those identified in Fig. 8.5. The study concluded that there is a set of different risk factors dependent on project type, and this will have an impact on the quantity surveyor's approach to modelling risk impacts, i.e. it is wrong to use a 'one that fits all' approach.

Expected monetary value (EMV)

One method of bringing probability and impact together is by the use of the expected monetary value (EMV) technique. With this approach, possible outcomes – the financial gain or loss – are weighted by the expected probability of each occurring and combined to produce an aggregate result.

This approach can be used in several areas, for example, to produce an overall project outcome by application to independent elemental costs, or to calculate a sensitive rate in production of a tender.

Monte Carlo simulation (quantitative risk analysis)

On most construction projects, it is likely that quantitative risk analysis is neither sensible nor needed due to the relative payback from the extra time and expertise required to complete such an appraisal. Simulation could be appropriate for large and complex projects as it is a powerful tool to understand the likelihood and consequences of interrelated risk events each of which may impact on the project outcome, in terms of cost, time or quality. This analysis technique removes the subjectivity from the assessment of probability and outcomes by randomly selecting values within the uncertainty range.

As far as cost risk is concerned, Fig. 8.6 shows notional costs for a hypothetical project which, for reasons of simplicity, has been reduced to four elements.

The costs shown in Fig. 8.6 have been produced following a workshop carried out in the early stages of the project's development. Discussions relating to the substructure element may have been recorded something like this:

> . . . 3.0 Substructure: The QS has produced a budget estimate amounting to £170,000. This includes the provision of £20,000 for excavation in bad ground (say hard rock removal). The architect advised that in a project constructed earlier in the year, on

	Least cost (a)	Most likely (b)	Highest cost (c)
Substructures	150 000	170 000	245 000
External walls	325 000	335 000	345 000
Roof	185 000	195 000	240 000
External works	155 000	215 000	235 000
Totals	815 000	915 000	1065 000

Fig. 8.6 Cost model showing a minimum, maximum and most likely cost (*Source:* Ashworth and Hogg 2000/Taylor & Francis).

an adjacent site, no rock was encountered. The engineers agree with the architect that rock is unlikely but are concerned that some piling may be required. No data from site investigations are available at present. The budget prepared by the QS is accepted as a reasonable provision for the substructures; however, a minimum cost and maximum cost are also identified.

Similar considerations have been made concerning each of the elements shown. A question to consider now is, 'Does the client have adequate finances to build the project?' Since this may depend on the amounts selected, which costs do we use? A negative client or client advisor may select the worst-case scenario in each element, resulting in a predicted total cost of £1,065,000. Whilst this approach to risk may be understandable in exceptional circumstances (where a 'fail safe' position is required), it is not in most construction projects. In the example used, there may be a 10% chance of the worst case in each element. The chances of each occurring simultaneously are thus $0.1 \times 0.1 \times 0.1 \times 0.1$ with a consequential probability of 0.0001 (i.e. 10,000 : 1).

Further consideration of the above cost model will reveal other weaknesses: there is an unmanageable range of 'what-if' scenarios; the values given are discrete (i.e. 'in between' values are not accommodated – the model does not allow a substructure cost of £160,000); no allowance is made for the fact that the minimum and maximum values are distinctly less probable than the 'most likely' value.

Simulation allows us to model each element in terms of cost likelihood (costs being used in the example, but this can be applied to schedule risk also). It allows for continuous values (as opposed to discrete) and accounts for the possibility of each value by using probability density.

Figure 8.7 shows triangular probability density distributions representing each of the four elements considered, where probability (P) is on the vertical axis and minimum cost (a), most likely cost (b) and maximum cost (c) are on the horizontal axis (Ashworth and Hogg 2000).

Following the construction of the elemental cost models in the form of probability density distributions, simulation software is used to randomly select elemental values that are collected to generate an estimate of total project costs. This exercise or iteration is repeated many times to produce, say, 500 estimates. Since, in each iteration, the selection of values is dependent upon each elemental probability density distribution, most of the elemental values will be selected about point 'b'. The frequency at which total project estimates will comprise elemental estimates tending toward values 'a' or 'c' is very low

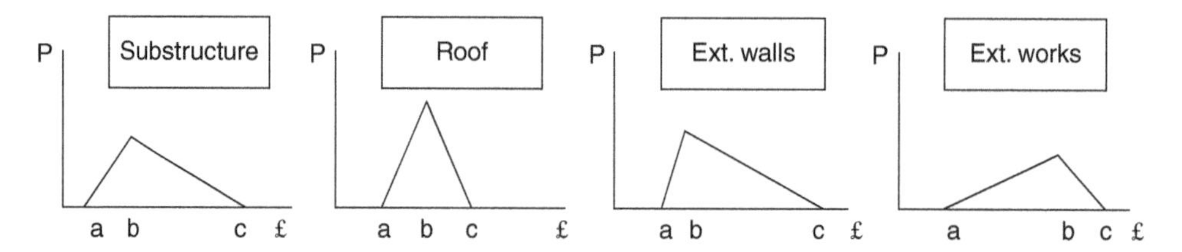

Fig. 8.7 Indicative probability distributions of the elements contained in the cost model (*Source:* Ashworth and Hogg 2000/Taylor & Francis).

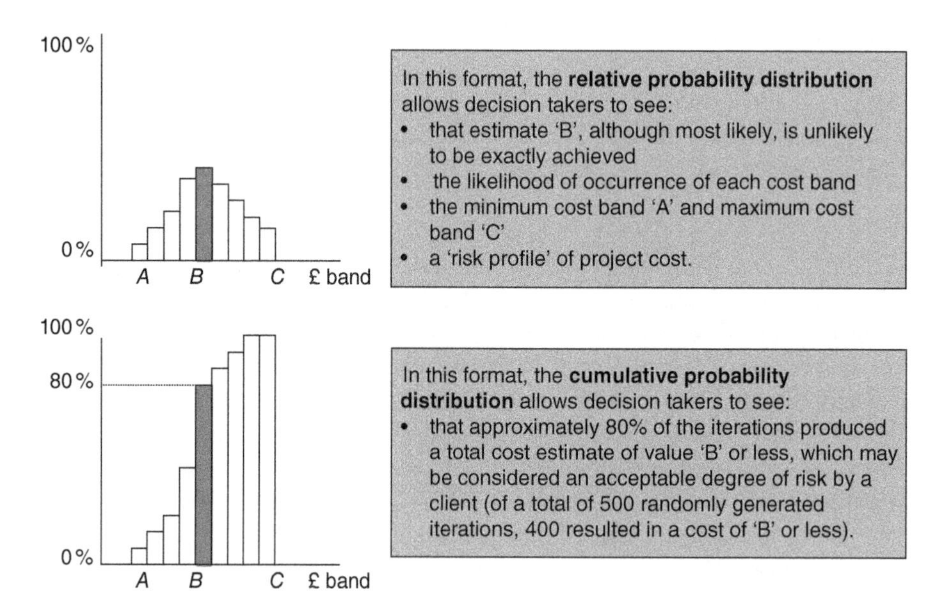

Fig. 8.8 An illustration of output from a simulation exercise (*Source:* Ashworth and Hogg 2000/Taylor & Francis).

and the likelihood of an estimate being produced from the sum of four minimum elemental costs (point 'a') or four maximum elemental costs (point 'b') is mathematically unlikely. The output of the simulation exercise can be presented as a relative (Fig. 8.8) or cumulative probability distribution. The information provides decision takers with a much clearer picture of the risks involved than by the provision of a single point estimate (Ashworth and Hogg 2000).

Information in this form allows clients and client advisors to see a full picture of possible project outcomes and therefore assists in decision-making. To illustrate the significance of this, consider the relative probability distributions of two projects that have been generated from a simulation exercise (Fig. 8.9). Both have similar 'most likely costs' (i.e. £5 million, which in normal conditions one would assume would be the basis of project estimates traditionally reported to clients), but the risk profile of project 'B' is considerably less attractive than project 'A' (Ashworth and Hogg 2000).

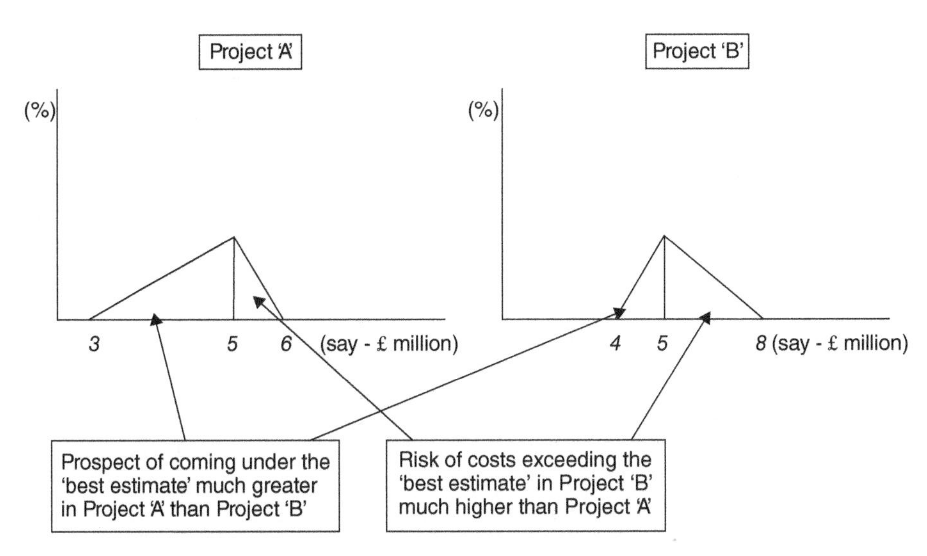

Fig. 8.9 Comparative simulation outputs (*Source:* Ashworth and Hogg 2000/Taylor & Francis).

Risk evaluation

Risk evaluation is informed by the outcomes of the risk analysis and enables the project team to implement the appropriate actions related to each of the risk events. As discussed, it is important to identify and focus on the significant risk items. Methods of deciding on the most important risks in a project have been outlined above.

Risk management plan

There are only a small number of risk response options available for consideration, which is helpful in simplifying the process. These may be categorised as follows (note, there are two parameters to risk, its likelihood or occurrence and its impact):

- A risk can be *shrunk* or reduced by, for example, establishing more and better information about an unknown situation (removing uncertainty) or redesign to reduce the risk exposure.
- A risk can be *accepted* by a party as unavoidable and any alternative strategy may be considered as being inefficient or impossible to adopt; this risk will be carefully monitored and controlled.
- A risk may be *distributed* to another party, for example contractors usually distribute construction risk by selecting reputable subcontractors to carry out the work
- A risk may be *eliminated* or avoided by the rejection of a project objective, which may lead to reappraisal of a particular part of the proposed works.

This may be presented as a convenient mnemonic: SADE – Shrink Accept Distribute Eliminate.

In so doing, there is no order of action suggested in the consideration of the four active risk management options. However, it is reasonable to consider the potential reduction of a risk before giving thought to further, perhaps more drastic, action since the level of the new risk may influence ensuing considerations.

There are two further passive alternatives in addition to the active options outlined above. It is possible to *monitor* risks without action (i.e., 'keeping an eye on the situation') or unintentionally *accept* the risk, the natural default situation if all other risk considerations are overlooked. Both alternatives should be considered as a poor response and both may result in disaster, particularly the latter.

To demonstrate the effect of the above risk management plan options, consider the following project scenario:

Project: the construction of a block of residential flats in which a double basement car park is required. The flats are positioned in a busy inner city location, surrounded on all sides by main roads/buildings in occupation. The ground is known to contain a significant amount of landfill, and existing services (including large Victorian sewers, gas mains and water mains) have been vaguely identified as close to the intended works. Clearly there is an abundance of potentially major risk events in constructing

Risk management option	Possible action	Possible secondary risks
Shrink the risk	Obtain more accurate information about the nature of the site and location of existing services. Relocate the car park. Direct a specific construction method, e.g. contiguous piling	Additional information may be inaccurate. Relocated car parking may produce additional problems. Selected construction method may cause problems to contractor resulting in additional risk events
Accept the risk	Include in the Schedule of Construction Risks to to cover the eventuality. Insure against the risk. (Insurance in this case is seen as the contribution to a wider contingency fund. Some may classify this as distribution)	Risk provision is inadequate. Risk acceptance by client promotes more 'carefree' attitude by contractor resulting in more risk eventualities. Insurance provision inadequate
Distribute the risk	Pass all associated risks to the main contractor	Excessive risk premium. Contractor loss resulting in claims or insolvency. The latter will result in significant problems to the client
Eliminate the risk	Abandon the basement car parking. (This may not be a true option without the abandonment of the total project since planners may require the facility)	Location of new car parking causes new problems. Piling (bringing new risk) is a requirement of the new substructure design since the poor ground overburden is no longer removed during basement construction

Fig. 8.10 Identified risk – earthwork collapse say (high probability, very high impact) (Updated from: Ashworth and Hogg 2000).

the basement car park, including: earthworks collapse; cutting through existing services; damage to existing buildings and roads; problems of construction due to confinement of site; and bad ground.

Appraisal of risk management plan options

It is important to recognise that when risk management action is taken, in each case (including that of the 'elimination' option) secondary risks should also be considered. These risks, which are identified in Fig. 8.10, arise as a result of the selected risk response strategy.

An opinion held by many employed in the construction industry is that the best way to manage risk is to pass it to another party. This is prevalent throughout the construction hierarchy: clients to contractor/consultants; contractor to subcontractor. This may be in the belief that such action results in the removal of the risk which, as shown in Fig. 8.10, is not the case. It is possible to enlarge risk by distribution and this should be appreciated. This can be shown by consideration of the design and build procurement option. The key aim of this method of procurement is single point responsibility; this includes the transfer of design risk from the client (and his consultants) to the contractor. Since the design process in design and build is likely to be carried out speculatively and be of a relatively short duration, it is realistic to presuppose that at the point of contractual agreement, the design will be less complete than at a similar stage with the traditional procurement option. If it is acknowledged that the design is less firm, it should also be evident that the risks relating to the design will be amplified. Thus, with design and build procurement, design risk is enlarged, not eliminated. This does not refute the existence of client advantages of this risk transfer but demonstrates the requirement to fully recognise the effects of risk allocation (Ashworth and Hogg 2000).

Considerations in risk allocation

As previously indicated in the consideration of the active risk management plan options shown in the above example (Shrink, Accept, Distribute, Eliminate), the most desirable method of management appears to be by distribution to a third party with acceptance by the existing owner being the least preferred. The effective distribution of risk is a key element of risk management and is a principal objective of contracts which should be arranged – from a client's perspective – to optimise exposure to risk.

When taking into consideration the allocation of risk to another party, thought should be given to the following factors:

- The ability of the party to manage the risk
- The ability of the party to bear the risk if it eventuates
- The effect that the risk allocation will have upon the motivation of the recipient
- The cost of the risk transfer.

There are many examples of inappropriate risk allocation within the construction industry that occur due to the strong desire to minimise risk exposure at all costs.

Willingness of a party to accept risk

The readiness with which a party may be prepared to accept a risk will depend on several key factors, including:

- Attitude to risk: a party who is risk averse is, in essence, less willing to accept risk than seek risk. This fundamental will also be translated into the assessment of a risk premium.
- Perception of risk: a party who has recently experienced a risk event on a previous project is quite likely to perceive the probability of a similar occurrence on a new project more highly than if the risk event had not occurred. This viewpoint may be translated into additional risk premium.
- Ability to manage risk: theoretically, a party unable to manage a risk due to lack of resources or experience should be less willing to accept a risk than someone with the necessary expertise.
- Ability to bear risk: theoretically, a party unable to bear a risk due to the lack of the necessary financial back-up should be unwilling to accept.
- The need to obtain work: this factor is likely to be the most significant of all as risk acceptance is market sensitive. When the construction industry is in recession, a party is more willing to accept risk as a necessary means of business survival. Alternatively, when work abounds, a full consideration of project risk can be allowed for.

Risk monitoring and control

During the risk management process, a list of identified risks will be produced and expanded during the analysis and evaluation stages to contain important information relating to each risk item. This 'database' is referred to as the 'risk register' and is used as a management tool and will normally include key details of each identified risk with possible reference to:

- Description of the risk
- The predicted probability and impact
- 'P/I rating' of the risk
- Details of the strategy to be adopted to control both impact and probability
- Identification of the owner of the risk (the accountable party)
- An action window which identifies the dates by with the action should be completed and reviewed.

RICS Management of Risk provides example risk registers.

The risk register acts as a control document and assists as a means of monitoring the management of risks throughout a project. For example, an identified risk in a refurbishment project may be the need to eradicate dry rot, which at the time of project design is not evident. Since, based on previous experience, the probability of its occurrence in the building is deemed high, and the potential impact large, as a precaution a risk allowance is made for timber treatment throughout the building. In addition, the client's consultants include the provision for the replacement of timbers and additional work to the structure, where found necessary. The time that this possible expenditure would occur, if required, would be between month 3 and 4 of the contract. These and other details are retained in a risk register. At the closure of the 'action window', i.e. at the end of month 4, no dry rot had been found. At this stage, all areas of the building have been fully exposed, and inspections carried out. No additional works are found necessary. This situation allowed the release of the risk allowance, and the removal of this item from the monitoring process.

In addition, risk registers may be used as a reference tool for future project evaluation, with many of the included risks being relatively common in occurrence and very similar in content in most project situations.

Risk reporting

The risk management process should be integral to both the projects and client's communication and reporting strategy. Effective monitoring and reporting will enable appropriate and timely decisions to be made to support the implementation of the risk register's actions. Regularly reviewing risks will ensure that the risk ratings can be amended to reflect reality and where the project changes and appropriate corrective actions taken. The risk register is a 'live' document and as the project progresses risk actions will be closed and where appropriate new risks identified, assessed, evaluated and actioned.

Response when a risk eventuates

If and when things ultimately go wrong a contracting party who has accepted the corresponding risk will be accountable for the related losses. Whilst this is understood, it is also quite possible that the party suffering the loss will be moved toward recovery in some way. In some situations, this may be manifested in contractual claims. This reality should be accepted and the mitigation of risk, through a well thought-out risk management plan, should be seen as desirable irrespective of who is in possession.

BIM and risk management

The benefits of BIM in relation to risk management are twofold; in that its implementation on a project has the potential to reduce the risk profile and also support the risk management process (Pittard and Sell 2016). The shared environment of BIM enables greater availability and improved quality of project information, which impacts positively on the effectiveness of the risk management process, in terms of identification. The facilitation of shared information promotes a collaborative culture in the awareness and understanding of the project risks. BIM integration with project management, risk management and 4D planning simulation software enables different scenarios to be investigated over the project life cycle to support risk assessment as qualitative analysis of the impact of the likelihood of risks on both cost and time can be ascertained (Papachatzi 2019). The greater understanding gained through visualisation of the construction process leads to more proactive risk mitigation. However, it is also important to add that poor implementation of BIM in itself will add risk associated with technology and skill-based issues.

References

Ashworth, A. and Hogg, K.I. *Added Value in Design and Construction*. Longman. 2000.

BSI. *BS ISO 31000 Risk Management – Guidelines*. BSI. 2018.

BSI. *BS EN IEC 31010 Risk Management – Risk Assessment Techniques*. BSI. 2019.

BSI. *BS 31100 Risk Management – Code of Practice for the Implementation of BS ISO 31000:2018*. BSI. 2021.

Lester, A. *Project Management, Planning and Control*. 7th Edition. Butterworth-Heinemann. 2017.

Lewis, H., Allen, N., Ellinas, C. and Godfrey, P. *Engaging with Risk*. London. CIRIA. 2014.

Odeyinda H., Kelly S., and Perera S. *An evaluation of budgetary reliability of bills of quantities in building procurement*. Proceedings of RICS COBRA. Paris. 2009.

Papachatzi, D. *Risk management in construction projects using BIM*. In: European Conference on Computing in Construction. 2019. Available at https://www.youtube.com/watch?v=-XAHCQ2aEqc (accessed 01/05/22).

Pittard, S. and Sell, P. *BIM and Quantity Surveying*. Routledge. 2016.

RICS. *Management of Risk*. 1st Edition. RICS. 2015.

RICS. *Cost Prediction*. Royal Institution of Chartered Surveyors. 2020.

9

Procurement

KEY CONCEPTS
• Procurement • Methods of price determination • Contractor selection • Procurement routes • Partnering

LEARNING OUTCOMES
After reading this chapter, you should be able to: • Identify the factors that influence procurement choice. • Understand the range of procurement routes and their advantages and disadvantages. • Appreciate good practice in the selection of contractors. • Identify the benefits of partnering.

COMPETENCIES
Competencies covered in this chapter: • Procurement and tendering.

Introduction

Procurement is the process that is used to deliver construction projects. The R.I.C.S defines procurement 'as the overall act of obtaining goods and services from external sources (i.e. a building contractor) and includes deciding the strategy on how those goods are to be acquired by reviewing the client's requirements (i.e. time, quality and cost) and their attitude to risk' (RICS 2014).

Willis's Practice and Procedure for the Quantity Surveyor, Fourteenth Edition.
Allan Ashworth and Catherine Higgs.

The procurement process has three phrases:

- A planning phase when decisions are made around the strategic approach to procuring goods and services,
- An acquisition phases when contracts are obtained, evaluated and awarded
- A contract management phrase when the contract is administered and compliance with requirements is confirmed (BS ISO 10845-1:2020)

Clients who have made the major decision to build are faced with the task of procuring the construction works that they require. This may be a daunting prospect, given the level of financial commitment and other risks associated with the venture, the complex nature of construction and the possible perception of the construction industry as one that frequently underperforms.

Approximately 50 years ago, clients of the construction industry had only a limited choice of procurement methods available to them when commissioning a new construction project. Since then, there have been several catalysts for change in procurement, such as:

- Government intervention
- Pressure groups being formed to create change for the benefit of their own members, for example, the British Property Federation
- International comparisons, particularly with the USA and Japan, and the influence of developments relating to the European Union
- The apparent failure of the construction industry and its associated professions to satisfy the perceived needs of its customers in the way that the work is organised
- The increase in Public Private Partnerships (PPP) projects
- The impact from research studies into contracting methods
- The response from industry, especially in times of recession, towards greater efficiency and profitability
- Changes in technology, particularly digital technology
- The attitudes towards change and the improved procedures from the professions
- The clients' desire for early contractor involvement and drive to support collaborative relationships
- The publication of headline reports: in 1994, the Latham Report, in 1998, the Egan Report and in 2009 the Wolstenholme Report

This has resulted in a significant shift in methods of procurement used by clients.

The *National Construction Contracts and Law Report 2018* (NBS 2018) shows a steady decline in the use of 'traditional' procurement by clients: from 59% in 2011 to 46% in 2017. Design and Build has seen a steady increase over this period with 37% clients reporting its use in 2017, supporting the desire for more collaborative relationships.

General matters

The wide range of procurement systems now available, and the understanding that procurement choice may have a significant bearing on the outcome of a project, signify both the opportunity and importance of meeting the procurement challenge with a well-considered strategy. The selection of appropriate contractual arrangements for any but the simplest type of project is difficult because of the diverse range of views and opinions that are available. Much of the advice is conflicting and lacks a sound base for evaluation. Individual experiences, prejudices, vested interests and familiarity, together with the need for change and the real desire for improved systems, have all helped to reshape procurement options available. The proliferation of differing procurement arrangements has resulted in an increasing demand for systematic methods of selecting the most appropriate arrangement to suit the particular needs of clients and their projects.

Whilst the main issue is that of satisfying the client's objectives, a matter examined in detail later, at an implementation level the following are the broad issues involved.

Consultants or contractors

These issues relate to whether to appoint independent consultants for design and management, or to appoint a contractor direct or an integrated team. The following should be considered:

- Single point responsibility
- Integration of design and construction
- Need for independent advice
- Overall costs of design and construction
- Quality, standards and time implications.

Competition or negotiation

There are various ways in which designers or constructors can secure work or commissions, such as invitation, recommendation, speculation or reputation. However, irrespective of the final contractual arrangements that are selected, the firms involved need to be appointed. Evidence generally favours some form of competition in order to secure the most advantageous arrangement for the client. There are, however, many different circumstances that might favour negotiation with a single firm or organisation. These include:

- Business relationship
- Early start on site
- Continuation contract
- State of the construction market
- Contractor specialisation
- Financial arrangements
- Geographical area.

Also, the advent, development and promotion of partnering has changed the view of some clients toward the need for competition (see the section on Partnering). In determining the need for competition, it must not be assumed that the choice between that and the option of negotiation is clearly defined, as each case must be decided on its own merits.

Measurement or reimbursement

There are in essence only two ways of calculating the costs of construction work. The contractor is either paid for the work executed on some form of agreed quantities and rates or reimbursed the actual costs of construction. The following are the points to be considered between the alternatives:

- Necessity for a contract sum
- Forecast of final cost
- Incentive for contractor efficiency
- Distribution of price risk
- Administration time and costs.
- Cost control.

Traditional or alternative methods

Until recently, most projects built in the UK have used single-stage selective tendering as their basis for contracting. With a wider knowledge of the different practices and procedures around the world, and some dissatisfaction with this uniform approach, other methods have evolved to meet changing circumstances and aspirations of clients. The following factors should be considered (these are examined in more detail later in this chapter):

- Support for development of collaborative practices
- Length of time from inception to completion
- Overall costs inclusive of design
- Accountability
- Importance of design, function, aesthetics, maintainability, sustainability and future flexibility in use
- Quality assurance
- Organisation and responsibility
- Project complexity
- Risk apportionment.

Standard forms of contract

A wide variety of different standard forms of contract are in use in the construction industry. The choice of a particular form depends on a number of different circumstances, such as:

- Client objectives
- Private client or public authority
- Type of work to be undertaken
- Status of the design
- Size of proposed project
- Method used for price determination.

Local authorities use different forms of contract than central government departments, while some of the larger manufacturing companies have developed their own forms and conditions. These often place a greater risk on the contractor and this is in turn is reflected in the contractor's tender prices. The different industry interests continue to develop a plethora of different forms for their own sectors. A trend toward greater standardisation would be welcomed. As long ago as 1964, the Banwell Report recommended the use of a single form for the whole of the construction industry as being both desirable and practicable. Standard contracts help simplify and speed up procurement procedures (HM Government 2022).

Although the general layout and contents of the various forms are similar, their details and interpretation may vary immensely. Different forms exist for main and subcontracts, for building or civil engineering works, and according to the relationship between the client, consultants and contractors, whether procured through a traditional, or more collaborative procurement routes. The different versions of the JCT (Joint Contracts Tribunal) forms of contract are used on the majority of building contracts. These were amended and published as a comprehensive set of forms in 2016.

There are alternatives to the contracts published by JCT. The New Engineering Contract (NEC4) published in 2017 is a suite of standard contracts widely used for major construction and infrastructure projects both within the UK and internationally. The FIDIC 2017 Rainbow Suite of contracts are also widely used internationally. The RIBA publish two standard forms of building contracts, typically used on UK projects for small value works. These alternatives have differing contractual approaches to that provided by the JCT suite of contracts and further meaningful commentary is considered to be beyond the scope of this text.

Methods of price determination

Building and civil engineering contractors are paid for the work that they carry out on the basis of one of two methods:

1) *Measurement.* The work is measured in place, i.e. in its finished quantities, and paid for on the basis of quantity multiplied by rate. The rate includes the associated costs for labour, plant and materials. Measurement may be undertaken by the client's surveyor; in which case an accurate and detailed contract document can be prepared. With this method, the risk relating to measurement is carried by the employer and that for the rate by the contractor. Alternatively, measurement may be undertaken by the contractor's surveyor or estimator, in which case it will be detailed enough only to satisfy the contractor concerned. With this situation, the risk relating to both measurement and rate is carried by the contractor.
2) *Cost reimbursement.* The contractor is paid the actual costs based on the quantities of materials purchased and the time spent on the work by operatives, plus an agreed amount to cover profit. Elements of measurement contracts may be valued on the same basis by the adoption of dayworks.

Measurement contracts

Measurement is the most often used method of price determination for contracts. The alternative forms of measurement contract used in the construction industry are as follows.

Drawing and specification

Drawing and specification is the simplest type of measurement contract and is really only suitable for small or simple project work. Each contractor measures the quantities from the drawings and read in conjunction with the specification prices them to determine the tender sum. The method is thus wasteful of the contractor's estimating resources and does not really allow for a fair comparison of tender sums. The contractor also has to accept a greater risk, since in addition to being responsible for the pricing they are also responsible for the measurements. In order to compensate for possible errors, contractors will tend to overprice the work.

Performance specification

This method of tendering requires the contractor to provide a price based upon the client's brief and user requirements alone. The contractor must therefore choose a method of construction and type of materials suitable to meet the project outcome. The contractor will take responsibility for the design and may select the least expensive materials and methods of construction that comply with the laid down performance standards. Some design and build contracts (see section on Procurement options, later in this chapter) may be based on performance specification.

Schedule of rates

In some projects it is not possible to predetermine the nature and full extent of the works. In these circumstances, a schedule is provided that is similar to a bill of quantities, but without the quantities. Contractors then insert rates against these items and these are used to calculate the price based on remeasurement. This procedure has the disadvantage of being unable to provide a contract sum, or any indication of the likely final cost of the project until the work is remeasured and the rates applied at the completion of the works. On other occasions, a comprehensive schedule already priced with typical rates is used as a basis for agreement. The contractor in these circumstances adds or deducts a percentage adjustment to all the rates. This standard adjustment can be unsatisfactory for the contractor, as some of the listed rates may be high prices and others low prices.

Bill of quantities

A bill of quantities provides the best basis for estimating, tender comparisons and contract administration. The contractors' tenders are therefore judged on price alone as they are all using the same measurement information. This is an efficient approach to the measurement of the works, since only one party is responsible rather than each individual contractor. Bills of quantities are referred to in JCT 2016 as the Contract Bills. As a negative, this type of documentation relies on the production of working drawings before tender stage and is time consuming to prepare. It is therefore not a practical approach where time is short. Also, the employer's acceptance of the risk relating to measurement and design contributes to uncertainty of final cost.

Bill of approximate quantities

In some instances, it may not be possible to measure the work accurately. In this case, a bill of approximate quantities is prepared and the entire project measured upon completion.

This is also a useful approach where an early start on site is required; approximate bills can be measured from incomplete drawings.

Cost-reimbursement contracts

These types of contract are not favoured by many of the industry's clients as there is an absence of a tender sum and a predicted final cost. This type of contract also often provides little incentive for the contractor to control costs. It is therefore only used in special circumstances, for example:

- Emergency work projects, where time cannot be allowed for the traditional process. For example, following a fire in the departure hall of a major airport, the key priority of the client would be to ensure passengers could use the airport without disruption as quickly as possible. Any possible additional cost arising from the nature of the contract would probably be insignificant relative to the possible loss in revenue.
- When the character and scope of the works cannot really be determined. For example, this is frequently the case with elements of measurement contracts for which the works are valued on a dayworks basis.

Cost reimbursement contracts can take several forms. The following are three of the types that could be used in the above circumstances. Each of the methods pays contractors' costs and makes an addition to cover profit. Prior to embarking on this type of contract, it is important that all the parties concerned are fully aware of the definition of contractors' costs as used in this context. Expenditure on the project needs to be carefully audited and the likely end cost reported regularly to the client.

Cost plus percentage

The contractor is paid the costs of labour, materials, plant, subcontractors and overheads, and to this sum is added a percentage to cover profits. The percentage is agreed at the outset of the project. A disadvantage of this method is that the contractor's profit is related directly to expenditure. Therefore, the more time spent on the works, the greater will be the profitability. In other words, lower efficiency leads to greater profit.

Cost plus fixed fee

In this method, the contractor's profit is predetermined as a fee and is agreed for the work before the commencement of the project. There is therefore a possible incentive for the contractor to attempt to control the costs because it will increase the rate of return. However, to counter this, it is also possible that, since the fee is fixed, the contractor can improve his profitability only by reducing management costs, precisely what the client is seeking. In practice, because it is difficult to predict cost accurately beforehand, it can cause disagreement between the contractor and the client's professional advisers when trying to settle the final account, if the actual cost is much higher than that estimated at the start of the project.

Cost plus variable fee

This method requires a target fee to be set for the project prior to the signing of the contract. The contractor's fee is made up of two parts: a fixed amount and a variable amount depending on the actual cost. This method provides an even greater incentive to the contractor to control costs but has the disadvantage of requiring the target cost to be fixed on the basis of a very rough estimate.

Contractor selection

There are essentially two ways of selecting a contractor: through competition or by negotiation. This will apply to any working arrangement, including strategic partnering, which in the first instance requires the appointment of a contractor partner. Competition may be restricted to a few selected firms or open to almost any firm that wishes to submit a tender. The contract options described later are used in conjunction with one of these methods of contractor selection. It is important that the tendering process is fair, transparent, non-discriminatory and recorded accurately. The Fraud Act 2006 defines fraudulent activities, whilst section 7 of the Bribery Act 2010 penalises companies that do not have adequate procedures in place to prevent fraudulent behaviour.

As a consequence of the UK's, post Brexit, independent membership of the World Trade Organisation (WTO) committed to the Agreement on Government Procurement (GPA), UK firms have retained access to public procurement opportunities in 48 countries (Cabinet Office 2020). Information on public works within the EU and other countries is published on the SIMAP portal using Tenders Electronic Daily (TED). European-derived legislation imposes restrictions on UK tendering arrangements for the procurement of public goods and works. Where government-related expenditure is involved above a prescribed contract value, currently just over £4.73 m, it is a requirement by The Public Contracts Regulations 2015 to publish contract notices on the UK *Find a* Tender service. However, it is anticipated as a result of the Green Paper *Transforming public procurement,* the UK public procurement regulatory framework will be simplified. The Sourcing Playbook (2021) sets out 11 key policies providing best practice guidance that applies for the sourcing and contracting of all public works projects.

The JCT Practice Note Tendering 2017, although not mandatory, provides guidance and suggested good practice for the selection of contractors and the awarding of construction contracts within the UK using selective tendering (see 'Selective Competition' below). This guidance updates previous JCT publications based on the 1997 CIB Code of Practice for the Selection of Main Contractors. The latter document identified the key principles of good practice to be adopted when appointing contractors in competition (either by single or two-stage tendering), which are still applicable today:

- 'clear procedures should be followed that ensure fair and transparent competition in a single round of tendering consisting of one or more stages'
- 'the tender process should ensure receipt of compliant, competitive tenders'; where contractors feel it necessary to attach conditions to tender submissions due to their inability to fully comply with the tender documents, they should be given the opportunity to withdraw the qualifications, but without amending the price. If variant tenders are to be considered, this must be stipulated within the Invitation To Tender.
- 'tender lists should be compiled systematically from a number of qualified contractors'.

- As stated, it may be necessary to consider European-derived procurement law when compiling lists of tenderers. Considerations to be made when selecting the preliminary list of firms include: the firm's financial standing and record; its recent experience of building over similar contract periods; the general experience and reputation of the firm for similar building types; the adequacy of its management; technical capabilities (for example BIM) and its capacity to undertake the project and provide resources in respect to health and safety requirements, approach to quality assurance and sustainable credentials. This can be done through a pre-qualifications questionnaire. PAS 91:2013 Construction Prequalification Questionnaires (BSI 2017b) provides a set of questions based on good practice within the sector. In 2021, the industry-wide pre-qualification system, the Common Assessment Standard, was introduced. Companies are able to demonstrate that they meet an agreed industry standard significantly reducing duplication and costs within the sector. Although, in some respects, the inclusion of some tenderers may appear to be automatic due to their size and previous record, consideration should be given to regional standing, resources and reputation which may differ from the national or another regional position. Generally, with a prequalified list of tenderers, there should be no doubt as to the ability of any of the tenderers to satisfactorily complete the contract.
- 'practices that avoid or discourage collusion should be followed'
- 'tender lists should be as short as possible'; the cost of tendering is usually between 0.5 and 1% of a company's turnover, with a success ratio of about 1 in 4, but varies depending on competition levels (Brook 2017) Therefore, increasing the number of tenderers on the tender list increases the potential costs of abortive tendering; this in turn will be reflected in a higher tender price. Figure 9.1 details the number of tenderers recommended in the JCT Practice Note. Whilst the rationale for this guidance is not in doubt, in practice, clients must consider the possibility of collusion and breaches in confidentiality. The risks of this are increased where the list of tenders is very small, three for example.
- 'suites of contracts and standard unamended forms of contract from recognised bodies should be used where they are available'
- 'confidentiality should be respected by all parties'
- 'sufficient time should be given for the preparation and return of tenders'; these times clearly depend on situation, such as project size and complexity, and procurement type.

	Contract Type	
	Construct only	Design and Construct
Tender list (number of firms)	3–6	Maximum 4
Tender period	Minimum 28 days	3–4 months when design not prescribed.

Fig. 9.1 Recommended practice (*Source:* Adapted from JCT Practice Note: Tendering 2017).

The recommendations are shown in Fig. 9.1. Insufficient time for the contractor to fully examine all aspects of the project and assess the impact on resources may lead to a higher price to cover risks. A lack of time may also lead to the contractor withdrawing from the process.

- 'sufficient information should be provided to enable the preparation of tenders'; this information is not just that relating to the project, specifications, drawings, etc. but how the tender process will be handled, managed and evaluated. This information will normally be contained in the Invitation to Tender. The JCT Tendering Practice Note provides a model form.
- 'conditions should be the same for all tenderers'; conditions refer to instructions provided, such as return date, site visit arrangements and the method for submitting queries.
- 'Tenders should be assessed and accepted on quality as well as price'; the basis of award of the contract is stated in the Invitation to Tender. Evaluation criteria such as quality, price, life cycle costs and environmental requirements must be stated with their associated weightings. This will inform the contractor the pricing strategy needed in order to win the bid. In the case of Public contracts, the basis of award should be the most economical advantage tender; in most cases this will be the tender that represents the best price/quality ratio. Table A4–A6 within BS ISO 10845-1:2020 *Construction Procurement Part 1* provides examples of commonly used quality criteria, their subject matter and qualitative indicators to be used in evaluation.

Consideration should also be given within the evaluation criteria, especially if BIM is implemented to, the tenderer's capacity and capability to meet information management requirements as identified in BS 19650-2:2018 *Organisation and digitization of information about building and civil engineering works, including BIM . . .* (2021b).

- 'tender prices should not change on an unaltered scope of works'
- 'there should be a commitment to teamwork from all parties'; this is very much the essence of the way ahead for improving the construction industry, central to the partnering approach and desired in all forms of contracting.

Sustainable procurement and tendering

BS ISO 20400 defines sustainable procurement as procurement that has the most positive environmental, social and economic impacts possible across the entire life cycle and that strives to minimize adverse impacts (BSI 2017a). Whilst the client's sustainable objectives to support the organisation's sustainability agenda will impact on the procurement strategy of the construction project, sustainable procurement is wider in that consideration is needed to how procurement decisions contribute to sustainable development in general (See Chapter 18). BS ISO 20400:2017 *Sustainable procurement guidance* provides guidance to organisations on how to integrate their sustainable practices into their procurement strategy and processes. With reference to a generic procurement process, the guide recommends where and how to enable sustainability practice. Annex C, though not construction related, provides useful examples of how a sustainable issue approach can be adopted and the potential of how such an approach to construction processes could potentially link to the UN Sustainable Development Goals. Sustainability requirements and evaluation criteria should be stated within the Invitation to Tender, as well as requiring contractors to demonstrate their experience and competence to fulfil sustainability practices throughout their supply chain.

Selective competition

This is the traditional and most popular method of awarding construction contracts. In essence a number of firms of known reputation are selected by the project team. The guidelines outlined above should be adhered to where relevant.

Open competition

Open competition, whereby details of the project are first advertised in local or trade publications or on websites inviting requests for tender documents, is not an efficient practice and, in consideration of the guidelines above, is not a recommended practice. Whilst the approach allows new contractors, or those who are unknown to the project team and the client, the possibility of submitting a price, the costs of tendering are high and the process of pricing the items lengthy. Where selective tendering is adopted with a list of six tendering contractors, the law of averages indicates that tendering costs for six projects will be absorbed in one successful submission. In open competition, this waste is much greater, depending on the total number of interested contractors, and is of course borne by clients. Some very reputable contractors may not be interested in tendering in such conditions.

Open tendering may relieve the client of the obligation of accepting the lowest price. This is because firms are generally not vetted before the tenders are submitted. Factors other than price must therefore be taken into account when assessing tender bids.

Negotiated competitive dialogue

The negotiated contract method of contractor selection involves the agreement of a tender sum with a single contractor. It is more likely to occur when the client and contractor have experience of working together and is an influencing factor in the increase in its use (NBS 2018). The contractor will offer a price using the tender documentation, and the client's surveyor then reviews this in detail with reference to their pre-tender estimate. The two parties then discuss the rates that are in contention, and through a negotiation process a tender acceptable to both parties can be agreed. Owing to the absence of any competition or other restriction other than the acceptability of price, this type of contract procurement is not generally considered to be cost advantageous. It can be expected to result in a tender sum that is higher than might have been obtained by using one of the previous methods, although substantiation of the order of this is difficult to assert. To protect the client's interest, it is normal to include the ability for the client to withdraw if agreement is unable to be reached. Because of the higher sums incurred, public accountability and the possible suggestion of favouritism, government departments do not generally favour this method.

A negotiated contract should result in fewer errors in pricing. It also accommodates contractor participation during the design stage, and this may result in savings in both time and money. It should also lead to greater cooperation during the construction period between the designer and the contractor. Where an early start on site is required, it clearly offers distinct advantages over a drawn-out competitive tendering approach.

Two-stage tendering

The main aim of two-stage tendering is to involve the chosen contractor on the project as early as possible. It therefore tends to succeed in getting the party who knows what to build (the architect or engineer) working in collaboration with the firm that knows how to build it (the contractor) before the design is finalised. The contractor's expertise in construction methods can thus be used in the architect's design. A further advantage is that the selected contractor will be able to start on site sooner than would be the case with the other methods of contract procurement.

At the first stage, an appropriate contractor must be competitively selected. This can be achieved by inviting suitable firms to price elements of work that are sufficiently designed and provision of a schedule of rates for other items. The contractor will also be required to state their overhead and profit percentages and preliminaries. The guidelines discussed earlier in the chapter are applicable to this stage in the process.

The prices established at the first stage will then form the basis during the second stage for the subsequent construction price agreement that will be achieved through negotiation.

Serial tendering

Serial tendering is a development of the system of negotiating further contracts, where a firm has successfully completed a contract for work of a similar type. Initially, contractors tender against each other, possibly on a selective basis, for a single project. There is, however, a legal understanding that several other similar projects would automatically be awarded using the same schedule of rates. The contractors would therefore know at the initial tender stage that they could expect to receive a number of contracts, which could provide them with continuity in their workload. As an alternative, a 'series' of projects may be awarded to a contractor who successfully tenders in competition on the basis of a notional 'master' bill of quantities that will include a comprehensive range of items and be used to price each future project in a defined series. In either situation, conditions would be written into the documents to allow further contracts to be withheld where the contractor's performance was unsatisfactory.

Serial contracts should result in lower costs to contractors who are then able to gear themselves up to such work by, for example, purchasing suitable types of plant, and who would generally benefit from economies through the increased total contract size. Serial contracts are appropriate to buildings such as housing and schools in the public sector. This method may also be usefully employed in the private sector in the construction of industrial units. It has been successfully used with industrialised system buildings.

This arrangement for letting contracts, although having several advantages including that of promoting a good working relationship, is not to be confused with the concept of partnering which is discussed in detail later in this chapter.

Procurement options

There are several procurement options available to the client and within each broad type there are several variants, each of which may be possibly refined to accommodate particular client needs and project specifics. For example, within a traditional arrangement, it is normal to have some of the works carried out under a cost plus or remeasurement arrangement and possible also to let a portion of the works on a design and build basis. An appreciation of the operation and application of each of the procurement options is essential to developing a sound procurement strategy.

Traditional

In this approach, the client commissions an architect to take a brief, produce designs and construction information, invite tenders and administer the project during the construction period and settle the final account. If the building owner is other than small, the architect, traditionally the first point of client contact, will advise the client to appoint consultants such as quantity surveyors, structural engineers and building services engineers. The contractor, who has no design responsibility (unless particular portions of the work are so identified), will normally be selected by competitive tender unless there are good reasons for negotiation. The design team members are independent advisers to the client and the contractor is only responsible for executing the works in accordance with the contract documents. Figure 9.2 shows the relationship of the parties.

The key feature of this form of procurement is the separation of design and construction. The client appoints a team of consultants, frequently led by the architect, to design the building and prepare tender documentation. The main advantages and disadvantages of this procurement option are as follows.

Advantages

- *A high level of price certainty for the client.* Since cost is known before construction commences and providing the design process has been completed fully in the pre-contract stage, a high degree of price certainty exists.
- *Ease of valuing design changes.* Cost management processes are aided as price information is established within tender documentation.
- *Competitive fairness.* Comprehensive tender documentation results in high tender quality as all contractors bidding on same basis.

Disadvantages

- *A relatively lengthy time from inception to start on site.* This is due to the sequential nature of the design and construction phases.
- *Problems relating to design error.* The risk relating to the design lies with the client. Post-contract design changes are frequent and may lead to resultant delays and disputes.

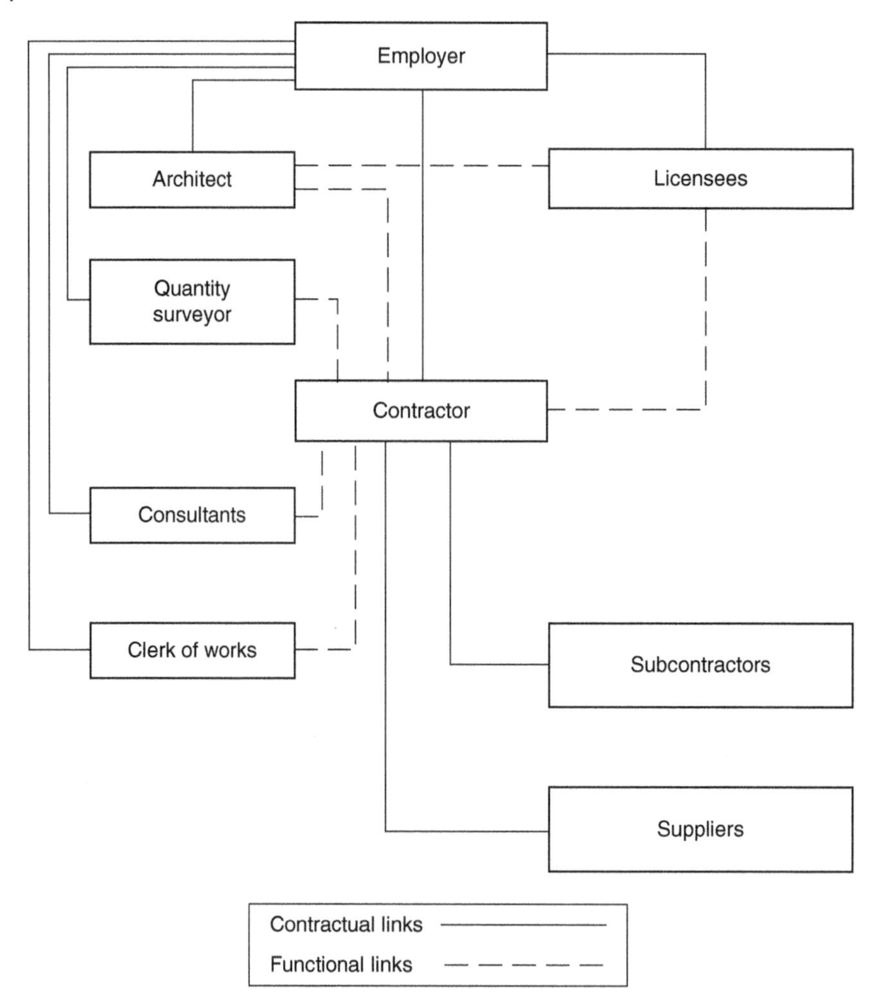

Fig. 9.2 Traditional.

- *Lack of involvement of the constructor in the design process.* As the contractor is not appointed at the design stage, there is reduced opportunity for a collaborative approach to buildability

Design and build

It has been suggested that the separation of the design and construction processes, which is traditional in the UK construction industry, has been responsible for a number of problems. Design and build (Fig. 9.3) can often overcome these by providing for these two separate functions within a single organisation. This single firm is generally the contractor. The client, therefore, instead of approaching an architect for a design service, chooses to go directly to the contractor. With this method of procurement, the contractor therefore accepts the risk for the design element of a project.

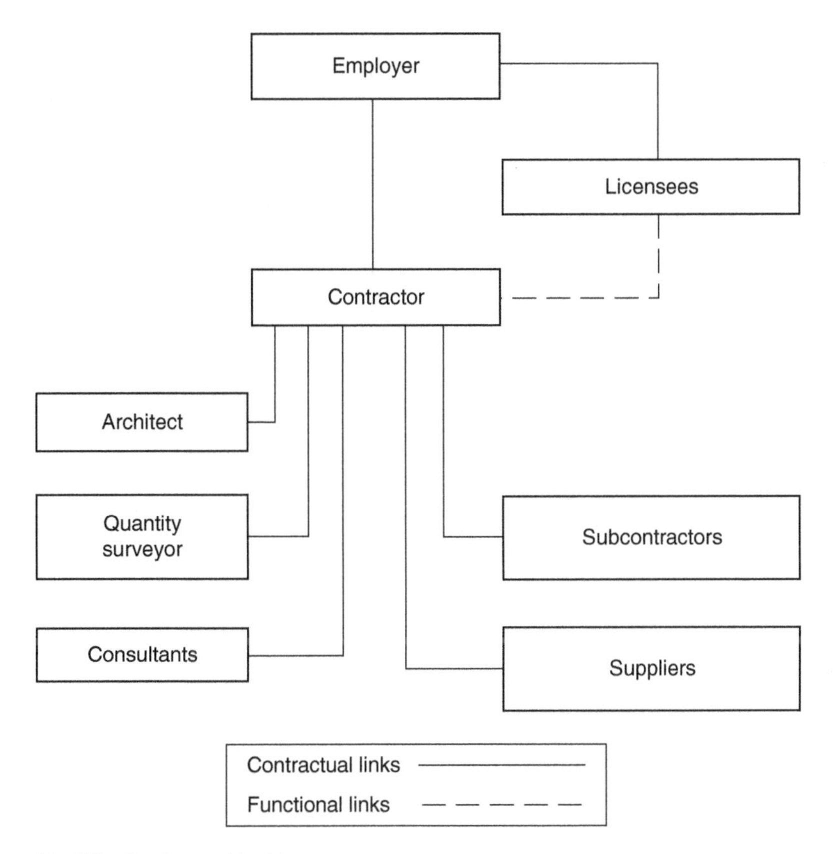

Fig. 9.3 Design and build.

It is common for the client to initially appoint design consultants to develop a brief, examine feasibility and prepare tender documents that will include a set of employer's requirements. These requirements will usually include a performance specification. Contractors are invited to tender on the basis that they will be responsible for designing and constructing the project and will submit a bid, which will incorporate design and price information. The contractor's proposals will be examined by the client and the project subsequently awarded. An issue to be considered in the early stages of the project is the nature of the employer's requirements, which may vary significantly in terms of detail. Clients may need to balance their conflicting desires to both direct the design and transfer full design risk to the contractor.

Develop and construct is an approach whereby the client's consultants prepare a concept design and ask the contractor to develop that design and construct the works. Frequently, the client attempts to transfer the risks associated with the early design by novation. With this approach, the client's architect and pre-contract designs are 'transferred' to the contractor who accepts the related liability as part of the contractual arrangement. This procedure is unpopular with contractors, who are inclined to believe that in such situations the allegiance of the architect remains with the client.

There are several advantages and disadvantages in the use of design and build that should be considered, including the following.

Advantages

- *Single point responsibility.* If problems arise during the works, the contractor will be motivated to reduce design problems and their mitigation when they arise.
- *Price certainty prior to construction.* Provided there are no client changes, a high level of price certainty exists.
- *Reduced project duration.* This is made possible due to the overlap of the design and construction phases.
- *An improved degree of buildability.* This may be achieved since the contractor has a greater opportunity to influence the design although, in practice, this will depend on the level of design direction contained in the employer's requirements.
- *Fitness for purpose.* The potential exists to extend design liability beyond reasonable skill and care to include fitness for purpose as the design and construction are provided as one service.

Disadvantages

- *Client's reduced ability to control design.* The scale of this concern depends on the design and build approach adopted.
- *Commitment prior to full design.* In essence, the client contracts to buy a building that is based on client requirements and yet to be fully designed.
- *Difficulty in comparing tenders.* The evaluation of the differing design alternatives contained within the contractors' proposals may add significant complexity and time to the normal tender review process.
- *Cost management difficulties.* A much reduced level of price information is available to the client's surveyor, creating significant cost management problems.

Management-based procurement strategies

There are several possible variants to a management-based procurement strategy. Each shares the main characteristic of the appointment of a party – usually a contractor – to manage the construction works (and in the case of design and manage, the design of the works also) in return for a lump sum or percentage fee. The scope of the management provision and methods of establishing contractual links in the construction supply chain provide differentiation between specific types. This approach to procurement is particularly suited to large, complex projects. In recent years management-based procurement strategies have been used less frequently, reflecting client's willingness to take on risk (RICS 2021).

Management contracting

Management contracting is a term used to describe a method of organising the project team and operating the construction process. The management contractor acts in a professional capacity, providing the management expertise and buildability advice required in return for a fee to cover overheads and profit. The contractor does not, therefore,

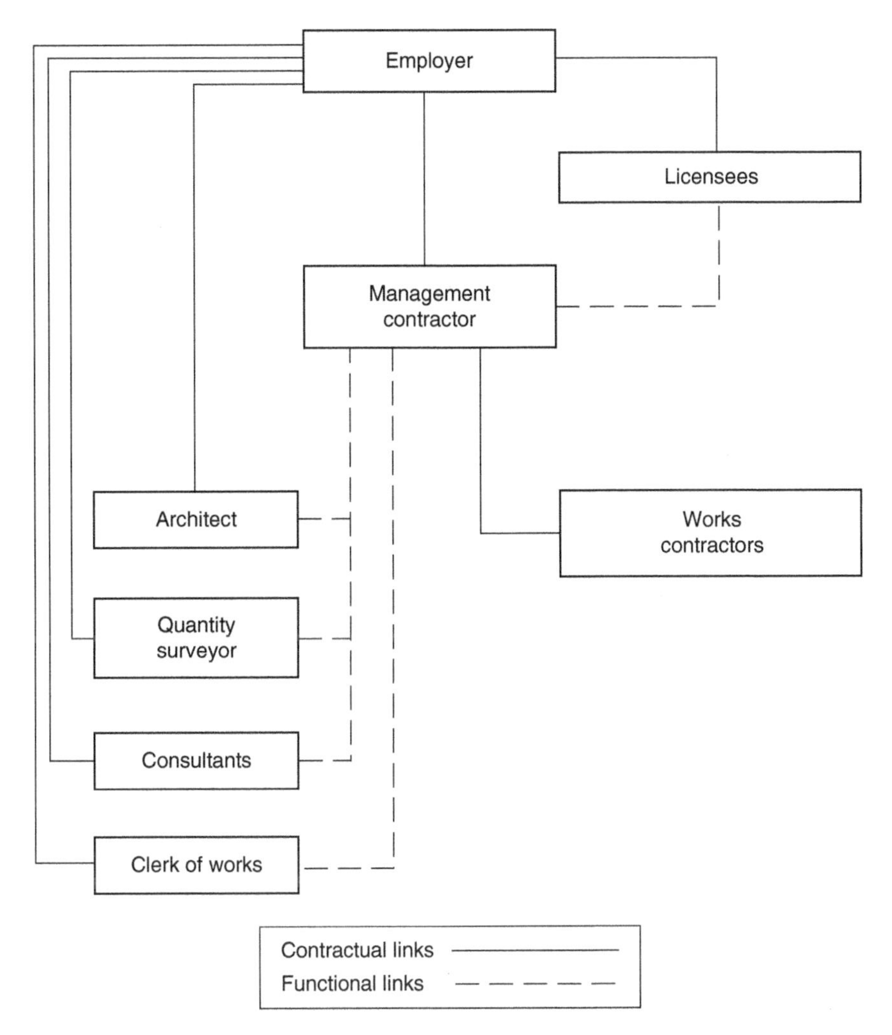

Fig. 9.4 Management contracting.

participate in the profitability of the construction work itself and does not employ any of the labour or plant, except for the possibility of the work involved in the setting up of the site and the costs normally associated with the preliminary works. The works are let on a package basis, usually by competitive tender (although other methods of procuring individual works packages may be adopted), and contracts are between the management contractor and works contractors. Figure 9.4 shows the relationship of the parties.

Advantages

- *Early involvement of the management contractor.* Because the contractor is employed on a fee basis, the appointment can take place early during the design stage. The contractor is therefore able to provide a substantial input into the practical aspects of the building technology process.

- *Reduced project duration*. Construction work can start as soon as sufficient work has been designed.
- *Accommodates later design decisions*. Some design decisions relating to work which sequentially may be at the end of the construction phase could be deferred due to work being let in packages.

Disadvantages

- *Commitment prior to full design*. Design is incomplete at the time of commencement and therefore aspects of price, quality and programme are uncertain when the client decides to proceed.
- *Increase in client risk*. The risks relating to additional costs arising as a result of the faults of the works contractors (e.g. delays, defective work, claims from other works contractors) lie with the client.
- *Cost management oversight*. As the final cost is not known until the last works package is procured, detailed cost reporting is required.

Design and manage

Design and manage is similar to management contracting in that a management contractor is paid a fee for managing the construction of the works; however, as the term may suggest, it extends the role to incorporate the management of the design of the project. The benefits and pitfalls of the system are similar to those experienced in management contracting. The major difference relates to the additional design responsibility of the management contractor. To this extent, the characteristics of this form of procurement bear similarity with design and build – the gain of single point responsibility and the loss of design control.

Construction management

Construction management offers an alternative to management contracting and shares a close similarity to management contracting; however, the main difference is that the individual work package contractors are in a direct contract with the client.

The client appoints a construction manager (either consultant or contractor) with the relevant experience and management expertise. The construction manager, if appointed first, would take the responsibility for appointing the design team, who would usually also be in direct contract with the client. The construction manager is responsible for the overall control of the design team and work package contractors throughout both the design and construction stages of the project.

Because of the direct contract arrangements, additional client involvement is required and this approach is therefore not recommended for those without adequate experience, resources and understanding of both the product of the construction industry and its processes (Hughes et al. 2015).

Contract strategy

In deciding on a contract strategy, recognition should be given to client priorities (and, where possible, desires) in terms of project duration, cost and quality objectives. In order to determine an appropriate method, it is necessary to match the needs of the client with the most suitable approach available. To enable this, an understanding of both the client's objectives and the operation and relative attributes of the available procurement/contract types and methods of appointment is required. The following points should be borne in mind.

Project size

Small schemes are not suitable for more elaborate forms of contractual arrangement. Such procedures are also unlikely to be cost effective. Small schemes will more likely use either selective or open tendering or a type of design and build. Medium to large projects can use the whole range of methods; with very large schemes, more advanced and complex forms of procurement may be necessary.

Client type

Clients who regularly carry out construction work are much better informed, develop their own preferences and will not require the same level of advice as those who build only occasionally. They may, however, need to be encouraged to adopt more suitable and appropriate methods, and the quantity surveyor will need to convince such clients that using such suggestions will lead to improved procedures. Experienced clients may have a working knowledge of the construction industry, may retain the services of in-house construction advisors and will be able to contribute to the process throughout. Inexperienced clients may have unrealistic expectations and the tendency to inappropriately interfere rather than contribute.

There is a range of client types which, although each client will be unique, may be categorised by key characteristics. The categories of client that are likely to be encountered in practice include: public bodies, including local authorities, who are experienced and have a large and wide portfolio of construction needs; large commercial developers who build for profit; and large and small companies who build to improve and extend their business and are thus owner occupiers. These different client types will have different procurement needs.

The priorities of these groupings will generally differ in terms of the balance of time, cost and quality objectives, accountability and certainty of output. For example, public sector clients are likely to be driven more by the need for low cost, cost certainty and accountability than are private sector clients.

Client procurement needs

Time

Project duration or completion dates may be critical to the success of a project; in some situations, if they are not met, it could lead to total failure in meeting a client's objectives. Whilst most clients are likely to have a desire for an early building completion, it is important to distinguish between this and true need since attempting to meet the objective or early completion is likely to have consequences on other project requirements. The choice of procurement strategy can have a large impact on the duration of a construction project.

Cost

If a limited capital budget is the prime consideration of the client, quality, in the form of a reduced specification, is likely to be restricted and project duration will be the optimum in terms of construction cost rather than client choice.

It is generally believed that open tendering will gain the lowest price from a contractor although, as previously stated, this brings several possible problems and is therefore not recommended. Negotiated tendering will be more expensive although it is difficult to ascertain an exact premium. Projects with unusually short contract periods tend to incur some form of cost penalty. The introduction of conditions that favour the client, or the imposition of higher standards of workmanship than normal, will also push up costs.

The degree to which cost certainty is required prior to commitment to construction or at project completion restricts procurement choice considerably. The risks associated with abortive design fees are also a factor of which clients should be aware.

Quality

The quality of a building is influenced by several factors, including: the briefing process; the suitability of materials, components and systems and their interrelationships within the total design; and the quality control procedures that are in place during both design and construction. The choice of procurement strategy can affect the design process and means of control by which the client and his advisors can monitor both specification and construction activity. It should be noted that quality is a subjective issue, sometimes difficult both to define and identify; it may not necessarily mean a more complex building or a higher specification.

Accountability

Whilst organisations receiving public funding will naturally be concerned with accountability since they are subject to public scrutiny, it is often assumed that accountability is less of a concern for companies in the private sector. This assumption may be misplaced as the majority of clients will have to justify expenditure.

Certainty of project objectives

Some forms of procurement (e.g. management contracting) incorporate an inherent facility to accommodate design development throughout a project and others are particularly unsuited to design changes (e.g. design and build).

Market conditions

The selection of the process to be used will vary with the general state of the economy and should also consider the predicted changes to it within the project's life. When there is ample work available, contractors may be reluctant to enter what they perceive to be unsatisfactory relationships. When the market is at a low point, contractors could be more willing to accept risk, a situation which should be understood by the client for its negative effects as well as positive, since liquidations and claims are more likely in such conditions. However, global market conditions, such as the impact of increasing energy costs and resource shortages, can significantly impact on the contractor's decision to tender, as seen in 2022, with the increases in prices of construction products and materials due to both scarcity and being energy-intensive products. The key to a successful procurement process in these circumstances is for the quantity surveyor to consider where possible the mitigation and sharing of the risk of inflation and advise the client accordingly (BCIS 2021).

The relative strengths and weaknesses of the available options may be evaluated in terms of the above factors. Since each strategy will contain a differing balance of the various attributes required by clients, a prioritisation of key objectives is necessary to enable the most suitable choices to be made.

Partnering

Many of the problems that exist in construction are attributed to the barriers that exist between clients and contractors. In essence, partnering is about breaking down these barriers by establishing a collaborative working environment that is based on mutual objectives, teamwork, trust and sharing in risks and rewards.

Within the UK construction industry, partnering activity was given significant impetus by the Latham Report (1994) and the report by the Construction Task Force (1998): *Rethinking Construction (The Egan Report).* Further attention was given to the need to increase the use of the partnering approach in 'Accelerating Change', (Strategic Forum for Construction 2002, chaired by Sir John Egan). One of the forum's six headline targets stated that '20% of construction projects (by value) should be undertaken by integrated teams and supply chains by the end of 2004, rising to 50% by the end of 2007'. The prominent recognition given to partnering by Egan was a clear indication of the strong belief that the partnering approach can make a major contribution to improving value within the construction industry. The Wolstenholme Report (2009) observed that the client's lack of awareness of the potential value of integrated supply chains and consultants' reliance on dated procurement methods was a key 'blocker' for the construction delivery model envisaged in the Egan Report. Recognising government's position as an exemplary client, the Government Construction Strategy (GCS) 2016–2020 was built on the success of GCS 2011–2015 identifying a principle objective to deploy collaborative procurement techniques which enabled early contractor and supply chain involvement.

Though dated, the following provides an authoritative definition of partnering. The version here (which, it is important to add, has been subsequently revised by the authors) provides a relatively simple view, which should be easily understandable by all:

> 'Partnering is a management approach used by two or more organisations to achieve specific business objectives by maximising the effectiveness of each participant's resources. The approach is based on mutual objectives, an agreed method of problem resolution and an active search for continuous measurable improvements.' (Bennett and Jayes 1995)

It can be seen from this definition that there are three main components to a partnering arrangement: mutual objectives; an agreed method of problem resolution; an active search for continuous and measurable improvements. Bennett and Pearce (2006) add 'feedback' as a fourth and important component as 'teams need to be guided by feedback about their own performance if they can deliver the substantial benefits that partnering can provide'.

Collaboration, cooperation, trust, respect and feedback are all key requirements that are stated within the Overriding Principles of working together in the JCT Construction Excellence Contract 2016, an appropriate form of contract to be used in partnering.

The nature of partnering is such that it may take several different forms depending on the situation and the objectives of the parties involved. However, it is possible to broadly classify a partnering arrangement as either 'project partnering' or 'strategic partnering'. Although the differences between these two, which relate to scale and level of relationship, are significant, the essence of the partnering concept is the same in both.

Project partnering (short term)

As the name suggests, project partnering relates to a specific project for which mutual objectives are established and the principles are generally restricted to the specified project only. The great majority of partnering opportunity is of this type since:

- It can be relatively easily applied in situations where legislation relating to free trade is strictly imposed
- Clients seeking to build on an occasional basis may use it.

Strategic partnering (long term)

Strategic partnering takes the concept of partnering beyond that outlined for project partnering to incorporate the consideration of long-term issues. The additional benefits of strategic partnering are a consequence of the opportunity that a long-term relationship may bring and could include:

- Establishing common facilities and systems
- Learning through repeated projects
- The development of an understanding and empathy for the partners' longer term business objectives.

The partnering process

Once the decision to partner has been made, the procedure of partnering principally involves a selection procedure, an initial partnering workshop and a project review.

The selection of a good client or contractor partner who is trustworthy and committed to the arrangement is fundamental to the process. Partnering can be used with traditional procurement methods in the initial selection stages. This is followed by a workshop that will be attended by key stakeholders and usually results in the production and agreement of a partnering charter to be signed by all participants.

During the project implementation stage, best practice includes workshops throughout the project to review performance. This will incorporate all relevant project matters including quality, finance, programme, problem resolution and safety. Benchmarking against Key Performance Indicators can act as a mechanism to do this. The ISO44001-2017 standard also provides guidance in developing and managing collaborative relationships between partnering organisations.

Advantages and disadvantages of partnering

The adoption of partnering – at a strategic level or for a specific project – is considered to bring major improvements to the construction process resulting in significant benefits to each partner. Further benefits are achieved if partnering is extended throughout the supply chain. These may include:

- Improvement in the delivery of the client's objectives
- Early contractor and supply chain involvement to improve buildability
- Reduction in disputes
- Reduction in time and expense in the settlement of disputes
- Reduction in costs
- Improved quality and safety
- Improvement in design and construction times and certainty of completion
- More stable workloads and income
- A better working environment to support innovation
- Improved transparency and trust between parties.
- Continuous improvement

However, in considering the use of a partnering approach, it is necessary to acknowledge the existence of some important disadvantages and concerns regarding its use. These may include:

- Initial costs
- Complacency
- Single source employment: strategic partnering could result in either party becoming very dependent on one client and thus be in a very vulnerable position should this source of work be threatened
- Lack of confidentiality of client/contractor processes and systems: disputes could occur if information, which may be regarded as commercially confidential, were to be withheld from one or more of the partners
- The lack of competition: the accepted route to securing a 'good value' price is through the competitive tendering process
- Absence of partnering through the supply chain: the benefits of partnering seem to be absent from the contractor–subcontractor relationship

As previously defined partnering is a management approach, a system of processes integrating activities of participants as objective targets within a behavioural relationship. To support this relationship, the objectives need to be clearly articulated within the contract (RICS 2021).

Alliancing	Long Term Partnering
Time bound, targeted end date	Open ended, regular reviews of performance
Performance focussed via discrete objectives (e.g. project milestones, client deliverables etc)	Performance focused via business delivery (growth, profit, cost base, etc)
Team formed from as many organisations as necessary	Primarily one-to-one relationships
Formal, legally binding contract underpins delivery of objectives	Joint agreement clarifies and supports relationship development
Vision, objectives and goals are clear at the outset	Vision is agreed, objectives are understood but may evolve, and goals emerge throughout the relationship

Fig. 9.5 Alliancing and long-term partnering (*Source:* ECI undated/The European Construction Institute).

Alliancing and long-term partnering

Long-term partnering is a distinct but complimentary process to alliancing (ECI undated). Alliancing is different, in that it has a set time period rather than the open-ended relationship of long-term partnering. Figure 9.5 shows the characteristics of each.

The Private Finance Initiative (PFI)

A Public–Private Partnership (PPP) refers to any alliance between public bodies, local authorities and central government and private companies to deliver a public project or service (Ashworth and Perera 2015). Up until 2012, the most common form of Public Private Partnership (PPP) introduced in the early 1990s was the Private Finance Initiative (PFI). PFI is a procurement method in which the private sector finances, builds and operates the provision of capital projects through long-term concession agreements, typically 25–30 years. It is a means of funding major capital investments without the need for government funding. It is expedited through private consortia who are given the opportunity to design, build and, increasingly, manage new projects. This one-stop approach, frequently contractor-led, is complex, incurs high costs in tendering and carries with it the risk of poor quality of service in facility operation. However, assuming the client has made clear the required outputs, it is seen by many to offer an attractive alternative on large government projects, with an inbuilt incentive to design in a manner that makes operation more efficient.

Projects include hospitals, schools, transport, defence, leisure, culture, housing and waste. However, since 2008, due to the weakening economy, the number of new PFI projects has decreased and adverse publicity about the performance of some PFI schemes (e.g., relating to NHS projects) raised further doubt about PFI activity. As a response to the criticism of PFI's lack of transparency of financial returns and procurement times, the new PF2 model was launched in December 2012. As of 31 March 2018, there were a total of

704 current PFI and PF2 projects with a total capital value of £57 billion (HM Treasury 2019). With continued government concerns of inflexibility, complexity and potential risk to finances, at the 2018 Budget it was announced that PFI and PF2 would not be used on any new contracts. However, the last PFI payment is due in 2049/50 on the Oxleas NHS Foundation Trust PFI signed in 1998 for 52 years (Booth 2018).

The role of the quantity surveyor

It is of fundamental importance to clients who wish to undertake construction work that the appropriate advice is provided on the method of procurement to be used. The active selection of an inappropriate form of procurement, or the passive acceptance of a regularly used and hence 'comfortable' practice, can have a major impact on the success of a project.

Advice must be relevant and reliable and should be given independently, without the intrusion of individual bias. The advice provided should exclude considerations of self-interest. A client should not be faced with unwittingly selecting an inappropriate form of procurement as a consequence of selecting a particular consultant or profession as the first port of call.

Quantity surveyors are in an excellent position as procurement managers with their specialist knowledge of construction costs and contractual procedures. They can appraise the characteristics of the competing methods that might be appropriate and match these with the particular needs and aspirations of the employer. Procurement management may be broadly defined to include the following:

- Determining the employer's requirements in terms of time, cost and quality
- Assessing the viability of the project and providing advice in respect of funding and taxation advantages
- Recommending an organisational structure for the development of a project as a whole
- Advising on the appointment of the various consultants and contractors in the knowledge of the information provided by the employer
- Managing the information and coordinating the work of the different parties
- Selecting the methods for the appointment of consultants and contractors.

References

Ashworth, A. and Perera, S. *Cost Studies of Building*. 6th Edition. Routledge. 2015.

BCIS. *Sharing the risk of Inflation in Construction Procurement*. BCIS News. 2021. Available at https://service.bcis.co.uk/BCISOnline/News/Article/3428?returnUrl=%2FOnline%2FNews&returnText=Go%20back%20to%20news%20summary%20page&sourcePage=Help (accessed 02/07/2022).

Bennett, J. and Jayes, S. *Trusting the Team*. Thomas Telford Ltd. 1995.

Bennett, J. and Peace, S. *Partnering in the Construction Industry – Code of Practice for Strategic Collaborative Working*. Elsevier. 2006.

Booth, L. *Goodbye PFI* House of Commons Library. 2018. Available at: https://commonslibrary.parliament.uk/goodbye-pfi/ (accessed 03/01/22).

British Standards Institution. *BS ISO 20400:2017 Sustainable Procurement Guidance*. BSI. 2017a.

British Standards Institution. *PAS91:2013 Construction Prequalification Questionnaire*. BSI. 2017b.

British Standards Institution. *ISO 10845-1, Construction Procurement: Part 1 Processes, Methods and Procedures*. BSI. 2021a.

British Standards Institution. *BS 19650-2:2018 Organisation and Digitization of Information About Building and Civil Engineering Works, Including BIM – Information Management Using Building Information Modelling*. BSI. 2021b.

Brook, M. *Estimating and Tendering for Construction Work*. 5th Edition. Routledge. 2017.

Cabinet Office. *Transforming Public Procurement*. Cabinet Office. 2020.

Cabinet Office. *The Sourcing Playbook: Government Guidance on Service Delivery, Including Outsourcing, Insourcing, Mixed Economy Sourcing and Contracting*. Cabinet Office. 2021.

CIB (Construction Industry Board). *Code of Practice for the Selection of Main Contractors*. Thomas Telford. 1997.

Construction Task Force. *Rethinking Construction (The Egan Report)*. Department of the Environment, Transport, and the Regions. 1998.

ECI. *Long Term Partnering: Achieving Continuous Improvement and Value*. European Construction Institute. Undated.

HM Government. *Government Construction Strategy 2016–2020*. London. Crown Copyrights. 2016.

HM Government. *The Construction Playbook*. Version 1.1 [online]. 2022. Available at https://assets.publishing.service.gov.uk/government/uploads/system/uploads/attachment_data/file/1102386/14.116_CO_Construction_Playbook_Web.pdf (accessed 26/09/22).

HM Treasury. *Private Finance Initiative and Private Finance 2 Projects: 2018 Summary Data*. London. Crown Copyrights. 2019.

Hughes, W., Champion, R., and Murdoch, J. *Construction Contracts: Law and Management*, 5th Edition. Routledge. 2015.

JCT Practice Notes. *Tendering 2017*. Sweet and Maxwell. 2017.

Latham, M. *Constructing the Team*. The Stationery Office. 1994.

NBS. *National Construction Contracts and Law Report 2018*. RIBA Enterprises Ltd. 2018.

RICS. *Tendering Strategies: RICS Guidance Note, UK*. 1st Edition. Quantity Surveying and Construction Group. 2014.

RICS. *Main Procurement Options*. RICS. 2021. Available at: https://www.isurv.com/site/scripts/documents.php?categoryID=65 (accessed 31/12/2021).

Strategic Forum for Construction. *Accelerating Change (The Egan Report)*. The Strategic Forum for Construction. 2002.

Wolstenholme, A. et al. *Never Waste a Good Crisis: A Review of Progress Since Rethinking Construction and Thoughts for our Future*. London. Constructing Excellence. 2009.

10

Contract Documentation

KEY CONCEPTS
• Contract documents • Form of Contract • Contract Drawings • Contract Bills • Contract Specification

LEARNING OUTCOMES
After reading this chapter, you should be able to: • Appreciate the purpose of contract documents, such as contract drawings, bills, and specification. • Appreciate the purpose of the three main sections included within the Form of Contract: Articles of Agreement, Contract Particulars and Conditions of Contract. • Understand the requirements of NRM2 in supporting the preparation of the Contract Bills. • Identify the course of action when discrepancies in documents arise.

COMPETENCIES
Competencies covered in this chapter: • Procurement and tendering. • Quantification and costing (of construction works).

Contract documents

The contract documents for a building project (JCT 2016, Section 1.1) comprise:

- Contract drawings
- Contract bills (or specification)

Willis's Practice and Procedure for the Quantity Surveyor, Fourteenth Edition.
Allan Ashworth and Catherine Higgs.

- Agreement
- Conditions of contract.
- BIM Protocol (if applicable)

In addition, immediately after the signing of the contract the architect will provide the contractor with other descriptive schedules and documents that are necessary for carrying out the works and any pre-construction information required under regulation 4 of the CDM regulations (2015). The contractor will also provide a copy of the master programme (JCT 2016, clause 2.9.1).

On civil engineering projects, using the NEC4 suite of contracts, the Form of Agreement will identify the documents which form the contract. These documents, such as drawings and specifications, are identified within the Contract Data (NEC 2017). The range of contract forms has been referred to in Chapter 9. When the choice of form of contract has been decided, the next step is to prepare the documents that will accompany the signed form of contract.

The contract documents for any construction project will include, as a minimum, the following information (Ashworth and Perera 2018):

- *The work to be performed.* This usually requires some form of drawn information, including plans, elevations and cross-sections. Additional details will also be prepared depending on the complexity and intricacy of the project. Increasingly this project information will be a BIM-derived information and data. This will provide information for the client and even a non-technical employer is usually able to grasp a basic idea of the architect's or engineer's design intentions. The drawings will also be required for planning permission and building regulations approval, where appropriate.
- *The quality of work required.* The quality and performance of the materials to be used and the standards of workmanship must be clearly conveyed to the constructor, the usual way being through a specification.
- *The contractual conditions.* In all but the simplest projects some form of written agreement between the employer and the constructor is essential. This will help to avoid possible misunderstandings. It is recommended that one of the standard forms of contract should be used. It is preferable to adopt the use of a standard form of contract rather than to devise separate conditions of contract for each project.
- *The cost of the finished work.* Wherever possible this should be predetermined by a firm estimate (tender) of cost from the contractor. However, it is recognised that on some projects it is only feasible to assess the cost once the work has been carried out. In these circumstances, the method of calculating this cost should be clearly agreed. Whilst value is now an important criterion, cost nevertheless forms a part of this calculation.
- *The construction programme.* The length of time available for the construction work on site will be important to both the client and the contractor. The client will need to have some idea of how long the project will take to complete in order to plan arrangements for the handover of the project. The contractor's costs will be affected by the time available for construction. The programme should include progress schedules to assess whether the project is on time, ahead or behind the programme.

Coordinated project information

The authors have included this and the following section within the chapter, due to the continued use of SMM7 within the industry. Research has shown that many problems on site are caused by poor or missing production information. Effective communication of high-quality production information between designers and constructors is therefore essential for the satisfactory realisation of construction projects (CPIC 2003). Shortcomings in drawn information have also been highlighted as a significant cause of disruption of building operations on site. Much of the site management time is devoted to searching for missing information or reconciling inconsistencies in that which is supplied. This, together with a lack of compatibility in project information generally between the drawings, specifications and bills of quantities, is a major concern. The difficulty is partially due to the fact that the information is sourced from several different professional disciplines involved in the project.

The industry's response, in 1987, to improve this situation was to produce a suite of Coordinated Production Information (CPI) conventions to support greater alignment for the production of drawings, specification and Bill of Quantities. This classification structure for the construction industry was the Common Arrangement of Works Sections (CAWS 2000).

Common Arrangement of Work Sections (CAWS)

CCPI as a working committee was reconstituted as the Construction Project Information Committee (CPIC) and CAWS was revised in 1998. The principles of these codes still apply; however, in response to the increased use of computers, *Production Information – A Code of Procedure for the Construction Industry* (CPIC 2003) was published to better reflect industry practice in the use of IT. In 2007, the BSI committee, recognising collaborative working, published a now superseded code of practice, BS1192:2007, to support projects where technology-enabled processes were used. The standard applied to all construction project documentation; it gave recommendations for structured names to convey information for effective information management and exchange.

The purpose of CAWS is to define an efficient and generally acceptable identical arrangement for all documents. The main advantages are:

- *Easier distribution of information, particularly in the dissemination of information to subcontractors.* One of the prime objectives in structuring the sections was to ensure that the requirements of the subcontractors should not only be recognised but should be kept together in relatively small tight packages.
- *More effective reading together of documents.* Use of CAWS coding allows the specification to be directly linked to other documents. This reduces the content in the latter while still giving all the information contained within the former.
- *Greater consistency achieved by implementation of the above advantages.* The site agent and clerk of works should be confident that when they compare the drawings with the contract bills they will no longer have to ask the question, 'Which is right?'.

Technology enabled information exchanges

Subsequently with the increased use of project-based technology-enabled processes, standards have been developed, which give recommendations for structured names to convey information for effective information management and exchange. The RIBA Plan of Works 2020 Toolbox includes a Design Responsibility Matrix, which details information exchanges to co-ordinate the production and use of information by cross-disciplined teams using the UNICLASS 2015 tables (RIBA 2020). These hierarchical classification tables provide a consistent structure and framework for information production and exchange. The UNICLASS system is compliant with ISO12006-2 Building Construction – organisation of information about construction works Part 2 – Framework for classification. This is mapped to NRM1 (Delany 2019). The NBS Digital Construction Report 2021 found that whilst Uniclass 2015 was the most commonly used classification structure (50% of those surveyed), a significant proportion were still using CAWs, 28% (NBS 2021)

Form of contract

This is the principal contract document. It is recommended that standard forms of contract are used as all parties are therefore familiar with responsibilities and procedures stated within. The form of contract, under JCT contracts, takes precedence over the other contract documents. The conditions of contract establish the legal framework under which the work is to be undertaken. Of concern. is typically, a third of contracts are signed after construction has commenced (NBS 2018). Whilst the clauses aim to cover for any eventuality, disagreement in their interpretation may occur. When disputes arise these should be resolved quickly and amicably by the parties concerned. Where this is not possible, it may become necessary to refer the disagreement to a form of alternative dispute resolution, adjudication or arbitration (see Chapter 15 Contractual Disputes). The majority of the standard forms of contract comprise, in one way or another, the following three sections.

Articles of agreement

This is the part of the contract which the parties sign. The contract is between the employer (building owner) and the contractor (building contractor). Articles 1 and 2 summarise the primary obligations under the contract, Articles 3 to 6 detail contractual and statutory appointments; Article 4 provides details on the quantity surveyor. This normally will refer to a company rather than an individual, and Articles 7 to 9 dispute resolution.

If the parties make any amendments to the Articles of Agreement or to any other part of the contract, then the alterations should be initialled by both parties.

Contract particulars

The Contract Particulars for the Conditions of Contract include that part of the contract which is specific to the particular project in question. It includes key information on, for example, the start and completion dates, the periods of interim payment due dates and the length of the rectification period for which the contractor is responsible. The Contract Particulars include recommendations on some of this information.

Conditions of contract

The Conditions of Contract include, for example, the contractor's obligations to carry out and complete the work shown on the drawings and described in the bills to the satisfaction of the architect (or contract administrator). They cover matters dealing with the quality of the work, cost, time, nominated suppliers' and subcontractors' insurances, fluctuations and VAT. Their purpose is to attempt to clarify the rights and responsibilities of the various parties in the event of a dispute arising.

Contract drawings

Ideally, the contract drawings should be complete and finalised at tender stage. Unfortunately, this is seldom the case, and both clients and architects rely too heavily on the clauses in the conditions allowing for variations. Occasionally, the reason is due to insufficient time being made available for the pre-contract design work or, frequently, because of indecision on the part of the client and the design team. One of the original intentions of the standard method of measurement was to only allow contract bills to be prepared on the basis of complete drawings. To invite contractors to price work that has yet to be designed is not a sensible or fair course of action. Tenderers should be given sufficient information to enable them to understand what is required in order that they can submit as accurate and realistic a price as possible. The contract drawings will include the general arrangement drawings showing the site location, the position of the building on the site, means of access to the site, floor plans, elevations and sections. Where these drawings are not supplied to the contractors with the other tendering information, they should be informed where and when they can be inspected. The inspection of these and other drawings is highly recommended since it may provide the opportunity for an informal discussion about the project with the designer.

Upon signing the contract, the contractor is provided with two further copies of the contract drawings (JCT 2016, clause 2.8). These may include copies of the drawings sent to the contractor with the invitation to tender, together with those drawings that have been used in the preparation of either the contract bills or specification. The contract drawings are defined in the Articles of Agreement as those which have been signed by both parties to the contract. It will be necessary during the construction phase for the architect to supply the contractor with additional drawings and details. These may either explain and amplify the contract drawings or, because of variations, identify and explain the changes from the original design. An information release schedule is to be provided and further drawings and details are expected to be provided.

Whilst drawings included in the traditional definition of contract documents are typically two dimensional, the use of Building Information Models (BIM) challenges this with BIM-based information and data replacing the traditional drawings and specification. This is evidenced in the NBS National BIM Report where 57% of those surveyed saw a BIM as contractually binding in the same way as the drawings and specifications (NBS 2018).

The legal and contractual requirements relating to BIM are discussed later in the chapter. When BIM is to be used in a project, the Information Protocol will specify the management, level, use and transfer of information within the *Information Model*. The Information Protocol will reflect the requirements of ISO 19650-1 and ISO 19650-2 (UK BIM Framework 2021).

Descriptive schedules

The preparation and use of schedules is particularly appropriate for items of work such as:

- Windows
- Doors
- Drainage
- Internal finishings.

Schedules provide an improved means of communicating information between the architect (or engineer) and the contractor. They are also invaluable to the quantity surveyor during the preparation of other documents. They have several advantages over attempting to provide the same information by way of either correspondence or annotations on drawings. Checking for possible errors is simplified, and the schedules can also be used to place orders for materials or components.

During the preparation of schedules, the following questions must be borne in mind:

- Who will use the schedule?
- What information is required to be conveyed?
- What additional information is required?
- How can the information be best portrayed?
- Does it revise information provided elsewhere?

The designer should supply the contractor with two copies of the descriptive schedules that have been prepared for use in carrying out the works. This should be done soon after the contract is signed, or as soon as possible thereafter should they not be available at this time as detailed in the information release schedule.

Contract bills

Some form of contract bills or measured schedules should be prepared for all types of building projects, other than those of only a minor or simplified nature. The bill, or schedules, comprises a coordinated list of items of work to be carried out, together with a description and the quantities of the finished work in the building. The bill may include firm or approximate quantities, depending upon the completeness of the drawings and other information from which it was prepared.

Purpose of contract bills

The main purpose of contract bills is for tendering. Each contractor tendering for the project is able to price the work on precisely the same information with the minimum amount of effort. This avoids duplication in quantifying the construction work and allows for the fairest type of competition. Contract bills are not appropriate for all types of construction work, and other suitable methods of contract procurement should be used. For example, for minor works drawings and a specification may be adequate, or where the extent of the work is unknown, payment may be made by using one of the methods of cost reimbursement.

In addition to tendering, contract bills have the following uses. These should be borne in mind during their preparation:

- Valuations for interim payments
- Valuation of variations
- Ordering of materials if used with caution and awareness of possible errors and future variations (the risks relating to this lie with the contractor)
- Cost control
- Planning and progressing by the contractor's site planner
- Final accounting
- Subcontractor quotations for sections of the measured work
- Cost information and data for cost forecasting purposes
- Taxation and grant purposes.

Preparation of contract bills

Contract bills are prepared by the quantity surveyor, adopting best practice procedures. The items of work are measured in accordance with a recognised method of measurement. Building projects are generally measured in accordance with NRM2. Other methods of measurement are also available to the construction industry for work of a civil engineering, highways and industrial engineering nature. Typically, the preparation of the contract bills is no longer a manual process. This will be discussed in more detail in Chapter 11. The contents of contract bills are typically as follows.

Preliminaries

The NRM2 provides a framework for this section of the bill. It consists of two sections:

- *Information and requirements.* This describes the building project particulars, including information such as the drawings, form of contract, the site details, employer's specific requirements and any specific limitations or restrictions.
- *Pricing schedule.* This is a schedule in which the contractor inserts costs relating to both employer's requirements and contractor cost items, such as accommodation, services, facilities and temporary work items.

The value of the preliminary items will vary by market conditions and the size and nature of the contract. For 2020, the mean preliminary percentages for each quarter ranged from 14% to 17% of the contract sum (BCIS 2022).

Measured works

This section of the bill includes the items of work to be undertaken by the main contractor or sublet to subcontractors. There are three principal breakdown structures for a bill of quantities: elemental, work sections and work packages. Appendix A1 of the NRM2 provides examples of each of these structures (RICS 2021). The choice of structure influences the grouping of the measurement and descriptions within the document; however, the NRM2 rules of measurement will be applied to each. Consideration needs to be given to ensure codification of items within the bill relate to that used for the cost planning stage of the project to enable effective reconciliation and management of costs. One of the benefits of using a measurement software package is the easy manipulation of different BQ formats and the ability to cross-reference items.

- *Elemental.* This groups the items according to their position in the building on the basis of a recognised elemental subdivision of the project, e.g. external walls, roofs, wall finishes, sanitary appliances. The group elements used are those as described in NRM1. In practice, contractors tend to dislike it since it involves a considerable amount of repetition in pricing.
- *Work section.* The items in the bill are grouped under their respective trades as defined in NRM2 Part 3. The advantages of this format are that similar items are grouped together, there is a minimum of repetition, and it is useful to the contractor when subletting.
- *Work package.* This groups the items into employer's or contractor's defined work packages.

Provisional sums

At the time of preparing the tender documentation, there may be works that cannot be entirely foreseen or detailed sufficiently to measure in accordance with the NRM2 tabulated rules of measurement. The quantity surveyor will therefore determine a sum of money to cover the cost of each of the items and specify where the provisional sum is 'defined' or 'undefined'. For defined provisional sums, the contractor will have made due allowance for the planning, programming and pricing of preliminaries for the related item.

Schedule of construction risks

At the time of preparing the contract bill there will still be a number of risks remaining to be managed by the employer and project team. The employer will wish to transfer this residual risk to the contractor. A Schedule of Construction Risks will therefore be included which describes each risk. The contractor is required to provide a lump sum fixed price for taking, managing and dealing with the consequence of the identified risk should it arise.

Credits

When the contract includes refurbishment and/or demolition of existing buildings, it is likely that old building materials, components and fittings arising from the work is of some value and the employer is content to pass ownership to the contractor to re-use. This section details such items, against which the contractor will enter the amount of credit that will be given for each item.

Dayworks

If required, a schedule of dayworks will be included in the contract bills.

Annexes

This section comprises information referred to in the contract bill but not included within it. Such information may include copies of letters from statutory undertakers and copies of quotations.

Summary

The summary of the bill of quantities may appear either at the start or end of the document. It consists of a list of each section of the bills, and against each the total price will be given. At the end of the summary, provision is made to total the list to calculate the total price. This is transferred to the form of tender and becomes the 'contract sum' referred to in the conditions of contract.

Form of tender

The form of tender is the contractor's written offer 'to undertake and execute the works in accordance with the contract documents for a contract sum of money' and will also state the contract period and any price adjustment. The tenders are submitted to the architect/contract administrator who will then make a recommendation regarding the acceptance of a tender to the client (see Chapter 11). If the client decides to go ahead with the project, the successful tenderer is invited to submit a copy of the priced bills for checking. The form of tender usually states that the employer:

- May not accept any tender
- May not accept the lowest tender
- Has no responsibility for the costs incurred in their preparation.

Contract specification

In certain circumstances it may be more appropriate to provide documentation by way of a designed specification rather than contract bills. The types of projects where this may be appropriate include:

- Minor building projects
- Small-scale alteration projects
- Simple industrial shed-type projects.

The specification provides detailed descriptions of the work to be performed, detailing the materials, workmanship and construction, to assist the contractor in preparing the tender. A specification is used during:

- Tendering, to help the estimator to price the work that is required to be carried out
- Construction by the designer to determine the requirements of the contract, legally, technically and financially, and by the building contractor to determine the work to be carried out on site.

Where the client wishes the contractor to take responsibility for the design, such in the Design and Build procurement method, a performance specification will be used, which will become part of the contract documents.

National Building Specification

The National Building Specification (NBS) is the recognised national standard specification system for the UK construction industry. NBS Chorus is a cloud-based platform that enables collaborative specification writing. It provides a library of specification clauses and construction standards from which relevant clauses can be selected and additional information to be inserted. The NBS Chorus thus facilitates the production of specification text specific to each project, including all relevant matters within a standardised structure.

NBS Source is a construction product library that integrates with NBS Chorus. Plug ins are also available to connect the specification information with CAD models to synchronise any changes made during the design process.

Schedules of rates

A schedule of rates is a compromise between a specification and contract bills. It is more like contract bills and does not include any quantities for the work to be carried out. Its main purpose is therefore to value the items of work once they have been completed and measured. A schedule of rates may be used on:

- Jobbing work
- Maintenance or repair contracts
- Projects that cannot be adequately defined at the time of tender
- Work required at very short notice
- Painting and decorating.

Master programme

It is the contractor's responsibility to provide the master programme. This shows when the works will be carried out. Unless otherwise directed by the architect, the type of programme and the details to be included are to be entirely at the discretion of the contractor. If the architect agrees to a change in the completion date because of an extension of time, then the contractor must provide amendments and revisions to the programme within 14 days (JCT 2016, clause 2.9.2).

Information release schedule

This schedule informs the contractor when information will be made available by the architect. The schedule is not annexed to the contract. However, the architect must ensure that the information is released to the contractor in accordance with what has been agreed in the information release schedule. In practice, the architect needs to coordinate the information release schedule with the contractor's master programme for the works and align with the Design Responsibility Matrix.

Discrepancies in documents

JCT 2016, clause 2.15 requires the contractor to notify the architect if any discrepancies between the documents are found. The form of contract always takes precedence over the other documents. The contractor cannot therefore assume that the drawings are more important than the contract bills. However, drawings drawn to a larger scale will generally take precedence over drawings that have been prepared to a smaller scale. JCT 2016, clause 2.17 expects the contractor to adopt a similar course of action should a divergence be found between statutory requirements and the contract documents. Any discrepancies or differences that result in instructions to the contractor requiring variations will be dealt with under Section 5. The contractor should not knowingly execute work where differences occur within the various documents supplied by the design team. However, it is not the contractor's responsibility to discover differences should they arise.

BIM and contracts

A BIM protocol is a supplementary legal agreement usually included as a schedule to a BIM-enabled contract (Winfield and Rock 2018). An agreed BIM protocol is a contract document. This schedule sets out the parties' rights, duties and risks allocation for the BIM process. *The Winfield Rock Report* identified that the JCT Design and Build (DB) was the most commonly used contract; 59% of those surveyed used it compared to use of JCT traditional (28%) and NEC Option C (23%) (Winfield and Rock 2018). The *BIM and JCT Contracts 2019 Practice Note* provides guidance on the provision within the JCT DB contract most impacted or relevant to projects implementing BIM (JCT 2019). Part B of this document provides a BIM Protocol checklist and Appendix 1, an example of common content of the Employer Information Requirement.

References

Ashworth, A and Perera, S. *Contractual Procedures in the Construction Industry*. 7th Edition. Routledge. 2018.

BCIS. *Study in Trends in Contract Percentages*. BCIS. 2022. Available at: https://service.bcis.co.uk/BCISOnline/ContractPercentages (accessed 22/01/2022).

BSI. *BS 1192:2007 Collaborative Production of Architectural, Engineering and Construction Information*. Code of Practice. 2007.

CAWS (Common Arrangement of Work Sections). *Committee for Project Information*. CAWS. 2000.

CPIC. *Production Information – A Code of Procedure for the Construction Industry*. Construction Project Information Committee CPIC. 2003.

Delany, S. *What is UNICLASS 2015?* NBS. 2019. Available at: https://www.thenbs.com/knowledge/what-is-uniclass-2015 (accessed 15/01/22).

Joint Contracts Tribunal (JCT). *Standard Building Contract with Quantities (SBC/Q)*. Sweet and Maxwell Thomson Reuters. 2016.

Joint Contracts Tribunal (JCT). *BIM and JCT Contracts 2019 Practice Note*. Sweet and Maxwell Thomson Reuters. 2019.

NBS. *National Construction Contracts and Law Report 2018*. RIBA Enterprises Ltd. 2018.

NBS. *Digital Construction Report 2021, Incorporating the BIM Report*. NBS Enterprises Ltd. 2021.

New Engineering Contract (NEC). *Preparing and Engineering and Construction Contract. Volume 2*. Thomas Telford. 2017.

RIBA. *2020RIBAPlanofWorktoolboxFeb2020xlsx.*RIBA. 2020. Available at: https://www.architecture.com/knowledge-and-resources/resources-landing-page/riba-plan-of-work (accessed 15/01/22).

RICS. *NRM2- Detailed Measurement for Capital Building Works: NRM2.* 2nd Edition. RICS Books. 2021.

UK BIM Framework. *Information Protocol to Support BS EN ISO 19650-2 – The Delivery Phase of Assets.* 4th Edition. The UK BIM Framework in association with the CIC. 2021.

Winfield, M. and Rock, S. *The Winfield Rock Report: Overcoming the legal and contractual barriers of BIM.* UK BIM Alliance. 2018.

11

Preparation of Contract Bills

KEY CONCEPTS

- Taking off process
- Standard methods of measurement
- Preparation of the contract bill
- Correction of errors in tender submissions
- Tender reporting
- E-tendering

LEARNING OUTCOMES

After reading this chapter, you should be able to:

- Appreciate the range and purpose of the standard methods of measurement used in industry.
- Understand how the quantity surveyor prepares the contract bills.
- Understand the role of the quantity surveyor in the tendering procedure.
- Appreciate the benefits of e-tendering.

COMPETENCIES

Competencies covered in this chapter:

- Procurement and tendering
- Quantification and costing (in construction work).

Appointment of the quantity surveyor

The appointment of the quantity surveyor is likely to have been made at an early stage when initial budget estimates were taken to test the feasibility of the project. This might be before any drawings are available, in order to provide some cost advice to the client. Only on very small projects will a quantity surveyor not be required at all. With the increased use

Willis's Practice and Procedure for the Quantity Surveyor, Fourteenth Edition.
Allan Ashworth and Catherine Higgs.

of alternatives to the traditional procurement route, contractors submit tenders in competition without quantities being provided. In these circumstances, quantity surveying firms are then sometimes involved in preparing quantities directly for contractors. Today, these can take several different formats. There may, of course, be special considerations that would justify such invitations for a larger contract and, equally well, similar considerations that could call for quantities being provided for a contract of a lesser value.

Receipt of drawings

Usually, the architect will send the quantity surveyor electronic versions of the drawings and any specification information. It is likely that access will be provided to a shared project collaboration area, such as sharepoint, a dropbox, or a cloud-based system. The format of the drawn information will be influenced by the architectural practice and the use of BIM on the project, so will vary from PDFs and CAD dwg files to BIM models. A timetable for the completion of the contract bills will be agreed, along with dates when any additional detailed drawings and information can be expected. If a BIM model is not used, it is likely that taking off drawings of some sections of the work, which are incomplete, will result in redundant and duplicate work. Therefore, if further drawings are to follow, it is helpful if the order in which they are being prepared can be agreed, having regard both to the architect's office procedure and the surveyor's requirements. One of the benefits of using technology, rather than manual processes, to both create and measure drawings is the ability to compare drawing and model revisions to flag and action differences in quantities generated.

Study of documents

As drawings are received, whatever the format, it is important to keep a record for audit trail purposes. Such a drawing issue register will identify the drawings that have been issued by the architect and structural engineers. The register should include issue date, drawing title, reference number, version, scale and brief particulars (Fig. 11.1). It is important to check what has been issued, as stated in email communication, with what has been received. Measurement software enables revised drawings or model files to be exchanged in place with superseded ones to ensure a current set of information. It is important to keep the drawing register up to date as the list of drawings from which the BQ is prepared is stated within the preliminaries section of the contract bills.

If the drawings are not electronic, except on the smallest jobs, they should be supplied in duplicates. It is advisable to number the sets so that it can be seen at a glance to which set a drawing belongs. On receipt of drafted rather than electronic drawings, the quantity surveyor will:

- See that all the necessary figured dimensions are given, both on plans and sections
- See that the figured dimensions are checked with overall dimensions given
- Insert any dimensions that can be calculated and may be useful in the measurement
- Confirm any errors in figured dimensions with the architect, so that the originals can be amended accordingly.

Project title: Southtown School					Project Nr: SS004
Consultant: Architect					Sheet Nr: 2
Drawing Register				Revisions	
Issue Date	Title	Nr	Scale		
11/7/22	Block plan	SS004/11	1:1000		
11/7/22	General plan	SS004/12	1:100		
11/7/22	Sections, elevations	SS004/13	1:100		
11/7/22	Classroom plans	SS004/14	1:20		
11/7/22	Cloaks, toilets plans	SS004/15	1:20		
14/8/22	Assembly hall plan	SS004/16	1:20		
14/8/22	Heating system detail	SS004/17	1:20		
14/8/22	Door detail	SS004/18	1:20		
14/8/22	Window schedule	SS004/20	N/A		
13/9/22	Drainage	SS004/25	1:200		
13/9/22	Classroom store fittings	SS004/27	1:20		

Fig. 11.1 Sample drawing register.

A careful perusal should be made of the specification or specification notes. By following through systematically the sections of the taking-off, gaps may be found in the specification which require information. These may be quite numerous when only notes are supplied, as 'notes' vary considerably in quality, thoroughness and extent. When a standard specification, such as NBS (National Building Specification) in one of its versions, is used then the opportunity can be taken to check that the correct alternatives have been chosen, that superfluous matter has been deleted, and that all the gaps have been filled in.

Schedules

Schedules are useful, both for quick reference by the taker-off and for eventual incorporation in the specification for the information of the clerk of works and site agent. They will normally be supplied by the architect with the drawings. Internal finishings should certainly be scheduled in a tabulated form, so that the finishes of each room for ceiling, wall and floors can be seen at a glance, with particulars of any skirtings, dadoes or other special features. Schedules for windows and doors would include frames, architraves and ironmongery. Those for inspection chambers would give a clear size, invert, type of cover and any other suitable particulars.

Some of the material on schedules may be otherwise shown on drawings, but the schedule brings the parts together and gives a clear view of the whole and is useful in informing the measurement approaches.

When BIM models are used, measurement software provides the facility to data drop information in a scheduling format that can be manipulated to schedule quantities by location. Different specifications in the schedule can also be highlighted on the BIM model giving a useful visual interpretation to aid greater familiarity with the specification used on the project.

Taking-off

Query sheets

After drawings and specification or the model have been examined, a first list of queries for the architect or engineer will be prepared. Queries will also arise during the measurement process. These queries will be either requests for information or technical queries. It is common practice for the company to have their own query sheet; however, these sheets will all be structured in a similar way (Fig. 11.2).

The query sheet will consist of four columns;

- Column 1: allows identification of the query for future reference.
- Column 2: is a description of the query and the clarification or confirmation required.
- Column 3: is a space for the response.
- Columns 4 and 5: gives the date and confirmation that the query has been actioned.

The queries will be shared electronically to the architect or engineer. A database will also be kept to manage this process. Measurement software packages will have integrated reporting tools to request further information for items on the drawings or model. It is important that a final check should be made when the take-off is completed to ensure all responses to queries have been actioned.

Division of taking-off

When measuring a project, the lead quantity surveyor in charge of the project will identify the measurement responsibilities of the team by subdividing the measurement by either elements of the building, trade or work package. The most common approach is by

Project Nr SS004		**Southtown School Queries**	Sheet Nr 2.	
Date: 11/7/22				
Ref	*Query*	*Reply*	*Date*	*Actioned*
1.	Finish to floor of entrance hall specified wood block, coloured as tile?			
2.	Should not dimension between piers on north wall be 5.08 not 5.03? (to fit the overall 56.85)			
3.	Specification for the dpc required.			
4.	Should brick facing to concrete beams be tied back?			

Fig. 11.2 Sample query sheet.

element, and allocation to the team will be influenced by experience. For example, it is common for graduate quantity surveyors to be given finishes. The amount of subdivision of the measurement will also depend on the size of the project, availability of staff and time allocated to complete the work.

When allocating the elements, consideration will be given to the relationships between each element. For example, the frame, with floor, roof, slab and beam casings would be the charge of one person and likewise the windows and external doors can hardly be subdivided.

The more the taking-off is subdivided the greater the risk of duplicating items or assuming that someone else will have measured specific items, so demarcation between elements must be clearly defined. There are certain items in which practice differs with different offices. For example, some surveyors measure skirtings with floor finishings and others with wall finishings. Some measure them net, others adjust for openings in the doors section. It is important, therefore, that clear rules are adopted.

Measurement software packages enable the lead quantity surveyor to allocate levels of authority within the system to reduce technical errors corrupting or deleting completed measurement sections.

Greater detail of the bill preparation process can be found in Willis' Elements of Quantity Surveying Chapter 20 (Lee 2020).

Methods of measurement

The quantifying of construction works is best done using an agreed set of rules or method of measurement. It is then clear to all users how the work has been measured and what has not been measured. NRM2 includes a list of 'works and materials' to be included in the first section of each work section.

The main aim of the different methods of measurements (Fig. 11.3) is to provide a clear set of rules that can be used for measuring a range of construction work, thereby avoiding ambiguity and disagreement in interpretation. The rules apply equally to work that is proposed or executed. Some words and phrases in contract bills have developed implied meanings, trade customs and practices. Standard phraseologies have also been developed to standardise and clarify meanings of bill descriptions.

Type	Method of measurement
Building work	SMM7 Standard Method of Measurement, 7th Edition NRM2 New Rules of Measurement 2, 2nd Edition
Civil engineering work	CESMM4 Revised Civil Engineering Standard Method of Measurement, 4th Edition
Construction work overseas	POMI Principles of Measurement International
Highway works	Method of Measurement for Highway Works
Rail Infrastructure	Rail Method of Measurement (RMM): Volume 2

Fig. 11.3 Methods of measurement.

Standard methods of measurement were first published by the RICS in 1922 and since then standard methods have been developed in partnership with industry stakeholders to reflect the diversity and changes within the construction industry.

Standard methods, such as SMM7 in 1988, developed a more detailed tabulated format to support standardised coding to be more readily suited for the use of information technology.

In recognition of the need to bring consistency to cost management of construction and maintenance work at any point in a building's life, in 2013 the RICS Quantity Surveying and Construction Professional Group, in consultation with industry, developed a suite of three documents supporting a more integrated cost management approach (see Chapter 5). Hence, the NRM suite is a set of rules for measurement to enable not only budget setting but also cost analysis. NRM2 Detailed Measurement for Capital Building Works replaced SMM7 and became operational in January 2013, providing guidance for the detailed measurement and description of building work for the purpose of obtaining a tender price. However, whilst there were minor amendments to the second edition of NRM2 published in 2021 in relation to risk profile of some items, it was noted findings from the industry research undertaken showed SMM7 was still used within the sector. The RICS therefore is making a conscious effort to better understand the reasons for the limited take up of NRM2 and continued use of SMM7 and incorporate their subsequent findings into a later edition (Thompson 2021).

Companies such as Network Rail have developed their own standard method of measurement as part of a suite of documents to provide an integrated measurement and valuation process for rail infrastructure works. This standard RMM2 is used through its supply chain (Doyle 2021).

Methods of measurement, such as CESMM4 Revised, have been amended to better reflect greater international use with the removal of references to British Standards, making it 'technical standards neutral'. References have also been removed to the ICE Conditions of Contract, being 'contract neutral' increasing its use with other conditions of contract (ICE Publishing 2021).

Countries in which quantity surveying has been long established have developed their own methods of measurement for building works, such as the Hong Kong Standard Method of Building Works. These have often been based on a UK SMM and have been adapted to suit local conditions, such as different methods of construction and their associated technologies.

Whichever the standard method of measurement rules used, they all include the following facets:

- Technical adequacy of the rules
- Lacking in ambiguity
- Ease with which the rules can be applied
- Measurement only of items of cost significance
- Consideration of a wide variety of applications.

Alterations in taking-off

Alterations are sometimes made by the taker-off after the dimensions have been squared and perhaps even later when they have been carried through to the bill stage. It is important that such alterations are at the same time taken through all the stages involved. If the taker-off marks the alterations with a pencil cross and hands the sheet personally to whoever is responsible for the next stage this should ensure that the corrections are made. Alterations should not be made to the taking-off by others. There may be a reason for an apparent error. Often, after making what appears to be a correction, it is found that the dimensions were right the first time.

During the preparation of the bill of quantities, if errors are found in the architect's and engineer's drawings, these should be corrected. The architect or engineer should be advised of these in time to allow the necessary corrections to be made prior to issuing them to the tenderers.

Standard descriptions

The descriptions in the contract bills are standardised both to ease the process of bill preparation and to overcome any misinterpretation during the pricing. Prior to the introduction of SMM7 in 1988, guidance was provided within the standard methods of measurement of standard phrases that could be grouped to form the description.

Reflecting the adoption of technology and the use of levels within the software description libraries, SMM7 as well as being a set of tabulated rules for measuring, also provided key phrases for writing the bills. Likewise, NRM2 Part 3 provides not just the rules of measurement for building components and items but the level of description required to accurately represent the quality of work to be carried out. This method has resulted in a much greater uniformity in the description of items within contract bills (see Fig. 11.4). This is helpful to contractors with their own processes of tendering, estimating and final accounting.

Everything that concerns price must be included in the contract bills. The unnecessary duplication of descriptions in specification and bill can be avoided by reference in the bill to clause numbers and headings of the specification, where there is a full specification prepared and issued with the bills. Where, however, the specification is not part of the contract, it is more satisfactory for the tenderer to have everything in the bills to save the extra effort spent in cross-reference.

Numbering items

Items in the contract bill can be serially numbered from beginning to end. Alternatively, each page can have items referenced from A onwards so that an item might be referenced as, for example, item 50 C or 94 B. The latter is the preferred method, since if new items have to be inserted at a late stage, the whole sequence of numbers does not need to be revised.

Schedule of basic rates (fluctuations option B)

If the contract alternative is used, requiring recovery of fluctuations to be based on the rise or fall in prices of materials, an appendix will be required to the contract bills, in which the contractor can set out the rates on which the tender is based. The appendix can contain a list of the principal materials (which the contractor can supplement if necessary) or it can

Item	Description	Qty	Unit	Rate	£ p
	Section 5: Excavating & Filling				
A	Site preparation; remove topsoil 150 mm deep	44	m^2		
	Excavation				
B	commencing 150 mm below existing ground level; bulk excavation not exceeding 2 m deep	7	m^3		
C	commencing 350 mm below existing ground level; Foundation excavation; not exceeding 2 m deep	11	m^3		
D	Disposal of excavated material off site	15	m^3		
	Retaining excavated material on site;				
E	topsoil; to temporary spoil heaps; 500 m average distance	7	m^3		
F	all other excavated material; to temporary spoil heaps; 300 m average distance	5	m^3		

Fig. 11.4 Example page from Bills of Quantities.

be left to the contractor to prepare the list. Alternatively, a fixed priced list could be given of the principal materials, but it is preferable to let contractors include their own rates, as one contractor may be better placed than another, for example by being able to purchase large volumes.

Schedule of allocation (fluctuations option C)

When the price adjustment formula for calculating variation of price claims is to be used, when the contract bills are complete and before they are sent out to tender, it is necessary to prepare an allocation of all the items in the bills. This is done by preparing a schedule of all work categories and then allocating all the bill items to the work categories. Preliminaries and certain provisional sums are included in a balance of adjustable items section.

When this allocation is complete it is included in the bills, and in submitting a tender a prospective contractor is deemed to have agreed to the allocation chosen by the quantity surveyor. It is unlikely that a tendering contractor will challenge the allocation, but if they do they must do so when submitting the tender. When the tender has been accepted it is a straightforward exercise to complete the schedule by filling in the appropriate amounts. The whole process is self-checking in that the end result must be the contract sum.

Completing the contract bills

When manual processes or excel sheets have been used, the draft bills should receive a final careful read through, an overall checking of the quantities and editing prior to their issue. An examination should be made to see that all gaps in the text have now been completed, such as cross-references to item numbers or summary pages. The bills should have their quantities compared by cross-checking against similar quantity related items. This aspect of work is usually carried out by the principal or a senior assistant whose experienced eye should be able to detect any major discrepancies if these are present. Someone not directly involved in the project will often be able to spot possible errors.

A final careful examination should also be made of the drawings and specification. All notes on the drawings should be examined in case any have been missed. The specification can be run through in pencil, clause by clause, to confirm that each has been included, either by the takers-off measuring or by those who drafted the preliminaries or preambles sections. At some stage between completion of the draft bill and passing of the proof the whole of the dimensions and the billing process should be examined to ensure that arithmetical checks have been carried out and the process audited. Finally the proof copy of the contract bills should be read through compared to the last draft to ensure all final edits have been incorporated.

The number of hard copies required of the finished document will depend on whether e-tendering is to be used and the number of tenderers on the project.

Computerised bill production

With the increased use of electronic drawings within the industry, the majority of bills of quantities will be produced using software that has the capability to import data from 2D, 3D, CAD and BIM models. These systems enable quantification directly from the drawings and module files, increasing the speed and accuracy of the measurement process. Such systems also offer the capability to transfer the measurement directly into cost management, whole life costing and e-tendering packages.

Copyright in the bills of quantities

Copyright is established by the Copyright Designs and Patents Act 1988 in every 'original literary, dramatic, or musical work'. It is not clearly established whether there is copyright in contract bills. The RICS has taken the opinion of counsel, and were advised that copyright existed, on the grounds that the bill was an original literary work within the meaning of the Act of 1911, which used the same wording. This is an expert opinion and it should be remembered that experts sometimes differ. This is a reason for actions in law. There are inevitably many clauses in a bill of quantities that are more or less standard and used in very similar form by many surveyors, but the quantities are undoubtedly original. There will be in all bills a number of items that are original and peculiar to that particular bill.

There might be those who would take advantage of an opportunity to reuse a surveyor's bill. The best protection is to have a specific reservation by surveyors in their agreements with their clients of the rights of reprinting and reusing bills. If, of course, the surveyor is advised, when instructed, that it is proposed to use the bill again as and when required, and the condition is accepted, there can be no redress. The difficulty is avoided if the RICS standard form of consultant's appointment (RICS 2019) is adopted without amendments.

In clause 13 of that agreement, copyright in all documents remains the property of the consultant.

The dimensions and other memoranda from which the contract bills are prepared are in a different category. These are the surveyor's own means to an end. It has been held that the surveyor is entitled to retain these documents unless, as is sometimes the case, the contract with the client provides otherwise.

Tendering procedure

Preliminary enquiry

The project team will usually prepare, often in consultation with the client, a list of firms to be invited to tender, in some instances by way of some form of pre-qualification questionnaire (PQQ). Guidance is provided in the *JCT Tendering 2017 Practice Note* (JCT 2017). Initially the firms selected will be sent an enquiry communication, normally via email, together with the questionnaire and a Project Information Schedule, in advance of any documentation to confirm that they are interested in competing for the works. A typical communication, based on the templates provided in Appendix A1 of the Practice Note, is shown in Fig. 11.5. The PQQ will establish the contracting organisation's capability to undertake the work, and questions may include financial standing,

Dear Sirs

Southtown Church of England Primary School

We are authorised to prepare a preliminary list of tenderers for the construction of the works described in the Project Information Schedule. Please can you indicate whether you would wish to submit a tender if selected to do so.

Should you wish to be selected, please complete the pre-qualification Questionnaire and return it to us, no later than 13:00 hours on 27th February 2023. Assessment of the responses will be carried out in accordance with the Information Schedule and we will inform prospective tenderers to be included on the list by 14th March 2023.

The is intended to be a single stage tendering procedure aligned to that described in the JCT Tendering Practice Note 2017. The tender process will be conducted electronically and the contract awarded on the basis of best value.

We expect all tenderers to submit a *bona fide* tender open for a period for acceptance of not less than 21 days.

The employer reserves the rights to postpone the intended closing date for bids and to accept any tender or no tender at all.

Yours faithfully

Quantity Surveyor

Fig. 11.5 Preliminary enquiry for invitation to tender (*Source:* Prepared in accordance with JCT Tendering Practice Note 2017 Appendix A1).

technical capability, management, supply chain information and insurance details. Alternatively, the contractor maybe asked to provide their current Common Assessment Standard certification. The Project Information Schedule, as detailed in Appendix A2 of the Practice Note, provides the details on which the contractor can make the decision to tender. The schedule will include the following:

- Title, description and location of the project
- Client and consultant details
- Anticipated cost range, start date and duration of project
- Tendering procedure details, and assessment, including evaluation criteria and weightings.
- The method for the correction of errors found in the contractors priced bill.
- The form of contract to be used
- If BIM is to be used the information management requirements (see Chapter 10).

On receipt of the confirmation from the contractor and the PQQs, the tenderers will be assessed against the pre-defined evaluation criteria and will be added to the shortlist of tenderers for the project.

Dispatch of finished bills

The contractors who are invited to tender should each be issued with the tender documentation. This will normally be in an electronic format; however, if a hard copy format is used, documents should be in duplicate and accompanied by separate envelopes for their return.

The tender documentation will reflect the chosen procurement method; for traditional procurement, these will be the drawings, bills of quantities and the specification. A Formal Invitation to Tender (Fig. 11.6) will be issued with the drawings and contract bills to the tenderers. It should state:

- The contract documents.
- Tender procedure information, including date and submission protocols.
- The basis for awarding tenders.
- Additional information to support the contractor, such as arrangements for visiting the site and whom to direct any queries that arise in their preparation of the bid.

It is also advisable to state that the client (employer) is not bound to accept the lowest or any tender or to pay any expenses incurred by the tenderer in preparing the tender. Figure 11.6 provides a worked example based on A3 in the *JCT Tendering Practice Note 2017*.

Tender queries

Once the bills are dispatched to the contractors for tendering purposes, a copy should be examined. Mistakes may still be found, even after careful checking has taken place, perhaps made in the rush to send out the bill in time. Queries may arise, too, from contractors tendering. Clarifications and corrections should be circulated to all contractors at the same time, in a timely manner for them to correct their copies before tenders are completed, and this can be via email or through the reporting tool of an e-tendering package. An

[13th April 2023]

Dear Sirs

Southtown Church of England primary School

You are invited to provide a tender for the works as shown and described in, and on the basis of the following:

- Drawings as listed on the attached schedule
- Specification and Bills of Quantities
- The Employer's Requirements
- The outline construction phase plan

The works are to be carried out in accordance with the Contract and conditions as specified in the Project Information Schedule issued with our Preliminary Enquiry dated 27th February 2023.

Please submit your tender on the attached Tender form in accordance with the tender submission instructions, so as to be received no later than 13:00 hours on 10th May 2023, together with the fully price document. Tenders received late will not be considered. Tenders will be assessed on the basis of best value and conducted in accordance with the principles set out in the JCT Tendering Practice Note 2017. The employer reserves the right to extend the submission deadline and to accept any tenders from those submitted or reject all. Please note that

- The site may be inspected by arrangement with [name] at this office
- Queries should be raised with [name] at this office

Yours faithfully

Quantity surveyor

Fig. 11.6 Formal invitation to tender (*Source:* Prepared in accordance with JCT Tendering Practice Note 2017 Appendix B1).

acknowledgement should be requested to ensure that all tenderers have incorporated the corrections. If the changes to the contract documents are significant, it may also be necessary to extend the tendering period. During the examination of the priced bills of the successful contractor it should be verified that these corrections have been made. A typical email is shown in Fig. 11.7.

Receipt of tenders

Delivery and opening

Tenders should be submitted by the time and date stated within the Invitation to tender. Within the Form of Tender (Fig. 11.8), contractors are normally advised that if their tender is received after the due date and time, it will not be considered, and therefore late submission is at the contractor's risk. However if tenders have yet to be opened, the *RICS Tendering Strategies Guidance Note* advices that, in the case of a private client, the client should be advised if they wish for the tender to be considered, then they need to provide explicit

20th April 2023.

Dear Sirs,

Southtown School: New Extensions

Will you please incorporate the following corrections in the contract bill:
Item 24D For 16 m^3 read '66 m^3'.
Item 35C For 'm^2 read 'm'.
Please acknowledge receipt of this communication.

Yours faithfully,

Quantity Surveyor

Fig. 11.7 Communication to contractors: corrections to the contract bills.

TENDER FOR: Southtown Church of England Primary School

To: LMN Chartered Architects

From: Contractor

We have examined the following documents referred to in the Invitation to Tender:

- Drawings as listed on the attached schedule
- Specification and Bills of Quantities
- The Employer's Requirements
- The outline construction phase plan

We offer to carry out the whole of the works as described in and in accordance with the documents referred to in this Tender for the sum of £2,726,517.00 (exclusive of VAT) within 65 weeks from the date of site possession.

We agree that if any errors are discovered prior to acceptance of the offer this will be dealt with in accordance with Alternative 2

We undertake that in the event of acceptance of this offer, we will execute a formal contract with the Employer within 21 days of being asked to do so.

This tender remains open for consideration for 35 days from the date fixed for submitting tenders.

Dated this 10th day of May 2023

Name: Contractor

Address . . . .

. . . .

. . . .

Signature . . . .

Fig. 11.8 Form of tender (*Source:* Prepared in accordance with JCT Tendering Practice Note 2017 Appendix B2).

instruction to do so (RICS 2014). Tenders received after opening will not be considered. The submissions will normally be opened in the presence of the client and the quantity surveyor and project manager/architect. It is essential that the opening is always witnessed and signed by at least two individuals.

Reporting of tenders

Each tender bid will be recorded on a Tender Opening Form, which will summarise as a minimum, the tender sum, contract period and qualifications. This quantitative data will be reported to the client and other consultants. Changes to the contract period stated in the preliminary enquiry will be regarded as a qualification to the tender. The contractor should be informed and asked to remove any qualifications; however, no amendment of the bid is permitted.

In considering the tenders received, factors other than price, qualitative data, may be of importance and these should form part of the assessment criteria provided to tenderers. The assessment of the bids in relation to this weighted criterion should be done independently of the examination of the priced documentation (JCT 2017). Whilst the quantity surveyor will focus on reviewing the validity of the tender bid prior to the tender analysis, they will also be expected to contribute to the decision-making related to the qualitative evaluation.

Examination of priced bill

Before acceptance of a tender, the tenderers whose offers are under consideration, if they have not already done so, are required to submit a copy of the priced bills to the quantity surveyor for examination. To aid comparison of tenders, the quantity surveyor may send an excel spreadsheet as part of the tender package, which will be completed in lieu of a priced bill. If an integrated measurement software package is used to create the bills, the contractor's priced bill will be submitted via the e-tendering tool.

If the project has been tendered using hard rather than electronic copies, the first check undertaken is an arithmetical check. Clerical errors can be made. It is important that the contract bills are as correct as they possibly can be prior to the signing of the contract. If, for example, an item has been priced at £0.50 per m and extended at £0.05, it will not be fair that either additional quantity or omission of the item should be priced out at the incorrect rate in adjusting accounts. All clerical errors should be corrected in the contract copy of the bills.

In addition to the arithmetical check, a technical check is also made of the pricing by examining the contractor's rates and prices in relation to the pre-tender estimate. Items may accidentally have been left unpriced or errors in pricing, such as items billed in square metres may have been priced at what is obviously a linear metre rate, or vice versa. An obvious misunderstanding of a description may be noted. Corrections should be made so that a reasonable schedule of rates for pricing variations results.

E-tendering tools aid the checking of the tenders, tenders can be compared with each other as well as with the pre-tender estimate (PTE). Items that are not priced or have the greatest variance between the PTE or other tenders can highlight possible errors.

A third reason for examining the priced bill is to ensure that the tenderer has not made such a serious mistake that they would prefer to withdraw the tender. Under English law, contractors may do this at any time prior to the acceptance of a contract. When such a serious error is detected, it is always advisable to bring this to the attention of the tenderer. The error will sooner or later be discovered, resulting in a risk that constant attempts will then be made to recover the loss to the detriment of the client's interest.

Ultimately, it is the aim of the quantity surveyor to ensure that the correct price for the contract is achieved; it is therefore important that any errors are bought to the attention of the bidding contractor and rectified.

Correction of errors

Two alternatives are described in *JCT Tendering Practice Note 2017*, Alternative 1 and 2, for dealing with genuine errors in rates and prices. The alternative to be adopted must be made prior to the tenders being invited and stated within the Project Information Schedule. Where overall price is dominant the contractor will be given the opportunity to either confirm or withdraw the tender and the bills will be endorsed to reconcile the error (Alternative 1). Notwithstanding the guidance provided in the JCT Practice Note, if the correction of the error does not bring the tender under consideration above the next highest, the architect may feel that the client should be advised to allow amendment to the tender.

Alternatively, the tenderer will be given the opportunity of confirming the original offer or amending it to reflect the correction of errors (Alternative 2). However, if the resultant tender means that the tender is no longer eligible for acceptance, i.e. is not the lowest or best value, then it will be withdrawn from the process.

When Alternative 1 has been stated as the method for correcting errors, the amount of the tender will of course not normally be altered, but documents will be endorsed to reconcile the error. Any difference will be shown as a rebate or addition as an addendum to the summary (see Fig. 11.9). This addendum will be used for interim valuations and adjustment of variations, to all rates except prime cost and provisional sums.

Figure 11.9 shows the summary detailing how alterations have been made correcting clerical and other errors found in the priced bills. The increased total means that all rates except PC and provisional sums (which the contractor has no power to reduce) will be subject to a percentage rebate. To calculate this percentage, PC and provisional sum amounts are extracted, as shown in Fig. 11.10.

The effect of this calculation is as follows. In the variation account all rates will be subject to an addition of 3.5% for water and insurances. The water and insurance percentage can be converted into a percentage on the contractor's own work instead of on the whole total as appears in this example. In that case, the two percentages can be combined into a single percentage. However, in this case, since the contractor has expressed water and insurances as a percentage of the whole, they are so treated.

Summary	£
Preliminary items	18662.00
Named subcontractors	55408.00
Groundwork	17479.06
	8131.16
In situ concrete	~~8031.16~~
Masonry	19083.69
Structural carcassing	4789.88
Cladding/covering	18103.97
	4392.36
	~~4393.36~~
	1884.38
Linings/partitioning	~~1884.28~~
Window/doors/stairs	8616.01
Surface finishes	1036.77
Paving/fencing	4302.74
Disposal systems	3728.98
Engineering services	16445.98
	£182064.98
	£~~181965.88~~
Water and insurance 3.5%	6367.57
	£188432.55
	£~~188333.45~~
Tender submitted £188 300.00	

Fig. 11.9 Corrected summary in contract bills.

Tender decision and reporting

It is normal practice that the design team will hold a Tender Review meeting to make the tender decision. This meeting will review the tenderer's performance against the quantitative data (price) and the qualitative data (non-price criteria). The use of relative price evaluation models where pro-rata scores is based on the lowest tender maybe used; however, issues do arise with combining price and quality ratings and therefore the domain of an experienced quantity surveyor. The architect will rely extensively on the quantity surveyor for advice on these matters. A report will be prepared for the client, setting out clearly the arguments in favour of acceptance of one tender.

The results of the tender are published to all the contractors on the tender list. It is normal practice to provide a separate list of those that tendered and a list of the tender sums and scoring against the criteria. The information is not linked but enables the unsuccessful contractors to see which organisations they were competing against and the tender sum range. This enables the contractor to review their bidding strategy.

			£
Provisional sums			10000
Mechanical services			14850
Electrical services			8223
Wood flooring			2200
External staircase			4050
Metal windows			6500
Water mains			200
Ironmongery			550
Sanitary fittings			890
Daywork			2610
			£50073
The rebate to be expressed as a percentage is:			
Errors	182064		
	181966	98	
Rebate in tender	188433		
	188300	133	
		£231	
(This total equals the total difference between £188 433 and £188 300.)			
The percentage is calculated as follows:			
Corrected total without insurance, etc.			182064
Less PC and provisional sums			50073
			£131991

$$\text{Percentage} = \frac{231}{131991} \times \frac{100}{1} = 0.18\%$$

Fig. 11.10 Calculation of percentage rebate.

Addendum bills

The lowest tender is sometimes for a higher amount than the client is prepared to spend. It is usually possible to reduce the tender sum by changing the design in terms of either quantity or quality. However, the correct application of cost planning procedures should avoid this problem. An addendum bill is prepared in a similar way to the variation account referred to in Chapter 13. Changes to the design are measured and priced using the rates from the contractor's priced bills. The addendum bill is prepared prior to the contract being signed and the contract sum is hence based on the revised tender figure. The bill modifies the original quantities, and the quantities so modified become part of the contract. The adjustments are mostly omissions, but balancing additions are also required. Where there are no rates or prices for these in the bills, they are agreed with the contractor through a process of negotiation. Sometimes, if a variation is complex but its value can be estimated fairly accurately, the adjustment can be made on a lump sum agreed by the parties concerned.

Preparation of contract bill of quantities

A fair copy of the priced bills is required for signature with the contract. If a blank copy has been sent to the contractor to be completed, this can be used as the surveyor's office copy after any alterations necessitated by the checking process are made. A corrected copy, for the contract, is made by the quantity surveyor.

The prices in the contractor's priced bill of quantities are confidential. They must be used solely for the purpose of the contract (JCT). Though they naturally contribute to the quantity surveyor's knowledge of current rates and prices, they can only be referred to for the surveyor's own information. They should not, for example, be discussed with another contractor.

Preparing the contract

Once the tendering procedure has been completed, it is necessary to complete a formal contract with the successful tenderer. The duty of preparing the contract by completing the various blank spaces in the articles of agreement is the responsibility of the architect, although the quantity surveyor is often asked to complete it. It is sometimes necessary to add special clauses to the conditions of contract and to amend other clauses. Where this is required, they must be written in, and the insertion or alteration must be initialled by both parties at the time the contract is signed.

Extreme care must be taken if clauses within standard conditions of contract are to be amended, as specific alterations can affect other clauses. It is not recommended. It is usual to seek legal advice if it is intended to make substantial amendments. Any portions to be deleted must be ruled through and similarly initialled. All other documents contained in the contract, i.e. each drawing and the contract bills, should be marked for identification and signed by the parties.

For the contract bills, this identification should be on the front cover or on the last page, and the number of pages can be stated. If the standard form with quantities is used the specification does not form a part of the contract. It will thus not be signed by the parties. If there are no quantities the specification is a contract document and must be signed accordingly. All the signed documents must be construed together as the contract for the project.

Contracts are either signed under hand, when the limitation period is six years, or when a twelve-year period is required the contract is completed as a deed, formerly under seal. In the latter case, it is important to ensure that this is duly recognised, as failure to do so could have serious implications.

Case law exists that illustrates the importance of ensuring that all the contract documents are in agreement with each other. In *Glesson* v. *London Borough of Hillingdon* (1970) EGD 495, there was a discrepancy between completion dates set out in the contract bills and completion dates in the Appendix to the JCT form of contract. Delays had occurred, and the question was raised with regard to the correct date to be used when calculating liquidated damages. The court held that under the relevant clause of the form of contract in use at the time (JCT 63), the date in the Appendix prevailed. Litigation would not have occurred had the contract documents been properly checked for any inconsistencies.

E-tendering

E-tendering is defined as the electronic issuing and receipt of any tender documentation as part of the procurement process (RICS 2010). E-tendering provides a framework where both consultants and tenderers are able to reduce costs, remove unnecessary administration and streamline the overall tendering process. The RICS e-tendering guidance notes advocate the importance of adopting standard practice both in terms of presentation and content. The use of standard practices will help consultants and tenderers benefit from a consistency in approach as well as the avoidance of ambiguities and technical incompatibility. A number of issues need to be examined in the tendering process from the initial preliminary enquiry through to tender acceptance or withdrawal. These include:

- Tendering methodologies
- Different electronic formats and their impact upon information exchange
- Benefits and constraints that different technologies provide
- Security issues, tendering procedures and workflow
- Assessment of tenders and notification of results.

The *National Construction Contracts and Law Report 2018* (NBS 2018) shows a steady increase in the use of electronic tendering, with 46% of clients always using it compared with 34% in 2015. Means of distribution includes web-based portals and cloud-based technologies. Measurement software packages often include integrated E-tendering tools to aid standardisation of pricing documentation, ease of distribution, management and receipt of tenders and efficient and effective tender analysis and reporting.

Online auctions

An alternative to e-tendering practices is the use of online auctions. Those used in the construction sector are normally reverse auctions, a form of live electronic negotiation allowing bidders to reduce their own bid in response to offers made by others. BS ISO 10845-2020 provides guidance on documents and procedures used in electronic auctions (British Standards Institution 2021). However, whilst online auctions are not restricted only to price-based bids, there is concern in the industry that such practices are contrary to Egan principles and best value cannot be obtained.

References

British Standards Institution. *ISO 10845-1, Construction Procurement: Part 1 Processes, Methods and Procedures*. BSI. 2021.

Doyle, W. 5 Tips for getting to grips with RMM. Rail diary. 2021 [online] Available at: https://www.raildiary.com/en/blog/2021/01/5%20Tips%20for%20getting%20to%20grips%20with%20RMM (accessed 23/01/2022).

ICE Publishing. Introducing CESMM4 Revised. YouTube video. 2021 Available at: https://www.youtube.com/watch?v=5YDCQB3yDuE (accessed 23/01/2022).

JCT Practice Notes. *Tendering 2017*. Sweet and Maxwell. 2017.

Lee, S. *Willis's Elements of Quantity Surveying*. 13th Edition. NY. Wiley Blackwell. 2020.

NBS. *National Construction Contracts and Law Report 2018*. RIBA Enterprises Ltd. 2018.

RICS. *E-tendering RICS Guidance Note*. The Royal Institution of Chartered Surveyors. 2010.

RICS. *Tendering Strategies. RICS Guidance Note*. 1st Edition. The Royal Institution of Chartered Surveyors. 2014.

RICS. *Standard Form of Consultants' Appointment Explanatory Notes*. The Royal Institution of Chartered Surveyors. 2019.

Thompson, S. NRM updated launched. *Construction Journal*. RICS. 2021. Available at: https://ww3.rics.org/uk/en/journals/construction-journal/nrm-an-updated-suite-in-the-works.html (accessed 23/01/2022).

12

Cost Management

KEY CONCEPTS

- Interim valuations
- Certificates
- Interim payments
- Retention
- Cost control and reporting

LEARNING OUTCOMES

After reading this chapter, you should be able to:

- Understand the importance of the interim valuation in providing cashflow to the contractor, their timing and certification.
- Understand the responsibilities of the quantity surveyor in interim valuations and post-contract cost control.
- Appreciate how the quantity surveyor ascertains the sum awarded in an interim certificate.
- Understand the role of retention in the payment process, the items subject to retention and the timing of retention monies release.

COMPETENCIES

Competencies covered in this chapter:

- Contract administration
- Contract practice
- Project finance (control and reporting).
- Quantification and costing (of construction works).

Introduction

The scope of the surveyor's involvement during the post-contract stage of a project will generally require the preparation of interim valuations, the preparation and agreement of the final account and the management of project costs throughout. Final accounts are

Willis's Practice and Procedure for the Quantity Surveyor, Fourteenth Edition.
Allan Ashworth and Catherine Higgs.

discussed in the next chapter. It is usual for both the client and the contractor to employ a surveyor or team of surveyors during the post-contract phase. The successful execution and completion of the post-contract procedures and the final account very much depend on cooperation between the client's appointed surveyor and that of the contractor. Whilst the responsibilities of these differ, there are areas of involvement common to both and it is important that each side has an understanding of the process and possible approaches. This is further underlined by the developments in procurement and contract choice, which continue to emerge.

Depending upon the size and nature of the project, the post-contract administration may be undertaken by site-based staff involved on a full-time or fractional basis. In any case, the duties to be performed will be somewhat similar. Likewise, the degree of involvement may vary according to the type of main contract. For example, if the contract is awarded on an approximate quantities basis requiring remeasurement on site, additional surveyors may be needed to carry out the site measurement. Alternatively, if the project is design and build, it is probable that the demands upon the time of the client's surveyor will be much reduced.

The issue of choice of contract and the impact it could have upon practice is somewhat difficult to overcome in a book of this type. There are now many contract options available and it is impossible to accommodate the specifics of each. Therefore, in the interest of clarity and pragmatism, the emphasis of this chapter is on the practical rather than contractual. Where reference to the contract becomes a necessary part of the explanation, the main form which will be assumed to be in use is the Standard Building Contract With Quantities 2016 (JCT 2016 Standard Building Contract SBC). Despite the decline in the use of this traditional approach over recent years, it is still considered to be the most commonly understood, and knowledge of it is a fundamental requirement for students and practitioners. Although only occasional reference to other forms of contract is made, much of the explanation and associated practice and procedure relating to JCT 2016 can be transferred to these with little or no amendment. However, although strong similarities may exist there are also subtle and distinct differences which the reader should be wary of and research further where practice demands.

Since the contractual basis of this chapter generally relates to JCT 2016 SBC, it accommodates the provisions of the Housing Grants, Construction and Regeneration Act 1996 (the Construction Act 1996), and amendments made by the Local Democracy, Economic Development and Construction Act 2009 (Construction Act 2009).

Valuations

The construction industry survives on cash flow and the role of the surveyor is of key importance in this regard. JCT 2016 makes clear the duty of the client's quantity surveyor in this respect (clause 4.9.2):

> 'Interim valuations shall be made by the Quantity Surveyor whenever the Architect/Contract Administrator considers them necessary for ascertaining the sum due in an Interim Certificate.'

Most construction projects encountered by the surveyor will have contractual provision for payments to the contractor for work done at regular intervals during the contract period. The amounts involved in the construction of major works and the duration of most projects warrant this approach. Therefore, the contractor has a regular cashflow, which is vital to profitability within the construction industry. This issue of regular payment is a major feature of the Housing Grants, Construction and Regeneration Act 1996 (the Construction Act) and the subsequent amendment in 2009, the contents of which are reflected in JCT 2016 SBC. Where contracts do not specify payment mechanisms, the relevant terms of the Scheme for Construction Contracts (England) Regulations 2011 apply.

Certificates and payments

In the course of a construction project for which interim valuations apply, an architect or contract administrator will be called upon to make decisions and issue a series of certificates that must be issued in accordance with the provisions of the particular contract. In so doing when acting in an arbitral situation, an architect is under a duty imposed by law to act fairly and impartially between the parties.

Where an interim certificate or payment notice is required by the contract, and certainly under JCT 2016, its issue is a condition precedent to payment. Interim certificates must be issued no later than five days after the monthly interim payment due dates, which are seven days after the interim valuation dates specified in the contract particulars. The employer must pay the contractor no later than 14 days from the interim payment due date. This monthly cycle of due dates continues up to the due date for the Final Payment.

Where the employer intends to withhold and/or deduct an amount from the payment due, the contractor should be provided with a 'Payless Notice' which should be issued no later than five days before the date of payment. This should show the amount to be withheld/deducted and details on how the revised amount is calculated. Whilst neither the surveyor nor the architect should exclude amounts from the certificate, when there are grounds for withholding, they should advise the employer of the entitlement to do so, including provision of the required detail. The quantity surveyor can issue the Payless Notice on behalf of the employer. The RICS provides sample notices within its document templates (RICS 2016).

Accuracy

The valuation for certificates should be made as accurately as is reasonably possible. In preparing or verifying the valuation, the surveyor has two opposing concerns:

- Under the contract the contractor is entitled to the value of work done, less a specified retention sum. If the valuation is kept low, the retention sum is in effect increased. To a contractor having a number of contracts in hand, these excessive amounts of 'retention' will mount up and demand additional capital, which may have serious consequences.
- The client must be protected against the possible insolvency of the contractor. When an overpayment resulting from an excessive valuation cannot be recovered from the contractor, as in the case of insolvency, the additional payment will be effectively 'lost' and

may become an additional expense to the client. In turn, this additional amount may become the liability of the surveyor.

It is reasonable to assume that, for the client's surveyor, the latter consideration will be of more importance than the former since self-preservation is a powerful incentive. Unfortunately, in cases where the financial reputation of the contractor becomes doubtful, the reasonable inclination of the surveyor is to become more cautious, which in turn may result in greater financial difficulties.

In determining the need for accuracy, the surveyor should appreciate that an interim valuation is not a final agreed payment for the work but merely a snapshot of the progress of the works which, depending on the work stage, contractors' programme, deliveries to site etc., may change significantly within 24 hours of the assessment.

Timing

If, as is usually the case, the contractor is entitled to payments at regular monthly intervals (in terms of JCT 2016), this should strictly be adhered to. Figure 12.1 provides an overview of the timing of the valuation date, certificates and payment. Interim valuation dates are established by reference to the first interim valuation date specified within the Contract Particulars. Subsequent valuation dates will occur on the same date in each month. For example, if the first valuation date is 13 September, the second will be 13 October, the third on 13 November. The contract recognises that these dates may be weekends or bank holidays, when this occurs the date will be amended to the nearest business day. It is prudent to agree all valuation dates before the project commences, as these impact the due dates, issue of certificate dates and payment dates. It is also important to ensure that the client's internal accounting processes can meet the payment periods stated within the contract. If they are unable to do so, payment contractual clauses could be amended; however, this

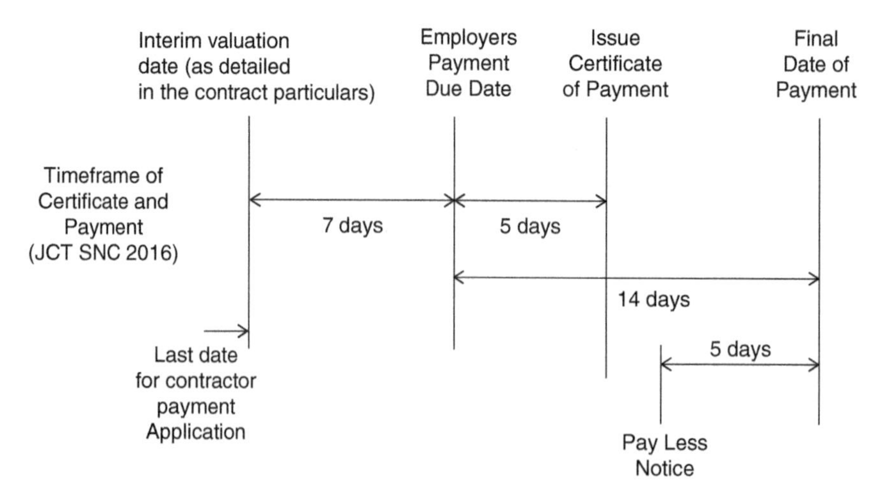

Fig. 12.1 Timings of valuation and payment dates.

should be done with caution. In any event, all parties should be aware of the implications of failure to certify and pay correctly in accordance with the conditions of contract.

In the past, the client's surveyor found it very helpful if the contractor submitted a statement with supporting invoices, delivery notes, etc. as a basis for the valuation. This approach saved a great deal of time and possibly was also more accurate since the contractor was able to provide more detail than may otherwise have been possible to establish during a site visit. JCT 2011 and subsequently JCT 2016 now reflect this previously accepted practice by means of clause 4.10. This allows the contractor to prepare and submit a Payment Application for interim payment, 'no later than the Interim Valuation Date' to the quantity surveyor. The application should state the sum due and the basis on which the figure has been calculated. It is important to emphasise that this action does not remove the responsibility for the valuation from the client's surveyor who, in normal circumstances, will be liable for its correctness and in the absence of the contractor's application is still obliged to prepare a valuation in accordance with clause 4.9.2 (Although the correctness of the certificate is the responsibility of the architect, it is normal to place reliance on the quantity surveyor's valuation and thus, in practice, liability for its accuracy is transferred). To verify the contractor's application, it is sensible for the client's surveyor to arrange a meeting on site with the contractor's surveyor and, where necessary, make any adjustment to the contractor's statement. Where such an adjustment occurs, the client's surveyor is obliged to provide the contractor with a statement showing the revised amount of valuation, together with supporting detail to allow the contractor to identify and understand the changes that have been made.

If an interim certificate is not issued or is issued late, then the contractor's interim Payment Application becomes an interim Payment Notice. Subject to any Payless Notice the employer must pay the contractor the sum stated in the Payment Notice. Where the contractor has not made an interim payment application, one can be made any time after the five-day period for issuing the certificate. In this case, the employer must pay within fourteen days of the submission rather than the due date.

Failure of the employer to pay the contractor may result in the application of one or more of the remedies available to the contractor:

- *Claim for interest on the late payment.* In addition to the amount due to the contractor, simple interest will be due on the amount of late payment, calculated at a rate of 5% above the Bank of England's base rate current at the due date, for the duration of the delay in payment.
- *Suspension of the work* (clause 4.13). If the employer fails to pay in accordance with the contract, the contractor may issue a notice to the employer, copied to the architect/contract administrator, to the effect that the works will be suspended after a further seven days, allowing time for delivery. Once the employer has paid the amount due, the work should recommence within a reasonable time thereafter.
- Determination in accordance with clause 8.9 of the standard form (JCT 2016).

The surveyor should be alert to the above conditions and advise the client and architect where necessary.

Extent of measurement

The extent to which measurement will be necessary when making valuations for interim payments will depend on the nature of the job and the stage it has reached. It may very often be possible to take the priced bill, identify the items that have been done, and build up a figure in that way. Some items will, of course, be only partly done, and in that case a proportion will have to be allocated. At a first valuation, for instance, there may be little beyond foundations, and to identify the appropriate items in the bill should not be difficult. If it is agreed that the foundations are two-thirds complete then the amount can be easily calculated. When, however, it comes to the superstructure, it may be necessary to take approximate measurements of such things as brickwork, floors and roofs. The surveyor should always bear in mind the value of the works being measured when determining the most appropriate approach. To avoid excessive labour in the measurement of relatively complex but low-cost items for valuation purposes, it may be more reasonable to agree with the contractor's surveyor that 'half the total value of the plumbing works' is a fair assessment of the works carried out. It should be remembered that detailed measurement for valuation purposes will likely have no further bearing beyond the interim valuation in hand and is therefore of passing value only.

Figure 12.2 indicates the apportionment approach to the preparation of an interim valuation, reflecting both a necessary pragmatism and a reasonable level of accuracy. Inaccuracy should not exceed ± 5% for contract sums less than £2.5 m and ± 2% for contract sums greater than £2.5 m (RICS 2015). In the example, concrete work has been collected into work between ground and first floor, first floor and second floor and second floor to roof. A tour of the site with the contractor's quantity surveyor will allow a prompt but reasonable assessment.

Interim valuation 5;
Date on site 18 December 2023

Bill Ref	Brief description (Concrete works)	Total £	Previous %	Amount £	Present %	Amount £
18/3 B–G	Reinforced Concrete in Cols					
	Say 40% GF–1st	24 000	100	24 000	100	24 000
	Say 30% 1st–2nd	18 000	75	13 500	90	16 200
	Say 30% 2nd–Roof	18 000	Nil	Nil	10	1800
18/4 A–F	Formwork to Cols					
	Say 40% GF–1st		100		100	
	Say 30% 1st–2nd		100		100	
	Say 30% 2nd–Roof		50		75	
18/5A–H	Reinforcement in Cols					
	Say 40% GF–1st		100		100	
	Say 30% 1st–2nd		100		100	
	Say 30% 2nd–Roof		50		75	
	Total carried forward					42000

Fig. 12.2 Apportionment of works for valuation purposes.

As indicated by the example, the usual method of valuation is to assess the total value of work performed at each interim valuation on a gross basis (the amount of work since the commencement of the project *not* as previous payment plus a little more). This approach negates any over valuation or under valuation errors that may have occurred in a previous valuation, although inclusion of columns showing the previous assessment may be a useful guide. There are various further shortcuts that could be taken; for example, annotating a copy of the bills of quantities with inserted columns for amounts of work carried out or by the generation of a spreadsheet that could also be used in the assessment of amounts for fluctuations.

When dealing with housing, or other projects for which there are a large number of similar units, it should be possible from the bill of quantities to arrive at an approximate value of one house at various stages, for example:

- Brickwork up to damp-proof course
- Brickwork to first floor level, with joists on
- Brickwork to eaves
- Roof complete
- Plastering and glazing complete
- Doors hung
- Plumbing and fittings complete
- Decoration complete.

The value for different types of the same size of house will not vary sufficiently to make a difference for certificate purposes. The work done at any time can be valued by taking the number of houses that have reached each stage, and pricing for half and quarter stages. With such projects, if the contractor's surveyor intends to submit a detailed payment application in accordance with JCT 2016, details of the approach should be agreed with the client's surveyor at the outset to avoid disagreement and abortive work.

On projects without firm bills of quantities, the value of the work carried out will be assessed through measurement of work on site or, where appropriate, by proportion of the individual items on the tender summary. It may be convenient to invest some time at the commencement of the project to agree the value of particular work stages as a basis for interim valuations.

Towards the completion of the project, it is a good idea to check what is left to finish, as a safeguard against error in a cumulative total. For the last two or three valuations, the contract sum might be taken as a basis, and deduction made of all provisional sums and percentage additions and of work not yet done. The various accounts against the first of these are added together with percentage additions pro rata, adjustment being made for the approximate value of variations and price adjustment. Again, once identifiable sections of the works are almost complete, a similar approach can be adopted whereby, for example, with reference to Fig. 12.2 above, concrete on each floor can be calculated as complete less minor amounts of work not yet done.

Preliminaries items

Each valuation will take into account the pricing of the preliminary bill or preliminaries items. There may be a number of items separately priced, or one total for sections or the whole in which case further analysis will be required. In addition, they may show an allocation between fixed and time-related costs or a combination of both in accordance with NRM2 (and SMM7).

Each priced item should be considered and a fair proportion of each included: fixed payments and time-related costs should be considered separately. Fixed payments occur where there is an identifiable item of work, for example a signboard with a single cost payable on construction at the beginning of the project and a further payment on removal at completion. Time-related items relate to items that attract costs on a regular basis during the works, for example site management; the amount for this which is included in the preliminaries could be divided by the period of the contract to give a suitable monthly sum.

Some preliminaries items may be allocated to more than one categories, such as the provision of internet. A proportion of the internet charges will be fixed (perhaps relating to set up costs) and some may be time related (monthly charges). Similarly, the price for provision of offices, mess facilities and storage units could be split into delivery cost (fixed payment), weekly rent (time related) and removal cost (fixed payment), and valuation made accordingly.

Even if it is anticipated that the contract period will be exceeded, the quantity surveyor should not make reductions to the time-related items until an instructed change to the project completion date. However, once instructed adjustments to the time-related costs should be adjusted to reflect the delay and to marry payments on account more accurately to the work carried out. If an extension of time award has been made, this could also affect the amount of the preliminaries allowance.

A cashflow forecast of preliminaries costs reflecting the apportionment to fixed and time-related costs, relating to the duration of the works may be prepared for agreement at the first interim valuation. Once this is accepted, it can be used for future valuations, not forgetting to monitor the progress of the works and adjust if necessary.

It should be noted that the use of either a time-related or a cost-related approach for the total preliminaries assessment at interim valuation stage is bound to be flawed and less accurate than the above method.

Subcontractors

The contractor is allowed to appoint subcontractors to carry out portions of the works, provided written consent is obtained from the architect/contract administrator in accordance with JCT 2016, clause 3.7. The payment of interim amounts to subcontractors is the responsibility of the contractor's surveyor. Section 2.8.14 of *RICS Subcontracting* (2021a) provides guidance on the main contractor's payment to subcontractors, with specific reference to the Construction Act. Applications for payment from subcontractors often form the basis of the main contractor's own application. Payments will be made to each subcontractor, generally in line with amounts certified to the main contractor. Reconciliation on a monthly basis will be required. The contractor's surveyor will analyse income received through certificate payments against costs of subcontractor payment and direct costs (see section 'Cost control and reporting' later in this chapter).

The client's surveyor should be aware that the interim accounts prepared by subcontractors should be considered as though the main contractor had produced them.

Named subcontractors

Named subcontractors are, as their title suggests, named by the architect/contract administrator. By providing a list of subcontractors and/or suppliers in the contract documents (at least three names must be provided for each portion of the works – JCT 2016, clause 3.8.1), from whom specified work or materials should be obtained, the architect/contract administrator is able to restrict and therefore control possible sources of supply. Once a subcontractor has been so named when finally selected by the contractor, they become subcontractors in exactly the same way as subcontractors chosen entirely at the discretion of the main contractor. No special requirements therefore arise regarding the client's surveyor's duties at interim valuation or final account stage.

In practice, named subcontractors may pose a problem to the contractor. The rates submitted by the contractor at tender stage for portions of the works covered by the named subcontractor provision are at the risk of the main contractor. It is therefore important that rates for the work to be carried out by one of the named subcontractors are adequate, and that they have been obtained from one of the named subcontractors. Failure to conform in this regard, for example by using rates from an unnamed subcontractor with whom work is regularly sublet, exposes the contractor to significant additional price risk. Where less than three names are provided in the list additional names may be added to the list by either party if approval is obtained from the other (clause 3.8.1).

Unfixed materials

Besides the value of work done, most forms of contract allow payment to be made to the contractor for unfixed materials and goods. The payment for such materials is an area of additional risk for the employer in that:

- At the time of payment, ownership of the materials may not be vested in the contractor but with subcontractors or suppliers. Therefore, in such situations, payment by the employer does not transfer ownership from the contractor to whom it did not belong in the first place.
- The materials, being unfixed, may be more easily lost, damaged or stolen. This remains the risk of the contractor, in accordance with clause 2.24 unless, of course, insolvency of the contractor occurs.
- In the event of contractor insolvency, unfixed materials, irrespective of true ownership, may be easily removed from site.

In consideration of the contract conditions, the surveyor should only include payment for unfixed materials stored on-site where:

- *The materials are correctly stored.* Storage on building sites may be less than ideal and wastage rates are generally considered to be too high. Payment for materials that, due to damage, cannot be incorporated within the works should not be allowed.
- *The materials are delivered reasonably, properly and not prematurely.* Materials delivered too far in advance of their incorporation within the works should not be paid for. There are several factors that should be considered when determining the reasonable timing of deliveries, including material availability, ownership of price risk (e.g. will the employer

gain benefit from early purchase and delivery by avoiding an imminent price increase) and works progress.
- *The materials are required for the project.* For example, materials being stored on a particular site for convenience of the contractor and eventual use on another site should not be allowed.
- *Adequate insurance* of the unfixed materials is provided by the contractor.

In practice, the surveyor will benefit by requesting a list of materials to be submitted by the valuation date, to be checked during the site visit. Where the contractor submits an application for payment, in accordance with the provisions in JCT 2016, this should be provided. The surveyor should bear in mind that consideration should be given for changes in material stock occurring between submission of the application and site visit. An item included as unfixed material in an application for payment prepared several days before the valuation date, may be incorporated within the works by the time of site verification. If this occurs, the value of the works will have increased without any consideration in the payment application, and therefore, if any adjustment is made, it should be an increase rather than decrease.

The risks associated with unfixed materials are increased where the materials are stored off-site. In an attempt to protect the employer, JCT 2016 makes special provision, in that payment depends on adherence to the following conditions:

- Materials or goods or items prefabricated for inclusion, which should be in accordance with the contract, should be identified in a list (Listed items), provided by the employer, that must be annexed to the contract bills. This provision ensures that the contractor knows, in advance, the extent of the materials off-site that will be allowed and may make a tender allowance for costs in connection with any item excluded from the list.
- Proof that the property is vested in the contractor should be provided, thereby allowing transfer of ownership upon payment of the certificate.
- The materials are to be set apart in the premises in which they are manufactured, assembled or stored and must be correctly labelled showing the name of the employer and the intended works.
- Proof of adequate insurance cover for the materials must be provided by the contractor, protecting the interests of the employer and the contractor in respect of the specified perils (clause 4.16.2.2).
- If stated as a requirement in the contract particulars, a surety bond for the value of such materials should be given.

The contract differentiates between uniquely identified 'listed items' (e.g. purpose-made glazing units) and those that are not uniquely identified (e.g. possibly, sanitaryware).

Problems have arisen over retention of title of goods when certifying payments for materials on- or off-site. JCT 2016 attempts to ensure that any unfixed materials included within an interim certificate become the property of the employer. However, despite this there may be co-existent supplier agreements by which the supplier retains the title of the goods until they are paid for, thus preventing the contractor from transferring ownership.

Valuing change

During the construction project, changes are inevitable. Such changes, as defined in clause 5.1, are *variations* and can be any alterations, modifications to the design, quality and quantity of work (JCT 2016). Consideration should also be made of the point in the project lifecycle, as changes later in the construction process may result in abortive work or less effective working practices etc. The opportunity for change decreases as the project progresses and, conversely, the cost of that change increases with time (RICS 2021b). Works associated with provisional sums and prime cost sums are included within the interim valuation provided an instruction has been issued. Pricing of variations and contractors' clams for loss and expense are discussed in Chapter 13 Final Accounts. Regarding payment for these at interim valuation, the surveyor should note that such amounts should be paid in full to the contractor once ascertained, without the deduction of retention. There may be some pressure on the client's surveyor to include payments on account prior to ascertainment; no such payment should be incorporated without the direction of the architect/contract administrator although, as later stated, in practice the surveyor's advice is usually sought in such matters.

At interim valuation stage, the assessment of the value of some variations, provisional items or sections and claims, may cause difficulty. For instance, where the foundations are 'All Provisional' it may not be possible to adequately remeasure the works prior to their completion on site and the contractor may need to be paid in total. In such cases, where the eventual construction varies substantially from the provisional measurement, the risk of under- or over-payment is increased. Similarly, a major variation requiring significant remeasurement may be completed on site before such work can be fully remeasured. The surveyor will need to use various methods of approximate estimating in order to make a fair allowance for such items.

Advance payment (JCT 2016, clause 4.7)

JCT 2016 makes provision for the employer to make an advance payment to the contractor to be supported by an advance payment bond provided by the contractor. The amount of the advance payment, and the timing of the payment, will be stated in the contract particulars. At interim valuation stage, where a prior advance payment has been made, it will be necessary to consider the reimbursement provisions. The amounts and timing of these reimbursements will also be stated in the contract particulars.

Price adjustment

If the *Option B Labour and material cost and tax* (JCT 2016) it is valuable to start checking the records of price adjustment at an early stage. At any rate, this should be done for the labour portion, when a running total can be kept month by month for inclusion in the certificate valuation. Materials are rather more difficult to keep up to date than labour, owing to the time lag in rendering invoices. In either case, increased costs, whether in respect of labour or materials, should not be included in interim certificates unless supporting details have been made available.

As stated elsewhere, by far the most common method of adjusting fluctuations is by way of the price adjustment formulae, Option C (JCT 2016). With this method, reimbursement of increased cost is automatic in each interim certificate by the application of the formulae to the current indices relevant to the proportions of the work categories carried out. This is discussed more fully in Chapter 13.

Retention

It is standard in building contracts to incorporate a provision whereby the client is entitled to retain part of the assessed value of the works at interim valuation stage. In JCT 2016, a rate of 3% of the contract sum will be used unless a different percentage is stated in the contract particulars.

Retention benefits the client as it provides a sizeable fund at the end of a contract that acts as an incentive to the contractor to complete the works and to make good defects where they occur. There are no perceived benefits to the contractor, however, and a retention bond, in accordance with clause 4.18 (JCT 2016) may be used as an alternative, requiring a surety to be approved by the employer.

Not all components of the interim valuation are subject to retention, those excluded are listed in clause 4.14.2 of the standard form.

When work is complete, part of the retention sum is released, the balance being held as security for making good of defects that may be found necessary within the rectification period. In compliance with the standard form, this retention release occurs in the first interim valuation issued on or immediately following the issue by the architect/contract administrator of the practical completion certificate. Hence, this provides a clear incentive for the contractor to reach this stage. In the standard form, half is to be released. The remainder of the retention is released under the next certificate issued after the Certificate of Making Good. Again, this provides a clear incentive for the contractor, on this occasion to promptly remedy any defects that may have occurred.

The rectification period, if nothing is stated in the contract particulars, is six months. The purpose of the rectification period is to provide a reasonable period during which defects may become apparent. Therefore, its length should be situation dependent rather than an automatic inclusion or default to six months.

Where part of the works is to be handed to the employer prior to contract completion, the release of retention for that section will be dealt with accordingly. This is incorporated within clauses 4.19 of the standard form.

Clause 4.17 of JCT 2016 covers the rules pertaining to the treatment of retention money. The employer's interest in the retention is as a fiduciary trustee, whereby the employer holds the money on behalf of the contractor, although without obligation to invest. This is intended to provide some protection to the contractor against default by the client. The contractor can direct the employer to place the retention funds in a separately identified bank account. Thereafter, the employer should issue a confirmatory certificate to the architect/contract administrator, copied to the contractor.

Liquidated damages

For many clients, it is vitally important that a construction contract is completed by a particular date and therefore most contracts incorporate a fixed date. Where the contractor fails to achieve this date, it is likely that the employer will suffer a loss. To compensate for such occurrences, most contracts, and certainly JCT 2016 under clause 2.32, allow the employer to deduct an amount, i.e. liquidated damages, calculated on the basis of a rate stated in the contract particulars (e.g. £10,000 per week or part thereof) from amounts due to the contractor. The delay may be calculated on the basis of the completion date stated in the contract particulars, or such extended period if an adjustment of completion date has been granted by the architect in accordance with clause 2.28 of JCT 2016.

The pre-ascertainment of the rate of damages is an important aspect of this provision. It must represent a reasonable pre-estimate of the loss that will be incurred by the client in the event of a delay and must thereby provide all parties with a known amount at commencement of the contract. The benefits of this are clear since, otherwise, actual damages would need to be calculated on each such occasion, a costly process and carrying with it additional risk to the contractor in that at the time of entering into the contract, the full contractual liability for delay is unknown. Once damages are fairly pre-ascertained, the amount paid to the employer in the event of a delay is limited to that stated in the contract particulars.

Although a delay in completion may have occurred, neither the surveyor nor the architect/contract administrator is entitled to deduct any liquidated and ascertained damages from the amount due to the contractor within either an interim or final certificate. In accordance with the contract, this entitlement falls to the employer. The architect/contract administrator must issue a Non-Completion Certificate to the employer, advising of the contractor's failure to complete the works by the completion date, as a prerequisite to any deduction. Although any action may lie with the employer and architect/contract administrator, it is good and proactive practice for the surveyor to monitor any situation where liquidated damages may arise and advise the architect/contract administrator and employer where necessary. This advice should include a calculation of the deduction that may be applied.

The employer is entitled to waive the right to deduct liquidated damages or reduce the amount to be applied by notification to the contractor in writing. Any deduction in the payment must be stated within the Payless Notice. In practice, this may be done in consideration of: the contractor's withdrawal of a contractual claim thus avoiding a lengthy dispute; or a reduction in the actual loss to the employer relative to the amount pre-ascertained. However, as stated above, the employer is under no such obligation.

Predetermined stage payments

Predetermined payments, based on stage payments, have been traditionally used on large multiple contracts such as housing; however, to use such a method on a complex building contract is much less feasible. Whilst predetermined payments are not featured within the JCT 2016 contract, clause 4.14 acknowledges the impact of any pre-agreed stage payments on the calculation of the Gross Valuation. That said, the process involved in stage payments may have some practical application within the preparation of an interim valuation, without any contractual basis.

Where stage payments are used, the contractor is paid an agreed amount when work reaches a certain stage of completion. For example, when the substructure is complete the payment of the cost of the substructures is due. Additional payments are then made upon the completion of each determined progress point such as external walls, upper floors and roof etc.

Use of stage payments allows both parties, contractors and clients, to know in advance the probable cash flow for the contract and to arrange their finances accordingly, although the completion of each stage is still dependent on contractor's progress. For contractors, it is often necessary to know in more detail the value of work done for their own internal purposes and for paying subcontractors.

Previous interim certificates

A careful check should be made to ensure that the figure shown as already certified is correct, as a slip here could lead to a serious error in the valuation. The architect/contract administrator should confirm the amount of the previous payment whether the previous interim certificate, payment notice or payless notice. The figure may be referred to as 'previous valuations', the architect/contract administrator being asked to verify before certifying.

Valuation forms

Valuation and interim certificate templates can be obtained from the RICS. These forms give all the information necessary to enable the architect to complete the certificate. However, nowadays it is more common that companies create their own standard valuation and certificate forms.

Care should be taken not to refer to the valuation as the certificate or to the surveyor as certifying. The surveyor only recommends, and it is for the architect/contract administrator to certify, who may take into account matters other than those within the surveyor's sphere, such as defective work.

Record keeping

The surveyor's copy of each valuation should be saved together with all correspondence relating to each valuation. This will include such things as the contractor's payment application, with any price adjustments and the rationale for change. All calculations made to support the valuation must be retained for future reference.

Valuation on insolvency

In the event of the contractor becoming insolvent, it is prudent for the client's surveyor to make a valuation of the work executed by taking the necessary measurements of the work up to the stage at which work ceases or is continued by another contractor. The valuation will include unfixed materials and goods, and the architect/contract administrator on behalf of the employer will be responsible for seeing that these are not removed from the site (see JCT 2016, clause 8.5.3). The purpose of such a valuation is that all parties are aware of the financial position, and some estimate of money outstanding to the insolvent contractor can be made.

Insolvency of the contractor is likely to severely disrupt the works and result in costs to the client. It is a complex area of involvement, which increases risk to the professional advisor. There may be several actions that the surveyor is able to take in such situations and further discussion on this aspect is contained in Chapter 14.

Cost control and reporting

It is vital to the client and, of course, the design team, that costs are effectively managed throughout the construction of a project. Clients generally desire the final cost of a project to be no more than the contract sum; for some clients this may be their most dominant concern throughout the project. It is the role of the client's surveyor to try to manage these costs by a process of monitoring design and site developments and advising the client and members of the design team of their likely impacts.

If working closely with the members of the design team, the client's surveyor should be aware at the earliest opportunity of proposed variations to the contract, including drawing amendments. Advance knowledge of proposed changes enables a full evaluation in terms of cost, quality and programming implications to be carried out in advance of their issue.

Although the initial estimate of variations to the contract is likely to be of a budgetary nature based on approximate measurements and notional rates or merely lump sums, it is important that such estimates are progressively updated as more detailed information becomes available in the form of firm measurements, quotations or daywork records. It is also necessary for the surveyor to review all correspondence and meeting minutes issued on the project in order to identify the potential cost implications of the issues contained therein. Similarly, it is also beneficial if the client's surveyor is aware of what is actually happening on site. Occasionally, changes may occur which are undocumented but for which the client may be liable.

Regular financial reports will be required to advise the client of the anticipated outturn costs. These are commonly produced at monthly intervals. As mentioned earlier, the report will be tailored to meet specific client requirements. Certain clients will only require a simple summary statement of the current financial position (see Fig. 12.3), others will require a detailed report identifying the cost implication of each instruction (whether issued or anticipated) and the reason for it. On a complex project, this may result in a lengthy document. In addition to the advice given in the regular financial statement, it may be necessary to provide the client and members of the design team with cost advice more promptly if a major issue arises.

The regular report will identify adjustments to the contract sum in respect of the following:

- Issued instructions
- Adjustment of provisional sums
- Remeasurement of provisional work
- Dayworks
- Increased costs (if applicable)
- Provision for claims and anticipated future changes.

SHOPS AND OFFICES DEVELOPMENT WILLOW CENTRE		
COST REPORT NO.6		
	Date	
SUMMARY	£	£
Contract Sum		4,268,751
Risks		120,000
Adjustments for:		4,148,751
Instructions Issued – Section 1	62,138	
Named Subcontractors – Section 2	(8,619)	
Provisional sum Expenditure – Section3	(4,603)	
Anticipated Variations – Section 4	30,190	
Ascertained Claims	–	
	79,106	79,106
Anticipated Final Account	£	4,227,857
Current Approved Sum		4,268,751
Balance of Risks	£	40,894
Notes: Costs exclude VAT, Professional fees,and direct client costs		

Fig. 12.3 Financial Statement.

An updated cash flow (see later in this chapter) may also be included with the report to identify for the client the actual current level of expenditure relative to that previously anticipated. This will allow the adjustment of the future budgetary provision. Care must be taken to save all supporting calculations for each report for future reference when it comes to reviewing and updating the costs.

Where significant cost increases occur, it is good practice, where applicable, to prepare an outline of possible cost savings to such budgetary excess, before advising the client. Almost inevitably, this information will be required but advance preparation for this request will be helpful.

The contractor's surveyor also submits regular financial reports to his senior managers and directors. These will show a different emphasis to reports prepared for the benefit of the client. The contractor will carry out frequent cost/value reconciliations which are likely to have three broad cost elements:

- *Actual value.* This is the real value of the works in accordance with the quantity of work performed valued at the rates estimated at the time of tender and contained in the contract documents.
- *Actual cost.* This is the real cost of carrying out the works on site, irrespective of the tender estimate upon which the contract sum is based. Predominantly, this will be work carried out by subcontractors.
- *External valuation.* This is the valuation that is agreed by the client's surveyor and is included in the interim certificate for payment by the client. In a perfect world, this

	Project A	Project B	Project C
Actual value	£150, 000	£160, 000	£140, 000
Actual cost	£140, 000	£150, 000	£150, 000
External valuation	£155, 000	£140, 000	£155, 000

Fig. 12.4 Cost reconciliation.

amount, subject to retention, should be equal to the 'actual value' whenever an interim valuation is carried out. In practice, however, the nature of the interim valuation process makes this a difficult objective to achieve.

The relationship of these three elements of cost, value and external valuation is of key importance to both the contractor and the client. Figure 12.4 considers three project situations that could occur at interim valuation and financial report stage:

- *Project A.* The contractor is in a profit-making situation since the cost of performing the works is less than the value of the works contained in the contract sum. Also, in the short term relating to the interim stage, the contractor's cash flow is positive. This project is therefore in a relatively harmonious position, albeit that the client is paying in excess of the true value of the works. This additional payment is in effect a 'hidden borrowing' by the contractor.
- *Project B.* As with Project A, the contractor is in a profit-making situation. However, in the short term, due to the shortfall between actual cost incurred and payment received, the contractor's cash flow is negative. This may prove to be a serious situation for the contractor, leading ultimately to liquidity problems.
- *Project C.* The contractor is in an overall loss-making situation since the cost of performing the works is more than the value of the works contained in the contract sum. This position may be disguised in the short term due to the excessive external valuation which, as for Project A, contains some 'hidden borrowing'. Unfortunately, whilst the contractor may be avoiding a short-term cash flow problem, if this cost/value balance continues in the long term, the contractor will suffer a loss on the project. In such a situation, in addition to the obvious problems to the contractor, others may also suffer negative consequences. The contractor will be motivated toward attempting to find legitimate opportunities to retrieve the loss from other parties including the client, the subcontractors and suppliers, the consultants and possibly employees.

Cash flow

A forecast of cashflow is normally requested by clients and provided by the surveyor. This should be prepared in association with the contractor since it will be greatly influenced by the intended programme of works. Software is available which will assist in the preparation of a cashflow forecast based on criteria specific to the project and the 'S' curve of expenditure shown in Fig. 12.5.

Cashflow forecasts are useful for two reasons: they may be used as a basis upon which to arrange project finance, and they can assist in monitoring the progress of the works.

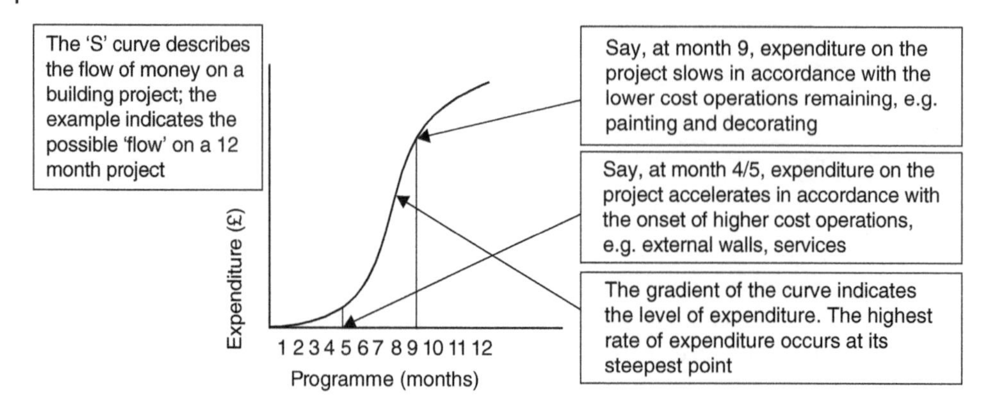

Fig. 12.5 Indicative 'S' curve.

With regard to the latter, it may be useful to have on file a note of the net amount of the contractor's work (excluding provisional sums), so that an eye can be kept on the proportion of this in each valuation. The value of certificates can be graphically represented to enable comparison with the anticipated cash flow. There may be reasons, such as site conditions or inclement weather, that result in a shortfall against that anticipated, but any serious departure from regularity should be looked into as it might be an early indication of delays and difficulties being experienced by the contractor.

BIM and valuing and managing change

The use of BIM integrated with cost and scheduling software enables the valuing of changes to the contract in the form of both variations and loss and expense claims (Pittard and Sell 2016). Variations to the BIM model can be amended and the impact on the costs and sequencing and programming can be automated. The benefit over traditional processes is that the data and information integrated within BIM can generate timely and effective information on the consequences of change. When a design change occurs, the updated 2D and 3D views of the portions of the model can be shared with the construction teams within the 'collaboration' facilities of the model. The 3D capability will visually identify what has been added, modified or deleted (Autodesk Building Solutions 2021). This not only supports the quantity surveyor in their cost management function but also enables the risk management activities associated with the changes to be analysed (RICS 2021b).

References

Autodesk Building Solutions. *Sharing Design Changes in BIM Collaborate Pro*. 2021. Available at https://www.youtube.com/watch?v=GF7VC8YdUyc (accessed 03/07/2022).

Joint Contracts Tribunal (JCT). *Standard Building Contract with Quantities (SBC/Q)*. Sweet and Maxwell Thomson Reuters. 2016.

Pittard, S and Sell, P. *BIM and Quantity Surveying*. London and New York. Routledge. 2016.
RICS. *Interim Valuations and Payment*. 1st edition. RICS. 2015.
RICS. *Payless Notice for use with the JCT Standard Building Contract With Quantities*. 2016. Available at: https://www.isurv.com/downloads/file/7452/jct_2016_pay_less_notice_for_use_with_standard_contract (accessed 20/02/2022).
RICS. *Subcontracting*. 1st Edition. Royal Institution of Chartered Surveyors. 2021a.
RICS. *Change Control and Management*. 1st Edition. Royal Institution of Chartered Surveyors. 2021b.

13

Final Accounts

KEY CONCEPTS

- Final accounts
- Variations
- Pricing variations
- Provisional sums
- Fluctuations
- Timing and completion of the final account
- Final account audit

LEARNING OUTCOMES

After reading this chapter, you should be able to:

- Identify the constituent parts of a final account, its timing and completion
- Understand the authorisation of variations and the methods of pricing.
- Understand the difference between a defined and undefined provisional sum.
- Identify the alternative fluctuation options.
- Understand the purpose of a final account audit.

COMPETENCIES

Competencies covered in this chapter:

- Contract administration
- Contract practice
- Project finance (control and reporting).
- Quantification and costing (of construction works).

Willis's Practice and Procedure for the Quantity Surveyor, Fourteenth Edition.
Allan Ashworth and Catherine Higgs.

Introduction

Many construction projects result in a final cost that is different from that agreed by the client and contractor at the commencement of the construction works. The calculation and agreement of this final construction cost, the final account, is usually of the utmost importance to both the employer and contractor. Therefore, parties to the contract need to ensure that the final account incorporates a fair valuation of the works carried out. This chapter deals with the principles of measuring for variation accounts and the practical implications of contract conditions covering the calculation and agreement of the final account. Whilst the RICS guidance document, '*Valuing Change*' (RICS 2010), is based on JCT 2011, the guidance provided on defining, valuing and other practical considerations associated with variations still applies.

Within lump sum contract arrangements, the price agreed by the client at commencement will usually require adjustment for several matters, including:

- Variations
- Provisional measurements
- Provisional sums
- Fluctuations
- Claims.

These are considered below. In addition to contractual provisions, there are other external factors that are likely to influence the project environment, the contract administration and the preparation of the final account. The degree to which this applies could depend on a range of factors, including:

- *Status of documentation at formation of the contract*
  The quality of the contract documentation will have a bearing on the proximity of the final project cost with that of the agreed contract sum. For example, a thorough pre-contract design process allied to accurate contract bills is likely to result in fewer post-contract changes than projects with incomplete designs and less accurate contract bills.
- *Skills of the contract administration and cost management team*
  An experienced contract administration and cost management team will be more able to maintain construction costs within the client's budget.
- *Market forces*
  In times of recession, contractors and subcontractors could be operating within particularly small profit margins. In such economic conditions, the tendency toward contractual claims may be more acute.
- *Client and contractor attitude*
  An adversarial contract environment is likely to be both a consequence of and a contributor to cost variations. In recent years, the construction industry has recognised this problem and has encouraged a change of attitude, for example via procurement routes that promote collaboration practices (See Chapter 9, Procurement).
- *Resources*
  Limited post-contract services, sometimes a consequence of clients wishing to reduce fees, is likely to reduce the extent and quality of post-contract control, resulting in a greater risk of cost variation.

Variations

General procedures

Once a contract has been signed, its terms cannot be changed unless the contract itself contains provision for variation, or the parties make a further valid agreement for alteration. The standard form contracts contain extensive machinery for variation; however, the only variations permitted are those that fall clearly within the contractual terms. If the desired change is not covered by those terms, it can only be effected properly by fresh agreement. In this connection, care should be exercised to ensure that the new agreement is itself a valid contract and that it is supported by consideration given by both parties. When making visits for interim certificate valuations or site meetings, the surveyor should be mindful of the variations that have arisen. It is often valuable to have seen the work in the course of construction.

The responsibility for issuing instructions is with the architect/contract administrator. The surveyor requires such instructions from the architect/contract administrator as an authority to incorporate the value of any resultant variation within the final account. Instructions from the architect/contract administrator may be issued for many reasons; however, the most common is to amend the design in some way. JCT 2016, clause 3.10, sets out the procedures for issuing instructions, though it is not necessary for the architect/contract administrator to issue an instruction using a standard form. It is, however, both usual and good practice, and an example of an instruction from the architect/contract administrator using JCT 2016 is shown in Fig. 13.1. Under clause 3.12, provision is made for dealing with instructions other than in writing and this is discussed later in this chapter.

The contract may also provide (e.g. JCT 2016, clause 2.14.3) that errors in the bill of quantities shall be treated as a variation and adjusted.

Drawing revisions

Many instructions from the architect/contract administrator will be accompanied by revised drawings. These should be dated on receipt and used to update a database of drawings which should be carefully maintained, as during the pre-contract stage. It is good practice to indicate that these drawings are post-contract, likewise drawings which are superseded by new drawing issues should be recorded and identified as such. This procedure will reduce the risk of using redundant information in the preparation of the final account.

There is some possibility that drawings will continue to be issued during the tendering period, i.e. once the contract bills have been completed and issued to the tendering contractors. If the changes contained in these revisions are relatively minor, it will be adequate for the client's surveyor to measure and price approximately and advise accordingly. However, if the changes are significant, it may be necessary to prepare addendum contract bills (see Chapter 11) and issue them as a further document to the tendering contractors; thus, the cost implications of the revisions will be included in the contractor's tenders.

Surveyors should also note the practice of design consultants to include on their drawings a schedule of revisions, indicating an outline of the changes contained in each updated drawing issue. This may be very helpful in identifying variations to design; however,

Architect's Instruction

Issued by:	Smith and Jones Architects		
Address:	Address		
Employer:	Willow Developments Ltd	Job reference:	456
Address:	Address	Instruction no:	10
		Draft no:	-
Contractor:	ABC Construct Ltd	Created date:	Date
Address:	Address	Sheet:	1 of 1
Works:	Shops and Offices		
Situated at:	Willow Centre		
Contract dated:	Date		

Under the terms of the above-mentioned Contract, I/we issue the following instructions:

	Office use: Approximate costs	
	£ omit	£ add
Construct inspection chamber and Install drainage pipework in Accordance with drawings 456/12B, 13C and 14D	ON ARCHITECT AND QS COPIES ONLY	
To be signed by or for the issuer named above — Signed [signature] Sub-total		

Amount of Contract Sum	£
±Approximate value of previous issued Instructions	£
Sub-total	£
±Approximate value of this Instruction	£
Approximate adjusted total	£

Distribution

- [x] Contractor | 2 |
- [x] Employer | 1 |
- [ ] Quantity Surveyor
- [ ] Structural Engineer
- [ ] M&E Consultant
- [ ] Clerk of Works
- [ ] Planning Supervisor
- [ ]
- [ ]
- [ ] Other
- [x] File | 1 |

Fig. 13.1 Architect's instruction © RIBA. Reproduced with permission by RIBA.

it should not be relied on solely since some drawing revisions may not be adequately scheduled. Each new drawing issue should be carefully examined to identify any amendment to the design.

A separate file for instructions from the architect/contract administrator should be kept with other variation-related correspondence. Such information may be found to explain the instructions and indicate the designer's intent, so helping when it comes to measurement.

Chapter 12 identified the benefits of using integrated BIM systems to both identify revisions and support the assessment of their impacts in terms of costs, time and risk.

Procedure for measurement and evaluation

Oral instructions

Instructions that have yet to be formally authorised by the architect/contract administrator are occasionally contentious and care needs to be taken to ensure that they are valid in terms of the contract. There are occasionally situations of convenience or perceived necessity whereby an instruction may be given orally. With regard to instructions other than in writing, the contract (clause 3.12) states that:

- The instruction shall be of no immediate effect.
- The instruction must be confirmed in writing by the contractor within seven days of issue and will take effect a further seven days from receipt by the architect/contract administrator, unless they dissent for any reason. This requirement on the part of the contractor is relaxed where the architect confirms an oral instruction within the seven-day confirmation period.

A contractor wishing to *ensure* payment for a variation arising from an oral instruction should therefore follow the contract accordingly. It is prudent to do so in a timely manner to avoid any disputes due to ambiguity of the verbal instruction. If neither party follows this procedure but the contractor complies with the oral instruction the architect/contract administrator may confirm any such instruction before issue of the final certificate.

Measurement

The evaluation of variations will usually include two distinct operations: measurement, either from revised drawings, model information or from site, and valuation. The procedures for both measurement and costing are outlined below.

Irrespective of who does the work, it is likely that the measurement of variations would be more easily carried out based on project information, whether drawings or model revisions. However, there are situations where it may be useful or essential to physically see the work carried out; for example, where provisional quantities form part of the contract and require site measurement, such as replastering in refurbishment projects. There are several approaches to site measurement, depending upon the circumstances. Traditionally, site measurements can be taken using a dimension book that will be 'worked up' later in office conditions. Alternative approaches could include the use of schedules of items that can be neatly pre-prepared in the office and used as a site tick sheet, or drawings that can be accurately dimensioned or annotated during a site visit. Where measurement occurs on site, it is not essential that the omissions be set down as the additions are measured. It is more pragmatic while on site to measure the additions, leaving the omissions to be evaluated in the office.

As a general principle, adjustment will be made by measuring the item as built and omitting the corresponding measurements from the original dimensions. However, there will be occasions when it might be easier to adjust a contract item by either 'add' or 'omit' only. For example, if all emulsion on walls is amended from two coats to three coats, an 'add' item of the contract quantity at the price for the additional coat would be a more efficient approach.

It is very important to keep omissions and additions distinct, and it is suggested that the words 'omit' or 'add' should always be written at the top of every page and at every change from omission to addition. Perhaps this advice sounds pedantic but should inversion between 'omit' and 'add' occur, the scale of error could be very large. When the measurement, calculation and billing process are separated and performed by various staff within an organisation it is easy to see how this type of error could result. Each item of variation should be headed with a brief description and instruction number. In accounts for public authorities, the references for instructions from the architect/contract administrator may well be required by the auditors.

Grouping of items within the final account

Before any bill of remeasurements is started, the surveyor should decide on the suitable subdivision into terms that will be adopted in the account. Since items may or may not correspond with the instructions from the architect/contract administrator they may be arranged in a different order: an instruction may be subdivided or several grouped together, if their subject matter suits. The arrangement of variations within the final account may reflect other objectives, for example where fluctuations require the categorisation of work or where capital allowances are being sought on some aspects of the completed building.

Quite probably, the adjustment of foundations will be the first variation for which measurements are taken. If there is a large difference between the contract provision for foundations and the as-built foundations, there may be some urgency for remeasurement and provisional agreement. In a situation of under provision, the contractor will require to be paid for the additional amount of work and the employer will need to accommodate the revised cost in the budget provision. Conversely, where the contract allowance is in excess of the as-built foundations, the consultant will need to ensure that overpayment does not occur and the only way to do this accurately is to carry out a remeasure. At this stage of the contract, it may not be known what other variations there will be; however, the foundation adjustment can be regarded as an item with which other variations will not interfere to any great extent. In the event that there is a minor change in plan for which an instruction from the architect/contract administrator is issued, any subsequent change to the foundations may be dealt with as a variation within a variation. Unless there is any special reason for distinguishing, the lesser variation will be absorbed in the greater. When it eventually comes to adjusting for the change in plan, this will be done for the superstructure only.

It might happen, however, that the complete value of the change in plan is separately required. For example, in the rebuilding of a fire-damaged building, for which reimbursement from insurers applied, it might be that a change in plan was being made at the client's request and additional expense. In that case, the foundation adjustment would have to be subdivided to give the separate costs required. If the variation is a completely additional room, then the foundation measurements for that room can easily be kept separate from those for the foundations generally, which is preferable as it gives more relative values.

As the list of variations develops, the surveyor will be able to decide on how to group them. For instance, there may be one instruction from the architect/contract administrator for increasing the size of storage tanks, another for the omission of a drinking water point and a third for the addition of three lavatory basins. Each of these will be measured as a separate item but, for convenience, the surveyor may decide to group these together as 'variations on plumbing'. It is good practice to highlight the 'reason for variation' in cost reports to ensure that the client can appreciate why costs have changed.

It will be convenient to group the very small items under the heading of 'Sundries', preferably in such a way that the value of each can be traced.

The role of the clerk of works

If there is a clerk of works, the client's surveyor should ensure that arrangements are made for records of hidden work to be kept in the form required. This will assist when examining accounts submitted by the contractor or otherwise prepared. Clerks of works vary, of course, in their ability and experience and it would be unwise to assume that the clerk of works knows exactly what is required without any guidance. Also, other duties may prevent the clerk of works from carrying out remeasurement work as and when it is necessary. The depths of foundations, position of steps in foundation bottoms, thickness of hardcore and special fittings in drainage (i.e. items of work which will become hidden) are all examples of items that the clerk of works may be asked to note and record. If these records are carefully kept and agreed at the time with the foreman, the client's and contractor's surveyors should have no difficulties arising from lack of knowledge.

The clerk of works may also play an important role in verifying materials, labour and plant records which relate to variations for which dayworks are to be utilised in their evaluation (see later in this chapter).

Where employed, a clerk of works may issue instructions on behalf of the architect/contract administrator. Such instructions will not take effect unless the architect confirms them in writing within two working days of issue.

Pricing variations

Pre-costed variations

The preparation and agreement of a final account can be a time-consuming and expensive process for both parties, so much so that in many cases it has been found that the measurement period often exceeds the contract period. An attempt to limit this problem is contained within JCT 2016. In addition, where a major and complex variation occurs, there could be considerable cost uncertainty at the time of issuing the required instruction to carry out the works. To overcome these problems, the principle of pre-costing variations as a Variation Quotation is incorporated within JCT 2016 through clause 5.3 and Schedule 2.

Under these conditions of contract, the architect/contract administrator has the power, if thought appropriate, to seek a firm price quotation from the contractor for variations before confirming the order. Notification of the intended variation is sent to the contractor, who is then required to price the instruction, breaking the cost down into the cost of the work and the cost of any concomitant prolongation and disruption. The price is then submitted for scrutiny by the quantity surveyor, who passes recommendations to the employer who, if the price is acceptable, provides the contractor with an acceptance in writing. This is followed by the issue of a confirmed acceptance from the architect/contract administrator. This procedure is subject to a strict timetable, and provision is made for adequate information to be supplied by both the architect/contract administrator and the contractor. If an acceptable price cannot be achieved using this procedure, then the traditional methods of agreeing the cost are adopted.

Under the terms of clause 5.3 and Schedule 2, this procedure is subject to the agreement of the contractor who, in addition, can be required to provide a method statement.

Pre-costing of variations may save a great deal of time and provide cost certainty with regard to the variation affected. However, there are drawbacks, not least of which is that pre-costed variations would be expected to be more expensive than those arrived at by traditional means:

- The pre-costing of variations lacks the element of competition provided in the traditional approach wherein tendered contract rates form the basis for pricing variations
- The costing of prolongation and disturbance, both totally unknown factors at the time of pricing, has to be something of a gamble, and contractors must err on the pessimistic side to safeguard their position
- The time frame is such that the contractor may have inadequate time to prepare (thus encouraging mistakes) and the client little time to consider (thus encouraging oversight)
- In the event that the employer does not accept a Schedule 2 quotation that has been prepared on a reasonable basis, a fair and reasonable amount will be added to the contract sum for the abortive work in connection with its preparation.

Where a Schedule 2 quotation is used, fluctuations if applicable will not be applied.

The pricing of measured work

The methods of pricing variations are described in the contract conditions (clauses 5.6 to 5.10, 'Valuation Rules', in JCT 2016) and follow a logical and hierarchical sequence in terms of their choice of application:

- Where applicable, the rates in the contract bills (or relevant schedule of rates) should be used. Clearly, this will apply to omission; however, additions may vary in terms of not only specification but also the nature of the work execution. For example, it is reasonable to expect that the rate for paint to ceilings 2.5 m above floor level will be less than the same painting specification at a height of 5 m, which will involve additional labour.

Likewise, there may be a significant change in the quantity of work resulting in lesser or greater efficiency.

- Where contract bill rates are not applicable, they may be adjusted to take into account the revisions in working conditions and quantity. In the 'paint to ceilings' example above, the new rate for valuation of the variation may be adjusted to reflect the additional labour involved. It may be possible to use other bills of quantities or schedule items of work in this agreement. For example, we may have an alternative painting specification, at both ceiling heights, showing a price differential of, say, £0.53/m^2 reflecting the additional labour involved at the greater height. It may therefore be reasonable to use this price differential in establishing a revised rate for the new item of work.
- Where variations do not resemble any contract bills item in terms of either specification or work context, a 'fair valuation' should be carried out. This may involve returning to the first principles of estimating whereby the rate for the item of work is considered in terms of material, labour and plant costs incorporating an addition for overheads and profit. This process may be avoided or reduced where the new item of work is relatively common and/or where there is available evidence of similar work, perhaps from a recent similar project.

Dayworks

Certain variations, which it may not be reasonable or possible to value on a measurement basis, may be charged on a prime cost basis. Daywork sheets, referred to as 'vouchers' in JCT 2016, clause 5.7, will be rendered for these by the contractor; they set out the hours of labour of each named operative, a list of materials and details of plant used. Section A12 of the BQ will state the notice period the contractor must give prior to carrying out any work on a daywork basis (NRM2 2021). It is the duty of the architect/contract administrator to verify the daywork vouchers, not later than seven working days after the work was carried out. If there is a clerk of works, it will usually be one of his/her duties to verify the dayworks voucher on behalf of the architect/contract administrator. The clerk of works' signature is not in any way authority for a variation, nor does it signify that the item is to be valued on a daywork basis instead of by measurement. When there is no clerk of works, the architect/contract administrator will be expected to sign the sheets. Neither the architect/contract administrator nor employer's surveyor, without being continuously on site – which is unlikely – can correctly guarantee that the time and material are correct but, if these appear unreasonable for the work involved, they can make enquiries to satisfy themselves.

JCT 2016, clause 5.7 provides for pricing daywork as a percentage addition on the prime cost or, if applied to labour, the All-Inclusive rate. The definition of prime cost is laid down in *Definition of Prime Cost of Daywork Carried Out Under a Building Contract* (RICS and the Construction Confederation), and these percentages, together with the All-Inclusive rates, are required by the contractor to be inserted in a space to be provided in the contract documents. These tendered percentage allowances will be used in the calculation of any dayworks arising during the project (see Fig. 13.2).

For work within the province of some specialist trades, different definitions of prime cost may be agreed, and these must be considered in the preparation of subcontracts.

The provision made for daywork within the BQ should be taken into account when considering the amount of any provisional sum for contingencies.

ABC CONSTRUCT LIMITED	
JOB No. 2436	DAYWORK No. 120
CONTRACT: WILLOW CENTRE	AUTHORISATION
W/E: DATE	AI No.
	CVI No. 102
	OTHER

DESCRIPTION OF WORK

Break out floor adjacent manhole. Insert 2 no. 100D 30 degree bends to form swan-neck. Install 100D vertical pipe within manhole with r/eye and rest bend. Adapt manhole benching, insert branch bend and Make good benching, brickwork and floor slab.

LABOUR

Name	Trade	M	T	W	T	F	S	S	Total	Rate	Add %	£
C Rose	B/layer			5					5	17.26	70	146.71
											Total Labour £	146.71

MATERIALS / PLANT / SPECIAL CHARGES

	No	Unit	Rate	Add %	£
Materials					
100D uPVC drain	1.5	m	11.92	10	19.67
...30 degree bends	2	No	29.00	10	63.80
...rest bend	1	No	35.50	10	39.05
...r/eye	1	No	58.00	10	63.80
...coupler	2	No	17.51	10	38.52
Site mixed concrete	1	Item	12.45	10	13.70
Cement and sand	1	Item	8.72	10	9.59
Plant					
Kango	5	Hrs	1.29	10	7.10
Angle grinder	5	Hrs	0.55	10	3.03
Task lighting	5	Hrs	0.48	10	2.64
Transformer and leads	5	Hrs	0.58	10	3.19
				Total materials etc £	264.09
				Total for sheet £	410.80

Signed by ABC CONSTRUCT LIMITED	Signed by Client's Representative
Date	Date

Fig. 13.2 Daywork sheet.

Generally, the attitude of the client's and the contractor's surveyors towards dayworks is very different. As far as the client is concerned, payment for work on the basis of cost reimbursement, which is what dayworks is, brings several disadvantages. With dayworks, the contractor will be reimbursed the full cost of all labour, materials and plant used, plus a percentage addition for overheads and profit. Therefore, the contractor has no incentive to complete the works in an efficient manner; in fact, as the cost of the variation work increases, so does the contractor's overheads and profit. For this reason, from the client's perspective, the use of dayworks will be resisted. However, they are likely to be to the advantage of the contractor. As with cost reimbursement contracts, dayworks can be expected to result in higher costs than with the methods of measurement and valuing previously outlined.

In practice, on most projects there are likely to be situations for which dayworks are the only fair method of valuation, i.e. where work cannot be properly measured; for example, work to remedy a design error, which has been discovered post-construction.

With reference to the above example of a completed dayworks voucher (Fig. 13.2), please note that the signatures affirm that the work recorded thus has been carried out, not that the work shall be valued on a dayworks basis nor, if priced, that the rates are correct in accordance with the contract. When calculating the amounts for inclusion into the final account, these should include the tendered allowance for overheads and profit, which may be a different amount for each of the three categories, materials, labour and plant.

Overtime working

Though there is nothing to prevent a contractor's employee from working overtime, subject to trade union control, this is normally entirely a matter for the contractor's organisation. No extra cost of overtime can be charged without a specific order. Where, therefore, overtime is charged on a daywork sheet, it will be corrected to the standard time rates unless there is some such special order.

It may be that, owing to the urgency of the job, a general order is given for overtime to be worked, the extra cost to be charged as an extra to the contract. Or the order may be a limited one with the object of expediting some particular piece of work. When a building operative paid, say, £13.00, per hour works an extra hour per day at time-and-a-half rate, i.e. £19.50, the half hour (£6.50) will be chargeable in such cases. As a matter of convenience, on the pay-sheet, if the normal day is 8 hours at £13.00 and 1 hour at £19.50, it will be 91/2 hours at £13.00. The half hour is not 'working time' at all and is therefore sometimes called 'non-productive overtime': the extra cost of payment for overtime work over normal payment. Any charges that are to be based on working time, such as daywork (where overtime is not chargeable as extra), must exclude the half hour. Where the extra cost of overtime is chargeable, the data will be collected from the contractor's pay sheets and, if necessary, verified from the individual operative's time sheets.

Provisional measurements

It often happens that work such as cutting away and making good after engineers is covered by provisional quantities of items such as holes through walls and floors or making good of plaster and floor finishings. The original bill may have been taken from a schedule supplied by the engineers, and the need for remeasurement on site must not be overlooked. These items should be distinguished in the contract bills by the label of 'provisional' where applicable to single items, or where an entire section is provisional, for example, foundations, the entire section would be marked as 'Substructures; ALL PROVISIONAL'. Such labelling would denote the need for the remeasurement of those items affected. It does sometimes happen that the provisional quantities reasonably represent the work carried out and can therefore be left without adjustment, but this should not be done merely to avoid what is certainly a rather laborious job. Nontechnical auditors are apt to frown upon such procedure. One of the few things they can do to check a technical account is to go through the original bill and see that all provisional items have been dealt with. An appendix to the variation account, showing how this has been done, can be of help to an auditor.

The main reason for the incorporation of provisional quantities within the contract bills is:

- To make a reasonable provision for work which will be required but for which exact detail is unavailable
- To obtain contract rates for provisionally measured items, which can be used in subsequent remeasurement.

The amount of work thus included should be a reasonable representation of the work involved and not a wild guess.

The valuation of remeasured approximate quantities may be based on the associated bill rates. However, where the approximate quantity allowance in the bills is not a reasonably accurate forecast, 20% or greater, adjustment will be necessary to accommodate the cost differential caused by the revised quantity. When the approximate and firm quantities are within 20%, the contract rate can be used without review (see the example in Fig. 13.3).

It should be noted that 'provisional quantities' are not to be confused with 'provisional sums', which are discussed later in this chapter. In terms of the contract, which in this matter follows the NRM2 Section 3.2.7 (2021), provisional quantities shall be used:

> 'Where work can be described and given in items in accordance with the tables but the quantity of work required cannot be accurately determined, an estimate of the quantity shall be given and identified as a "provisional quantity"'.

Provisional sums

Provisional sums are provided at tender stage for items of work for which there is little information or for work to be executed by a statutory body, e.g. electricity supply. Their inclusion provides a tender allowance for known work that cannot be properly measured or valued until later in the project. In terms of the contract, as with provisional quantities discussed above following NRM2, provisional sums are used 'where building components/items cannot be described ... they should be given as a provisional sum and identified as for either "defined" or "undefined" work as appropriate'. Examples of these are provided below.

Scenario: The refurbishment of 65 houses for a local authority

Treatment of existing plasterwork:

The existing plasterwork is of variable condition throughout the properties and therefore it is not cost effective to remove and replace all plaster throughout the scheme, nor is it possible to prepare a detailed schedule of the location and extent of plaster to be hacked off and replaced. Not only is the latter unforeseeable at design stage, it is practically impossible due to access to individual properties, the effects of damage arising from other refurbishment works (e.g. chasing for new wiring) and concealment due to wall coverings, furniture, etc.

Therefore, a sensible approach in pre-contract measurement is to approximate the amount of replastering per house (marking such quantities as '(Provisional)' and to later remeasure the exact amount of work carried out.

Extract from notional final account

BQ Ref		Quantity	Unit	Rate	£	
	House type: C: 23 Williams Street **Adjustment of provisional quantities** **Omit**					
18/3 B	Hack off wall finishes with chipping hammer, plaster walls. Remove and load into skip. (Provisional)	38	m^2	23.25	883	50
					883	50
	Add					
18/3 B	Hack off wall finishes with chipping hammer, plaster walls. Remove and load into skip.	45	m^2	23.25	1046	25
					1046	25

Fig. 13.3 Example of provisional quantities.

Defined work

Defined work relates to work that is not completely designed but for which particular information is available, which with reference to NRM2 incorporates:

- the nature and construction of the work
- a statement of how and where the work is fixed to the building and what other work will be fixed
- a quantity or quantities which indicate the scope and extent of the work
- any specific limitations identified (Section 2.5.5 of NRM2).

For instance, for an ornamental steel canopy at the entrance to a hotel: the nature and construction are given in the description of the work; the type and extent of work are clearly stated; the location of the canopy is given; and the method of fixing shall be by 'bolting to steel columns'.

Undefined work

Undefined work is identified as necessary at the time of tender; however, in addition to the lack of full design information upon which to measure, the particular details outlined above for defined work are unavailable. For instance, it may be known that landscaping work for the hotel project is required, but no further detail is available. A provisional sum calculated based on a previous project is included in the contract bills.

The important distinction to be made between defined and undefined provisional sums relates to programming, planning and pricing of related preliminaries. In the case of the steel entrance canopy, the contractor is required to allow for such in the tender submission. Provided no changes to these details occur, no adjustment will be made to the preliminaries, nor will the contractor be able to make reference to programme or planning matters in connection therewith. However, with regard to 'undefined work' due to the lack of detailed information available at tender stage, no such allowance is made. When the design of the related work is finalised and priced, the contractor will be entitled to include costs relating to preliminaries items and will also consider planning and programming matters.

Loss and expense

If the contractor has been delayed or progress of the works is disrupted by the 'Relevant Matters' listed in Clause 4.21 (JCT 2016), the contractor is entitled by written application to claim for loss and expense for any resultant additional costs. Relevant matters, specific to the quantity surveyor include:

- Errors and discrepancies in the contract bills and drawings.
- Work associated with approximate quantities items where the quantities in the contract bills do not reasonably reflect the work.

Within 28 days of the initial application of a claim, the quantity surveyor must notify the contractor of the initial ascertainment, identifying the difference between this and the contractor's assessment. Payment for loss and expense, once ascertained is included within each interim payment. The final account will therefore include any payments that have not previously been ascertained.

Fluctuations

Fluctuations are an allowance for building cost inflation that may or may not be reimbursed to the contractor, subject to the provisions of the contract. The amount of this cost factor depends on levels of inflation existing during the contract period. At times, this may be negligible, but at other times, for instance, as in 2020–2022, both material and labour costs rose significantly due to the Covid-19 crisis, oil prices, tariff on imports, sterling exchange rates and restrictions on the movement of foreign construction labour

(BCIS 2022). It is therefore important to do a risk assessment to identify whether fluctuations should be included within the contract.

The fluctuations provision to be used is stated in the contract particulars together with all relevant information to support the applicable fluctuation rules. JCT 2016 specifies three choices of calculating price changes: Options A, B and C. Only option A is included in Schedule 7. Option A provides for fluctuations relating to contribution, levy and tax, option B provides for labour and materials cost and tax fluctuations and option C for a formula adjustment. JCT 2016 also provides the flexibility of using alternative fluctuation provisions, but these need to be clearly stated.

Whilst it is the QS's responsibility to assess the price adjustments associated with fluctuations, it is very much seen as a specialist area (RICS 2016).

Option A – contributions, levy and tax

In situations where the client wishes to distribute the risk of building cost inflation to the contractor, option A is used. The choice of this contract option will require consideration of several factors including contract duration, market conditions and levels of present and predicted building cost inflation. Although such a contract choice will transfer the main burden of inflation risk to the contractor, option A does provide for the reimbursement of some inflation costs to the contractor caused by statutory matters such as changes relating to contributions and taxes on labour. These are identified in JCT 2016. In addition, clause A12 of schedule 7 provides for a percentage to be inserted in the contract particulars by the contractor, to be applied to all fluctuations to allow for cost inflation relating to some preliminaries items, overheads and profit.

Option B – labour and material cost and tax

Labour

The traditional way of adjusting fluctuations in the cost of labour and materials, although rarely encountered in practice today, is by way of a price adjustment clause (JCT 2016, option B). Under such a clause any fluctuation in the officially agreed rates of wages, or variation in the market price of materials, is adjusted. JCT 2016, clause B13, provides for a percentage to be inserted in the contract particulars by the contractor, to be applied to all fluctuations to allow for cost inflation relating to some preliminaries items, overheads and profit.

The checking of wages adjustment should be fairly straightforward from an examination of the contractor's pay-sheets. Wage rates are officially published by the construction industry joint council or other wage-fixing body so there should be no doubt as to the proper amount of increases or decreases, or the dates on which they came into effect. To these increases will be added allowances for increases arising from any incentive scheme and any productivity agreement and for holiday payments as set out in the contract. These increases will apply to workpeople both on- and off-site and to persons employed on-site other than workpeople.

Materials

The adjustment of materials prices is more difficult. The contractor will produce invoices for those materials from which the quantities and costs can be abstracted and the value will be set against the value of corresponding quantities at the basic prices. Prices must be strictly comparable. If the basic rate for eaves gutters is for 2 m lengths, an invoice for 1 m lengths cannot be set against that rate. The 1 m length rate corresponding to the 2 m length basic rate must be ascertained. There is also the difficulty of materials bought in small quantities, when again the price paid is not comparable with the basic rate. JCT 2016, option B, says '...if the market price...increases or decreases' and these are significant words. When in doubt the applicability of the contract wording must be considered.

Invoices should be requested for *all* materials appearing on the basic list. Claims for materials not appearing on the basic list are to be excluded from the calculation. It is therefore important the most cost significant materials are included on the list. The surveyor is responsible for seeing that fluctuations in either direction are adjusted. This is another case where non-technical auditors would be apt to worry if all items did not appear in the account.

The surveyor should also see that the quantities of the main materials on which price adjustment is made bear a reasonable relation to the corresponding items in the bill of quantities and variation account. An approximation, for instance, can be made of the amount of cement required for the concrete and brickwork, and any serious discrepancy should be investigated.

Option C – formula adjustment

Unlike Option A and B method, in which a calculation of actual amounts of increases and decreases is made, the formula method uses indices to calculate the amount of reimbursement. The BCIS provides monthly indices known as the Price Adjustment Formulae Indices (PAFI) As such, on a particular project, this calculation will therefore be technically inaccurate since it is a generalised approach for use across the construction sector. However, the benefits of its use are:

- The results are likely to be more predictable, particularly in unstable economic conditions
- The protection against price fluctuations is more comprehensive. With the alternative methods, there is likely to be a considerable shortfall in overall recovery on the contract
- The formula method greatly simplifies the administration of price fluctuation provisions
- The simplified process facilitates prompt payment of fluctuations in interim valuations
- The scope for dispute is greatly reduced
- Contractors can quote competitively on current prices with the confidence that reimbursement will be in terms of current prices throughout the contract
- The formula method can benefit from the application and use of information technology.

The JCT Formula Rules 2016 explain the formulae and provide information and assistance to those using them (JCT 2017). The formulae are of two kinds: the building formula and specialist engineering installations formulae.

The building formula uses standard composite indices (each covering labour materials and plant) for similar or associated items of work, which have been grouped into work categories. For example, the breakdown of concrete work categories reflects the differing material input and weighting of the index calculations. Despite the general heading of 'concrete', each of the items is subject to different cost influences – e.g. labour balance, cement, aggregates, fuel, timber, steel – and therefore justifies a different category. Rationalisation within the process can be seen by consideration of the work group for 'concrete: in situ'. The weighting of the two main material components, cement and aggregate, will differ for 1:2:4 concrete and 1:3:6 concrete; however, no differentiation is made in the calculation of fluctuations by the formula method. In reality, an increase in the cost of cement will have a greater impact on 1:2:4 than 1:3:6 concrete, but this detail is sacrificed for ease of application (i.e. to limit the work categories to a manageable number).

The formula is applied to each valuation, which will need to be separated into the appropriate work categories. In practice, this is done at the commencement of the contract by annotating the contract bills.

There are alternative applications of the formula available. Each of the work categories indices may be applied separately (Section 2; Part I of the rules). This provides the most sensitive possible application of the formula. Alternatively, the work categories may be grouped together to form weighted work groups (Section 2; Part II of the rules). Clearly, as explained above, the fewer work groups used, the less sensitive will be the indices to changes. It must also be practicable to analyse the tender and the value of work carried out in each valuation period into the selected work groups. This entails less work in separating the value of work carried out in every valuation.

These alternative uses are described in detail in the Formula Rules 2016 mentioned above, which also give notes on the application of the formula at precontract, interim valuation and final account stages.

The specialist engineering installations formulae (Section 2; Part III of the rules) cover electrical installations, heating, ventilating and air-conditioning installations, lift installations, structural steelwork installations, and catering equipment installations. These formulae use separate standard indices for labour and for materials, the respective weightings of which are to be given in the tender documents, except for lift installations and the shop fabrication for catering equipment installations where the weightings are standardised. In each case, the formula is expressed in algebraic terms and has been devised in conjunction with the appropriate trade association. It is intended that these specialist formulae will normally be applied to valuations at monthly intervals.

Completing the account

In terms of the contract, time limits for all parties to adhere to are provided:

- Clause 4.25.1: 'not later than 6 months after the issue of the Practical Completion Certificate or last Section Completion Certificate, the Contractor shall provide the...Quantity Surveyor with all documents necessary for the purposes of the adjustment of the contract sum'

- Clause 4.25.2.2: 'not later than 3 months after receipt... the Quantity Surveyor shall prepare a statement showing all adjustments...' The client's surveyor should prepare the final account and submit it to the contractor, via the architect/contract administrator.

These timescales will be affected by the completeness of the information provided by the contractor and the communication process between the parties.

As stated, whilst it is the client's surveyor's role to prepare the final account, the information required to do so will be provided by the contractor. Normally, subject to the considerations relating to variations discussed above, this will effectively include the preparation of the variation account This process is greatly aided by the Schedule 2 quotations received by the contractor during the course of the works.

In practice, the agreement of the final account will be helped by a good and collaborative working relationship between the surveyors involved. The contractor, having examined the account, is fairly certain to have some criticism. Unless the criticisms are of a minor nature and can be settled by correspondence, it is recommended the surveyors meet to resolve outstanding matters.

Audit

It is frequently assumed that the main role or function of an audit, whether it is of a company balance sheet, profit and loss account or the final account of a construction project, is to detect errors or, more importantly, fraud. This is an incorrect assumption and forms only subsidiary objectives. An audit of a final account, or any account, involves the examination of the account and the supporting documentation and, more importantly, the designated procedures involved. This enables an auditor to report that the account has been prepared to provide a true and fair view of the account. The auditor will:

- Compare the final account with contract bills
- Examine the records available
- Discuss aspects with relevant staff
- Examine the procedures used
- Prepare a report on the findings.

The auditor will use the skill and diligence that is normally expected and the process will involve an examination of all transactions involved. A technical audit, perhaps using the skills and knowledge of a quantity surveyor, may sometimes precede the more usual audit process.

The extent of the examination of the final account will depend on the auditor's experience and assessment of the internal controls that have been adopted. The audit will set objectives and incorporate what is known as the audit plan. The process is not intended to repeat that which has already been carried out by the quantity surveyor. It is chiefly concerned that:

- The processes used are correct
- The processes have not been departed from without good reason
- The accounts are free from error

- The final account is a fair and true record
- Sound accounting principles have been adopted and used throughout.

The following are the essential features of the audit plan:

- Critical examination and review of the system used for the preparation of the final account and the methods of internal control adopted
- Critical examination of the final account in order that a report can be made to the client as to whether the accounts are a true and fair record
- To ensure that the accounts have been prepared on sound accounting principles and in accordance with professionally accepted procedures.

The auditor will give particular attention to those types of transactions that could offer particular facilities for fraud. The amount of checking that will need to take place will depend on the quality and reliability of the system used. A good system usually relies on the collusion of two or more individuals and this provides some safeguard on which the auditor can base judgements.

The majority of errors that are discovered will be the result of miscalculation, carelessness or ignorance. However, on occasions, what may appear to be nothing more than a simple clerical error might ultimately be found to be fraudulent manipulation. The auditor must be able to show that reasonable skill and care have been exercised in designing the audit trail. Areas that appear dubious will be checked and spot checks made generally where it is considered necessary.

The culmination of the auditor's work results in a written report, sent to the client, indicating the opinion regarding the reliability of the outcomes and processes used. It will also offer suggestions of how these might be improved. The auditor's report will comment on the following.

The tender

The auditor should make sure that any relevant standing orders have been complied with and that the tenders were received and opened in the prescribed manner. Where the lowest price has not been accepted, the auditor will look for careful documentation as to the rationale of contractor selection. The auditor will need to be satisfied that appropriate arithmetical and technical checks were performed on the contractor's price and that errors were corrected within the agreed rules. Good tendering practices should be in evidence.

Interim payments

It is usual for clients to maintain a contracts payment register to record all of the payments made in respect of a project. It will be closed on the agreement and payment of the final account. The auditor will need to scrutinise the payments made to ensure that they are in accordance with the contract and the certificates from the architect/contract administrator.

Variations

The auditor will need to establish that the correct protocols were applied when issuing and authorising variations as described in the contract documents. The various conditions of contract lay down precise rules of instruction and valuation. These should have been followed unless there are good and documented reasons to the contrary.

The final account

The settlement of the final account is a lengthy and complex process and frequently extends beyond the period stated in the contract. The auditor's final task is to ensure that the amount of the final payment added to the sums that have already been paid equals the final account for the project. In order to reduce any possible delays in making the payment to the contractor, the auditor should examine the account as speedily as possible. Contractual claims may remain to be agreed for some time and the auditor may need to include provision for the auditing of these at some later date.

Where the client has been partially responsible for either the purchase of construction materials or the execution of a specialist portion of the works, the auditor will need to be satisfied that these have not been included within the final account.

Liquidated damages

Liquidated damages become due where the contractor fails to complete the works on time, unless there are agreed reasons that are beyond the control of the contractor. It is a rare occasion when an auditor will need to verify such an amount since extensions of time are frequently granted to cover for such a delay. However, the auditor should be satisfied that such a waiver was only granted for adequate and carefully documented reasons.

Fees

An auditor will also be involved in verifying the fees paid to the consultants. Commonly, these will be by agreement between the client and the consultants, frequently following a fee bidding process. They are usually influenced by the tender sum and the additional work involved during the progress of the works.

Timing and resources

It is not intended that final account preparation should commence only upon the completion of the construction of a project. Whilst this may be the case in some situations, it is neither correct nor good practice. If we assume a project duration of 18 months, there will be ample opportunity for final account measurement during the works. For example, at completion of the substructures, it will be in the interest of all parties to remeasure and, if possible, finalise the account for this element of the works. At this time, events are fresh in the mind and uncertainties relating to payment and final project cost can be reduced by prompt agreement. Continuation of such action throughout the course of the works will reduce the time required for completion of the final account at the end of the contract period. Again, this is advantageous to all parties. In the case of the client's internal financial arrangements it may be very important to achieve an early completion. A further factor on lengthy projects that should be borne in mind relates to the turnover of surveying personnel. In the event that final account measurement is left until the end of the project, it is quite possible that either (or both) the contractor or project consultant no longer employs the surveyors responsible for the earlier stages of a project.

In recognition of the factors outlined above, it may be beneficial to maintain a 'running' final account for which measurements and agreements occur throughout the project. Unfortunately, this philosophy may go awry, for example due to lack of available resources or where disputes arise. This is regrettable and will almost certainly lead to delays in the completion of final accounts, some of which are legendary in their duration, despite the obligations stated in the contract.

Despite the view given above that it is expedient to prepare aspects of the final account throughout the progress of the works, it is important to add a note of caution. It is not advisable to start too soon on measurement of variations when future developments, which cannot be foreseen, might affect the surveyor's work. One might, for instance, measure a number of adjustments of foundations or drains, only to find later that the whole of one of these sections must be remeasured complete.

References

BCIS. *BCIS Quarterly Briefing*. BCIS. 2022. Available at https://service.bcis.co.uk/BCISOnline/Briefings/EconomicBackground/3408?returnUrl=%2FBCISOnline%2FBriefing&returnText=Go%20back%20to%20briefing%20summary&sourcePage=Help (accessed 20/03/2022).

Joint Contracts Tribunal (JCT). *Standard Building Contract with Quantities (SBC/Q)*. Sweet and Maxwell Thomson Reuters. 2016.

Joint Contracts Tribufnal. *FR2016 Formula Rules 2016*. Sweet and Maxwell Ltd. 2017.

RICS. *Valuing Change*. RICS. 2010.

RICS. *Fluctuations*. 1st Edition. RICS. 2016.

RICS. *NRM2- Detailed Measurement for Capital Building Works: NRM2*. 2nd Edition. RICS Books. 2021.

14

Insolvency

KEY CONCEPTS
• Insolvency • Role of the quantity surveyor in relation to insolvency • Termination of the contract • Completion of projects impacted by insolvency • Performance bonds

LEARNING OUTCOMES
After reading this chapter, you should be able to: • Appreciate the frequency of insolvency within the construction sector. • Appreciate the role of the quantity surveyor in relation to insolvency. • Understand the provisions within the contract relating to the termination of the contract from both contractors and employers' insolvency. • Identify the actions that are taken following insolvency on a project. • Appreciate the purpose of a performance bond.

COMPETENCIES
Competencies covered in this chapter: • Corporate recovery and insolvency.

Introduction

Insolvency is a generic term that covers both individuals and companies. Bankruptcy applies specifically to individuals and liquidation to companies. The law relating to bankruptcy and that of insolvency is governed by a suite of Acts; the Insolvency Act 1986

Willis's Practice and Procedure for the Quantity Surveyor, Fourteenth Edition.
Allan Ashworth and Catherine Higgs.

(as amended), the Insolvency (England and Wales) Rules 2016, the Corporate Insolvency and Governance Act (CIGA) 2020 and the Insolvency (England and Wales) Amendment Rules 2021, as well as the Companies Act 2006. Unlike death or illness, the effect of these do not by themselves have the effect of terminating a contract.

The construction sector is responsible for more than its fair share of bankruptcies and liquidations. Limited liability companies become insolvent and then go into liquidation. The quarterly statistics published by the Government's Insolvency Service often reports that the sector is having the highest number of insolvencies compared to other industry groups. Whilst the construction industry group did see fewer insolvencies in the four quarters ending Q3 2021 compared to the same period in 2020, the sector represented the highest proportion, 17% of all insolvencies (Gov.uk 2021).

A company becomes insolvent when the value of everything that it owns comes to less than the value of its debts. An example of a statement of a contractor's assets and liabilities, following liquidation, is illustrated in Fig. 14.1. A company may enter voluntary liquidation or, more commonly, have been forced in this direction by a single creditor. The single creditor is usually a financier, such as a bank, who refuses to extend a loan or chooses to call in a debt. Even a profitable company may suffer in this way because of a shortage of cash to pay its bills. Hence, it suffers a cash-flow crisis. For companies to weather the business cycle, it is essential that they are careful with whom they do business, employ sound financial disciplines and verify and monitor companies. In times of recession, new as well as long-established companies cease to trade and more companies begin to pay their bills late, causing other firms to fail. But it is not just in times of recession that insolvency occurs; it is a common feature throughout the business cycle.

The kind of firm that goes into bankruptcy or liquidation ranges from the single trader, subcontractors, suppliers, general contractors through to, on occasion, the national contractor. Employers and consultants are also not immune from these statistics. When this happens, a knock-on effect is created and other firms teetering on the edge often go out of business as a result. The reasons for construction company failures include the following:

- A recognition that the construction industry is a risky business and often the risks involved are not fully evaluated
- Construction projects are used as economic regulators, with price fluctuations depending considerably on the state of the market
- Competitive tendering has been cited to explain the high incidence of contractor and subcontractor failure, although this has not been verified
- Insolvency has a knock-on effect on other, often smaller, firms, who are more vulnerable.

Avoiding the possibility of contractor insolvency in the first place is the preferred route to be followed, although this can never be completely assured. The JCT Practice Notes: Tendering 2017 recommends the questionnaire contains requests for information to assess the financial standing of the company prior to inviting a firm to submit a tender for a construction project:

- Report and audited accounts for last [2/3] financial years
- Statements of turnover for that period in respect of the divisional activities most closely related to the project

V R Mitchell & Sons (Ltd) Statement of Affairs 19-06-23			
	Book value		*Estimated to realise*
Assets			
Land and buildings		330 000	
Less Norwich Union Group	48 000		
Less National Westminster Bank	223 860	–271 860	58 140
Goodwill	5 700		0
Plant and equipment – owned	195 000		45 000
Value subject to HP	40 180		
Less outstanding HP	37 550		2 630
Motor vehicles	36 500		27 250
Stock – Chair frame department	13 250		13 250
– Builder's yard	90 000		35 000
Work in progress	372 000		185 000
Debtors Chair frame dept.	16 500		16 500
Trade contract	274 650		190 000
VAT	3 000		3 000
Cash ABU plc	13 500		
Rogers Plc	4 000		17 500
Available assets			593 270
Liabilities			
Preferential creditors			
PAYE	55 200		
NHI	47 750		
Holiday pay	19 250		122 200
Amount available to unsecured creditors			471 070
Unsecured creditors			
Redundancy pay	23 800		
Trade and expenses	685 550		
Sale contractors	395 500		1 104 850
Deficiency regarding creditors			–633 780
Share capital			
20 000 Ordinary shares of £1.00 each fully paid			20 000
Total deficiency			–653 780

Fig. 14.1 Statement of assets and liabilities.

- Contact details for bank references
- Alternative demonstration of financial status (where necessary for a true and fair view) (Section A3) (Joint Contracts Tribunal 2017)

The general reasons for any company failures are shown in Fig. 14.2, indicating the owner's viewpoint and those of the creditors. The perception of these two different groups regarding the financial failure of a company is markedly different in many respects. This may be a recognition that firms do not have clear perceptions about themselves.

	Owner's opinion %	Creditor's opinion %
Business depression	68	29
Inefficient management	28	59
Insufficient capital	48	33
Domestic or personal factors	35	28
Bad debt losses	30	18
Competition	38	9
Decline in assets values	32	6
Dishonesty and fraud	0	34
Excessive overhead expense	24	9

Fig. 14.2 Reasons for company failures.

The role of the quantity surveyor

It is important for all concerned that quantity surveyors carry out their work to the highest professional standards and integrity. In this context they should only work for bona fide employers. They should ensure that procurement practices are fair and reasonable and that risks are allocated to those parties who are best able to control them. Only financially stable construction firms should be recommended and, where necessary, performance bonds used. Interim payments should always be in accordance with the contract provisions, remembering also that these represent the lifeblood of the industry. Quantity surveyors should, as far as possible, seek to protect employers from undue loss and expense resulting from additional payments and extensions of time. One of the important roles of the quantity surveyor is financial propriety between all parties involved in the construction project. It is therefore important for the quantity surveyor to remain current in financial news within construction journals.

Prior to a contractor going into liquidation, there are often signs that the firm is in some sort of financial difficulties, including:

- Complaints from subcontractors and suppliers about the non-payment of their accounts
- Pressure to maximise interim payments
- Contractors requesting early release of retention
- Work is not making reasonable progress
- Changes in key members of staff
- Less frequent deliveries to site

The role of the liquidator

The commonly held view about receivers and their advisors is that they are like an undertaker employed to perform the last rites for an ailing construction firm. However, the true role adopted by the receiver is to preserve and salvage any or all the parts of the firm.

This is based on the fact that where insolvency occurs all those involved are losers; there are no winners. The types and values of assets which can be recovered vary considerably depending on whether the company is a contractor or developer, manufacturer or even a consultant. Contractors have become wise enough to separate the company's activities into groups such as plant hire, house building, general contracting, etc., so that if hard times occur then the whole business might not be lost. When insolvency becomes a possibility then speed becomes of the essence. The whole assets of the contractor must be maintained and not traded against favours elsewhere.

In the construction industry, when receivers are appointed, a firm of specialist quantity surveyors arrives at the same time. For each contract there are usually three choices:

- *Complete the project.* This is desirable and the more advanced a project is towards completion, the better is the likelihood of this occurring.
- *Abandon it.* This is the least desirable option, because there are usually assets still tied up in it.
- *Sell* the project to a third party.

In each of the above cases, the quantity surveyor, who must be an authorised insolvency practitioner, calculates the amount of work done, the materials on site, what is owed to subcontractors, suppliers and other creditors, amounts outstanding in retention etc.

For example, a contractor could be 60% of the way through a £10 m project, but of the £6 m work completed, the contractor might only have received £5 m owing to work being completed since the last interim valuation and outstanding retention sums. Coupled with this there is a good possibility that future profits may exist in the remaining work that needs to be carried out. However, life is never down to simple arithmetic and other factors will need to be taken into account, such as the amounts owing to the different creditors, the lowering of site morale and motivation, and the fact that the project might have been financially frontend loaded.

The law

The machinery to wind up a company can be set in motion either by the debtor themselves or by the creditors. The High Court or courts with the appropriate jurisdiction make the receiving order and require the debtor to submit to the official receiver or liquidator a statement of assets and liabilities (see Fig. 14.1). A creditor's meeting is usually held to decide whether to accept any suggested scheme or whether to declare the firm insolvent.

On completion of the liquidation, the official receiver will eventually turn all the assets into cash and then divide this amongst the creditors, paying them at a rate of so many pence per pound that is owed. Certain creditors may have priority, e.g. those who hold preference shares and the Inland Revenue, but most creditors rank equally.

Insolvency is broadly concerned with a firm or company's inability to pay its debts. Within the Insolvency Act 1986 (as amended) and Insolvency Rules 1986 (as amended) various situations can arise:

- *Voluntary liquidation.* This can be the result of either the shareholders or creditors making a special resolution to wind up the company, which is then accepted.

- *Compulsory liquidation.* This occurs when a company is regarded as not being able to pay its debts and is the result of a court order for a company to be wound up.

Creditors

Creditors may be secured, i.e. those who hold a mortgage, charge or lien upon the property of the debtor, or they may be unsecured. Unsecured creditors may be ordinary trade creditors, preferential creditors or deferred creditors. Preferential debts include the costs of insolvency proceedings. The proceeds of liquidation are distributed to the creditors and then shareholders.

Termination of contract (contractor insolvency)

All of the standard forms of construction contract incorporate provisions aimed at regulating the events of a contractor's insolvency. It usually results in one of two courses of action being taken. Either the contractor's employment is terminated and a new contract (probably) made with a new contractor, or alternative ways are examined to enable the same contractor to complete the works. On those occasions where the personal skill of the contractor is the essence of the contract, completion may only be acceptable under those terms. This situation is unusual in the construction industry and will usually only be applicable to specialist firms.

It should be recognised that insolvency does not of itself have the effect of terminating the contract, since the liquidator will normally have the power, after obtaining leave, to carry on the business of the debtor as may be necessary for the beneficial winding up of the company.

A liquidator has the power to disclaim unprofitable contracts which were still uncompleted at the commencement of the insolvency. The liquidator may decide to disclaim when most of the more financially fruitful work has been completed, or when some retention, for example sectional completion, has been released.

Where the work is reasonably far advanced, a liquidator is not likely to disclaim it if this can be avoided. Where a relatively small amount of work needs to be completed up to completion, then this will enable outstanding retention monies to be released. When a liquidator does disclaim a contract at this stage, the employer is likely to be in the most favourable position, since the retention sums can be used as a buffer when inviting other firms to tender for the outstanding work that needs to be completed.

Where the construction work is at a relatively early stage on a project when insolvency occurs, it is the employer who will want to ensure that the contractor completes the contract as originally envisaged. Under these circumstances, the employer is likely to have to pay higher prices to have the work completed by another firm, especially where the original contract was awarded through some form of competitive tendering. However, there is much less inducement on the part of the liquidator to want to carry on with such a project, especially if the rates and prices in the contract documents are economically unfavourable.

Where an employer is faced with poor performance from a contractor who is continuing with the project, there is still the right within the contract to determine the contractor's employment; for example, where the progress of the works is not being maintained.

Provision in the forms of contract

Section 8 of JCT 2016 makes provisions for insolvency. Similar clauses are found in the other forms of contract. The contractor is to inform the employer if it is intended to make a composition or arrangement with creditors, or the contractor becomes insolvent. More usually, where the contractor is a company, the employer must be informed if one of the following occurs:

- The contractor makes a proposal for a voluntary arrangement for a composition of debts or scheme of arrangement in accordance with the Companies Act 1989 and the Insolvency Act 1986
- A provisional liquidator is appointed
- A winding-up order is made
- A resolution is passed for voluntary winding-up, other than in the case of amalgamation or reconstruction
- Under the Insolvency Act 1986 an administrator or administrative receiver is appointed.

The forms of contract allow the employer in the above situations to:

- Ensure that the site materials, the site and the works are adequately protected, including offsetting any of the costs involved against monies due to the contractor
- Employ and pay others to carry out and complete the works
- Make good defects
- Use all temporary buildings, plant, tools, equipment and site materials owned by third parties subject to their agreement.
- Purchase materials and goods that are necessary to complete the works
- Have any benefits for the agreement for the supply of materials or goods assigned from the contractor
- Make payments to suppliers or subcontractors for work already executed and deduct these amounts from any monies due to the contractor
- Withhold the release of any retention monies
- Calculate the costs involved of the contractor's insolvency and show these as a debt to the contractor.

Where the temporary buildings, plant, tools, equipment and site materials are owned by a third party, the costs for these and the use of the contractor's facilities, described above, must be paid for by the employer. However, such costs will be offset against any payments that are made to complete the works. The legal title to the goods and materials must be clear (*Dawber Williamson Roofing* v. *Humberside County Council* 1979), so it is important to establish that ownership is transferred once delivered to site.

Factors to consider at insolvency

Until the contractor's legal insolvency becomes a fact, the employer is in a rather frustrating position. The employer must continue to honour the contract, otherwise the contractor may pursue a claim for breach of contract. Any interference on the part of the employer may also exacerbate the unfortunate situation.

The following are some of the principal actions that an employer should take immediately following the insolvency of a contractor.

Secure the site

The construction site should be secured as quickly and as expeditiously as possible. This is to prevent unauthorised entry or vandalism but also to avoid the possibility of materials or equipment being removed by the contractor or subcontractors. Although this would be in breach of the contract, it is clearly more beneficial to the employer to retain these rather than to have to go through the courts for their return when their whereabouts may have become unknown or even unrecoverable.

Materials

Materials on site for which the employer has paid belong to the employer unless their title is defective. Furthermore, they cannot be removed from site without the employer's written permission (JCT 2016 clause 2.24). Similar conditions exist in all the major forms of contract. Materials on site that have not been paid for can be used by the employer to complete the works. No payment is made directly since they will help to reduce the total costs of achieving completion and therefore reduce the financial indebtedness of the contractor. Materials off site for which the employer has paid are also in the ownership of the employer, providing that the provisions of clause 2.25 (JCT 2016) have been complied with. A list of the materials on site should be quickly made with a note of which of these have been paid for by the employer and legal ownership established.

Plant

The contractor's directly owned plant and other temporary structures on site can be used freely to secure completion. Hired plant and that which belongs to subcontractors cannot be used without express permission and payment. However, no plant passes directly to the ownership of the employer. Hired plant that is temporarily supporting the structure will obviously be retained by the employer, but at the usual hire rates.

Retention

Retention by the employer is one of the main buffers the employer has to face any possible loss. Any balance as a result of underpayment to the contractor or for work completed since the last interim payment will be used for similar effect.

Other matters

It will be prudent to ensure that insurances always remain effective. If a bond holder is involved, then these must be kept informed of what is happening and any progress that is being achieved.

Completion of the contract

An early meeting of everyone concerned should be arranged to consider the best ways of completing the works that are outstanding. It is usual for the liquidator to attend this meeting, but where this is not possible then the liquidator should be kept fully informed of all the decisions that are made.

If the liquidator decides to disclaim the contract there are several different ways of appointing a replacement contractor. It will be necessary, at a later date, to show the liquidator and the bond holder that due care was taken to complete the project as economically as was originally intended. The bulk of the work involved in drawing up a new contract will be performed by the quantity surveyor.

Where the amount of work carried out from the original contract is small, then the contract bills may be used to invite new contractors to tender for the work. Adjustments for the work completed will, of course, need to be made. If the contract is in its early days, then it is likely that priced documents will be available from the previously unsuccessful firms. Where the work is significantly advanced, then new documents or addendum bills of quantities will need to be prepared. The time delay involved in this may be unacceptable to the employer and alternative methods may need to be adopted. A new contract may, for example, be negotiated with a single contractor, perhaps on the basis of a single percentage addition to the original contract. Again, adjustments for the work completed will need to be made. The main advantage here is that the new contractor can start work on site immediately, with a minimum of delay, if this is possible. A third alternative is to pay for the remainder of the work on a cost plus basis, which may not be favourable to the employer. It is likely that the original contract was the most competitive and any new arrangement will therefore be more expensive.

With each of these methods, an allowance in the form of a provisional sum needs to be made to cover any remedial works or making good defects that have been left by the insolvent contractor. Whilst some of these defects may be apparent, others may not show for some time. Where the work is piecemeal and almost at practical completion, it may be better to employ a contractor who is suitably able to carry out such work; for example, if the outstanding work includes macadam hard standings, then it may be more appropriate to employ a firm who specialises in this kind of work.

An alternative method of securing completion is to award the new contract to a contractor for a fee. The selected contractor would then complete the works on the same basis as the original contractor, using the contract rates and prices and accepting total responsibility for the project, including its defects. The negotiated fee might be based on a percentage or lump sum to cover the higher costs involved in completing the project and the responsibility, for example, for defects.

A further suggestion is to offer the contract to a new firm on the same basis as that for the original contract, using the same rates and prices. This may appear, at first, to be an unattractive option for three reasons, which no contractor will accept. Firstly, the insolvent contractor was probably selected on a lowest price tender that no other firms could match, hence the prices could be uneconomical for other firms. Secondly, costs will need to be expended on rectifying the partially completed works. Thirdly, the new contractor will be responsible for the first contractor's work, including the making good of any defects.

However, the attractiveness of this proposal lies in the fact that retention sums are outstanding and these may provide more incentive than the three disadvantages combined.

The employer's loss

The main sources of loss to the employer are the probable delays in completion and the potential additional costs that might be incurred. If the original contractor had not become insolvent then the loss due to delays would normally have been recovered by way of liquidated damages. The employer is still entitled to show this as a loss, using the basis of calculation shown in the appendix to the form of contract. However, in practice, because of the contractor's insolvency, there may be insufficient funds to make this payment.

A second source of loss involves the additional payments that the employer may need to make to complete the project. This will include the temporary protection and extra site security, such as erecting additional hoardings, fencing and locking sheds. Additional insurances are likely to be required, especially to cover the period prior to appointing the new contractor. Additional fees will also be necessary for the services involved in placing and managing the new contract.

Expenditure involved

It is necessary to calculate a hypothetical final account, which would have been the amount payable to the original contractor had the insolvency not occurred. For most of the items involved this entails a translation of the actual final account by using the rates and prices from the original contract. A careful note must be made of dayworks which, in the original contract, may have been described as measured works. Claims for loss and expense that might have arisen may be a matter for professional judgement. In the case of fluctuations, had the original contract been on a fixed price basis then no account of these need be considered for the original contractor. The additional fluctuations due to the later completion can be genuinely deducted as some compensation against liquidated damages. Remedial work charged by the new contractor for the unsatisfactory work of the original contractor will be excluded from the hypothetical final account. Examples of typical calculations are shown in Figs 14.3 and 14.4.

Termination of contract (employer insolvency)

Whilst a greater number of contractors become insolvent than employers, the insolvency of employers is something that does occur from time to time during the execution of a contract. As the contractor is a supplier of goods and services, the Insolvency Act as amended by the CIGA prevents the operation of termination clauses as long as payment is received during the insolvency period when they can use the statutory right to suspend works under Section 112 of the Construction Act. Subject to CIGA, when insolvency of the employer arises, the employer must inform the contractor in writing of its occurrence

Contract 1		
Amount of original contract		1 356 000
Agreed additions for variations, prime cost sums, dayworks, fluctuations, claims[1]		82 800
Amount of the final account had the original contractor completed the works		1 438 800
Amount of completion contract	838 200	
Agreed addition for second contractor only	63 050	
Amount paid to the original contractor[2]	495 000	
Additional professional fees	12 500	1 408 750
Debt due to the original contractor		£30 050

[1]This includes all variations priced at the original contractor's rates and prices.
[2]The total amount of the first contractor's certificate, plus work done since then, was estimated to be £570 000. The original contractor receives in total £525 050.

Fig. 14.3 Examples of financial summary.

Contract 2		
Amount of original contract		2 292 000
Agreed additions (as above)		111 150
Debt due to the original contractor		2 403 150
Amount certified to original contractor	240 100	
Amount of completion contract	2 163 050	
Agreed lump sum fee	225 000	
Additional professional fees	36 000	2 664 150
Debt payable by the trustee in bankruptcy[1]		£261 000

[1]If there was a bond for 10% of the contract sum, then the employer should receive £229 200 towards the above debt. If the final dividend of five pence in the pound was paid to creditors at some later date, the employer would receive a further £1590 (£31 800@5p). The ultimate employer's loss is therefore £30 210.

Fig. 14.4 Examples of financial summary.

(Clause 8.10, JCT 2016). The following are the contractual consequences of insolvency of the employer:

- The contractor can remove from the site all temporary buildings, plant, tools, equipment, goods and materials. This must be done reasonably and safely to prevent injury, death or damage occurring.
- Subcontractors should do the same.
- The contractor prepares an account setting out the following:
  - The total value of work properly executed up to the date of determination, calculated in accordance with the contract provisions
  - Any sum in respect of direct loss and expense incurred either before or after determination
  - The reasonable costs associated with the removal of temporary works, plant and materials

  - The cost of materials or goods properly ordered for the works which the contractor has already paid or unable to cancel. These items become the property of the employer
- Within 28 days of submission to the account, the employer must pay the contractor without deduction of any retention (clause 8.12.5).

Construction contracts involve giving credit to the employer by the contractor in respect of work carried out and goods and materials supplied to the site prior to the issue of an interim certificate. The payment of the certificate is not made in full but is subject to retention. Also, a contractor has no lien on the finished work and the employer's insolvency may place the contractor in some real difficulty until matters are finally resolved. This difficulty can result in the contractor's own insolvency. A prudent contractor will, prior to signing a contract, wish to establish the financial standing of an employer.

Under JCT 2016, the employer's interest in retention monies is fiduciary as trustee on behalf of the contractor. This is to protect the contractor in the event of an employer's insolvency. It is now strongly recommended that the full provisions of the contract are invoked. A separate bank account should have been used for the depositing of retention monies.

The employment of the quantity surveyor and architect may not necessarily end at the employer's insolvency as will also be subject to CIGA. However, if they are to continue with their work they will require some assurances from the liquidator that they will be paid for their services in full.

Insolvency of the quantity surveyor or architect

Insolvency of architects or quantity surveyors is, fortunately, an unusual occurrence, but in common with other businesses does occasionally occur. Where an architect's or surveyor's practice becomes insolvent then the employer must within 21 days nominate a successor (article 3 and 4, and clause 3.5, JCT 2016). The contractor can raise reasonable objections and if these are upheld then the employer must nominate other firms. Under renomination, the firms appointed cannot overrule anything that has previously been agreed and accepted. These include matters relating to certificates, opinions, decisions, approvals or instructions. The new firms can make changes, as necessary, but these will always constitute a variation under the terms of the contract.

Performance bonds

A performance bond is a written undertaking by a third party, given on behalf of the contractor to an employer, wherein a surety accepts responsibility to ensure the due completion of the contractual works. Local and central government have frequently requested a contractor to take out a bond for the due completion of the work. The form of the performance bond, its value and period of validity will be stated in the contract particulars. The value often is in the order of 10% of the contract sum. Main contractors may also require performance bonds to be provided by their own subcontractors as a form of guarantee or insurance.

Bond holders are frequently banks or insurance companies. They are responsible for paying the amounts involved, up to the bond limit, should the contractor default, as in the case of insolvency. However, they have no control over the way in which the work is carried out or the project completed.

When an employer declares that a contractor has failed to perform the works adequately, the surety has normally three courses of action to follow:

- Pay damages up to the full value of the bond
- Engage another contractor to complete the work
- Make such arrangements that the contractor is able to finish the works.

Other bonds may be used which guarantee performance of the contractor. In these instances, the surety will satisfy themselves that the bid tendered is responsible, practical and complete. The charge for the bond is included within the contractor's tender and is thus ultimately borne by the employer as a form of insurance.

Wherever possible, standard forms of bond should be used where each party is then clearly aware of its implications.

References

Gov.uk. *Commentary – Company Insolvency Statistics*. Crown. 2021. Available at https://www.gov.uk/government/statistics/company-insolvency-statistics-july-to-september-2021/commentary-company-insolvency-statistics-july-to-september-2021 (accessed 23/03/2022).

Joint Contracts Tribunal (JCT). *Standard Building Contract with Quantities (SBC/Q)*. Sweet and Maxwell Thomson Reuters. 2016.

Joint Contracts Tribunal (JCT). *JCT Practice Notes: Tendering 2017*. Sweet and Maxwell. 2017.

15

Contractual Disputes

KEY CONCEPTS

- Reasons for disputes
- Litigation
- Alternative Dispute Resolution (ADR)
- Expert witness
- Claims

LEARNING OUTCOMES

After reading this chapter, you should be able to:

- Identify the area of possible disputes.
- Understand dispute procedures, including litigation and alternative dispute resolution procedures.
- Identify the main features of both adversarial and non-adversarial ADR procedures.
- Appreciate the role of the quantity surveyor in providing expert evidence.
- Understand the reason for claims and how they are addressed in the contract.

COMPETENCIES

Competencies covered in this chapter:

- Conflict avoidance, management, and dispute resolution procedures.
- Contract practice.

Introduction

From time to time, quantity surveyors find themselves involved in contractual disputes either in litigation in the courts, in arbitration or other forms of alternative dispute resolution cases (ADR). Their involvement is sometimes as witnesses of fact: that is someone who

Willis's Practice and Procedure for the Quantity Surveyor, Fourteenth Edition.
Allan Ashworth and Catherine Higgs.

was there at the time as project surveyor or manager. However, more often they are involved if appropriately qualified as expert witnesses, adjudicators, arbitrators themselves or as neutrals or mediators in ADR cases. Each of these situations is considered in this chapter, as alternatives to litigation in the courts.

Conflicts are a common feature of the construction industry. They occur daily and fortunately many are solved amicably between the parties involved using negotiation, without the need to resort to one of the above scenarios. However, the number of reported legal cases suggests that a certain litigious nature drives the construction industry. The costs associated with resolving these disputes, i.e. differences of opinion or interpretation are often high and damage the image of the industry. In the light of this, those responsible for drafting forms of contract for use by the industry have increasingly included provision for the parties to undertake some process or processes of ADR before proceeding further. Thus, the Standard Building Contract (Clause 9.1, JCT 2016) includes specific provision for mediation. This has been further reinforced in the courts in *Ohpen Operations UK Ltd v Invesco Fund Managers Ltd (2019) EWHC 2246 (TCC)*, the court stated that there is a clear, strong policy in favour of enforcing ADR provisions and encouraging parties to attempt to resolve disputes before litigation (Jackson 2020).

Why disputes arise

The construction industry is a risky business. It does not build many prototypes, with each different project being individual in so many respects. Even apparently identical building and civil engineering projects that have been constructed on different sites create their own special circumstances, are subject to the vagaries of different site and weather conditions, use labour that may have different trade practices and result in costs that are different. Even the identical project constructed on an adjacent site by a different contractor will have different problems and their associated different costs of construction. Disputes can therefore arise, even on projects that have the best intentions. Even when every possibility of disagreement has been potentially eliminated, conflicts can still occur; human nature will come to the fore leading to frustration, entrenchment of positions and a loss of objectivity impacting on the ability to agree a resolution. Some of the main areas for possible disputes occurring are:

General

- Adversarial nature of construction contracts
- Poor communication between the parties concerned
- Inappropriate use and understanding of forms of contract and warranties
- Fragmentation in the industry
- Poor quality project information

Employers

- Poor briefing
- Changes and variation requirements
- Changes to standard conditions of contract

- Interference in the contractual duties of the contract administrator
- Late payments.

Consultants
- Design inadequacies
- Lack of appropriate competence and experience, especially in relation to change management
- Late and incomplete information
- Lack of coordination and unclear delegation of responsibilities.
- Cost decisions
- Lack of appropriate risk management plan

Contractors
- Inadequate site management
- Poor planning and programming
- Poor standards of workmanship
- Disputes with subcontractors
- Delayed payments to subcontractors
- Coordination of subcontractors.

Subcontractors
- Mismatch of subcontract conditions with main contract
- Failure to follow and adopt agreed procedures
- Poor standards of workmanship.

Manufacturers and suppliers
- Failure to define performance or purpose
- Failure of performance.

The annual Arcadis *Global Construction Disputes Report* evidence-based survey identified the most common dispute causes in the UK for 2020 (see Table 15.1). Disputes were valued at approximately £27.7 million: the highest since 2013, and an average length of time to resolve disputes at just less than 10 months (Arcadis 2021). The report attributes the change in rankings to the imposed UK COVID-19 rules and the uncertainty around which party was liable for the time and cost risk. The *National Construction Contracts and*

Table 15.1 Dispute causes (*Source:* Arcadis 2021/Chamber of Commerce Amsterdam).

2020 Rank	2019 Rank	Most common dispute causes
1	2	Owner/contractor/subcontractor failing to understand and/or comply with its contractual obligations
2	*	Errors and/or omissions in the contract document
3	1	Failure to make interim awards on extensions of time and compensation

*New ranking in 2020

Law Report 2018 using a different set of criteria identified the most common issue in disputes as extensions of time (52% of respondents). Of note to quantity surveyors were the following cited causes; valuation of final accounts (45%), valuation of variations (42%) and valuations of interim valuations (14%) (NBS 2018). As the Arcadis study, the value of disputes were significant, with 12% of disputes being valued more than £5 million and the greatest frequency (74%) being between the client and the contractor.

Litigation

Litigation is a dispute procedure that takes place in the courts and is conducted in accordance with the Civil Procedure Rules. It involves third parties who are trained in the law, usually solicitors and barristers, and a judge who is appointed by the courts. This method of solving disputes is often expensive and can be a very lengthy process before the matter is finally resolved, sometimes taking years to arrive at a decision. The process is often extended, upon appeal, to higher courts involving additional expense and time. Also, since a case needs to be properly prepared prior to the trial, a considerable amount of time can elapse between the commencement of the proceedings and the trial.

The Pre-Action Protocol for Construction and Engineering Disputes sets out the process to ensure that proceedings, if they result in litigation are conducted efficiently. It ensures that both parties are informed of the nature of each other's cases, exchanged supporting information, have formally met, narrowed the area of dispute and if able to, settle cases early, fairly, and inexpensively without recourse of litigation (Ministry of Justice 2017). An objective of the protocol is for parties to attempt to resolve the dispute and engage constructively with ADR to do so. Where this has not occurred, the courts have been seen to penalise the parties in their judgements (RICS 2022).

Where the matter is largely of a technical nature, such as a building case, procurement claim, fire claim or professional negligence, the case may be referred in the first instance to the Technology and Construction Court (TCC). The amount of money involved in the case will determine whether it is heard in the County Court or High Court. TCC cases at the High Court are dealt with by High Court Judges, whereas circuit judges preside over the county court claims. The Technology and Construction Court Guide is intended to provide straightforward, practical guidance on the conduct of litigation in the TCC (HM Courts & Tribunal Services 2015). Construction cases accounted for 214 out of 472 (45%) TCC cases in England in Wales from October 2019 to September 2020 (Crown 2021). A database of the cases and the decisions of the High Court are made available to the public (Bailii 2022).

Alternative dispute resolution

Alternative Dispute Resolution (ADR) refers to dispute resolution processes that are alternatives to litigation. These processes as classified by Hughes et al. (2015) can be described as either adversarial or non-adversarial. The former referring to adjudication and arbitration and the latter referring to more conciliatory processes such as mediation, described later in the chapter. These non-adversarial ADR processes provide a dispute resolution mechanism that concentrates on resolving disputes by consensual, rather than adjudicative, methods.

To quote from the Centre for Dispute Resolution's introduction to ADR, 'These techniques are not soft options, but rather involve a change of emphasis and a different challenge. The parties cooperate in the formulation of a procedure and result over which they have control' (Gould et al. 2010). The main features are easily captioned under:

- *Consensus.* A joint objective to find the business solution
- *Continuity.* A desire to find a solution in the context of an ongoing business relationship
- *Control.* The ability to tailor a solution that is geared towards a business result rather than a result governed by the rule of law, which may be too restrictive or largely inappropriate
- *Confidentiality.* Avoiding harmful washing of dirty linen in public.

The advantages of ADR can be summarised as follows:

- *Private.* Confidentiality is retained
- *Speed.* A matter of days rather than weeks, months or even years
- *Economic.* Legal and other costs resulting from lengthy litigation are avoided.

However, none of these advantages will be achieved unless one vital ingredient is present. There must be goodwill on both sides to settle the matter on a commercial rather than a litigious basis. If this goodwill does not exist, then the parties have no option but to resort to arbitration or the courts, without wasting further time and resources.

Suitably qualified quantity surveyors have a role to play in such ADR processes. They are obvious candidates for the post of neutral in construction disputes, provided they have good negotiating skills and have undergone the necessary qualifications.

Adjudication

The Housing Grants, Construction and Regeneration Act 1996, Section 108, as amended by The Local Democracy, Economic Development and Construction Act 2009 impose statutory adjudication in most construction contracts (there are some exceptions identified in Sections 104 and 105, e.g. oil and gas contracts and private homes). This scheme has revolutionised dispute resolution for construction contracts (which include contracts between employers and construction professionals) because if one party to the contract wishes a dispute to be heard they can serve a Notice of Adjudication, detailing the dispute, on the other party. Within seven days of this Notice, an adjudicator, agreeable to both parties should be appointed. The adjudicator has 28 days from the date of the referral notice to make the award. The responding party's response to the dispute must be made within this 28-day timeframe. However, this timescale can extend by consent of the adjudicator. If required and in agreement of both parties, the deadline of the adjudicator's decision can also be extended.

Whilst there is no specific legal requirement for the content of the adjudicator's decision, it is recommended that it includes against each issue the findings and determination (RICS 2017a). The award will be binding and must be paid forthwith and will stand unless overturned in a later arbitration or court case. One of the benefits of adjudication is that it can often lead to a settlement without the matter going any further. This is because a party that has lost in an adjudication will think very carefully before proceeding with very expensive litigation or arbitration. They might well lose again, with the additional penalty of paying the other side's costs. Whilst few adjudications lead on to full arbitration or litigation, many cases in the Technology and Construction Court now deal with the enforcement of adjudication awards;133 out of 605 cases in 2019/2020 (Crown 2021). The judges are determined to uphold decisions unless it can be shown that there is no contract in fact or a major breach of natural justice has occurred.

Of note for the quantity surveyor is accounting issues constitute many of the disputes heard at adjudication. The Adjudication Society survey found 75% of cases related to payment; 39% payment/payless notices, 18% interim value of work, 16% relating to the final account and 2% relating to liquidated damages (Milligan and Jackson 2019). Due to the nature of these adjudication disputes and the need for the adjudicator to be a subject matter expert, suitably qualified quantity surveyors will often act in the role of adjudicator. The RICS offers a construction adjudication service and the RICS can be stated as the nominating body within the contract particulars.

For contracts that do not comply with The Housing Grants, Construction and Regeneration Act, Part 1 of the Scheme for Construction Contracts (England and Wales) Regulations includes provision for adjudication. *The Guidance Note: The Scheme for Construction Contracts* provides practical guidance on adjudication (CIArb 2020).

Arbitration

The use of arbitration by the UK construction sector has declined in recent years due to the introduction of construction adjudication. Arbitration is a judicial process regulated by statute by the provisions within the Arbitration Act 1996 (currently under review), used to resolve disputes in building contracts as a quicker and more effective alternative to litigation. Unlike litigation, the use of the arbitration process to resolve disputes comes about following the agreement of the parties, either when the dispute arises, or more often as a term of the original contract. For instance, the JCT forms of contract all provide that if a dispute arises between the parties to the contract, then either party can call for arbitration. When such clauses exist in contracts the courts, if asked, will generally rule that arbitration, having been the chosen path of the parties, is the proper forum for the dispute to be heard and will stay any legal action taken in breach of the arbitration agreement under section 9 of the Arbitration Act 1996.

The JCT publish arbitration rules for use with arbitration agreements referred to in the JCT contracts. They set out rules concerning the Notice of Arbitration, arbitrator's powers, conduct of arbitrations and various types of procedures: short procedure with hearing, without hearing (documents only) and full procedure with hearing. Each procedure contains strict timetables (JCT/CIMAR 2016). Alternative rules used within the sector include the RICS *Fast Track Arbitration Rules* for claims less than £100,000 and the international ICC *2021 Arbitration Rules*.

Arbitration and the law relating to it are subjects on their own. The traditional advantages of arbitration over the courts are fourfold:

- Arbitration proceedings are quicker than the courts
- Arbitration is cheaper than litigating in the courts

- The parties get a 'judge' of their choosing, a person knowledgeable about the subject matter in dispute, but with no knowledge of the actual case, rather than a judge imposed on them. Even if parties cannot agree on the name of the arbitrator, they may be able to agree on the expertise and qualification
- Unlike proceedings in the courts, arbitration proceedings are confidential, unless an appeal in court is made and the case is reported.

Regarding the first two of these traditional advantages, provided the proceedings are kept simple then arbitration can still prove quicker and cheaper than proceeding through the courts. However, whilst arbitration was developed as a speedier, cheaper and less formal alternative, this is not always the case (Hutt and May 2022).

However, the third and fourth advantages, choice of judge and confidentiality, still exist and are considered by many to be of overriding importance. Hence the continued popularity of arbitration references, although statutory adjudications under the Housing Grants, Construction and Regeneration Act 1996 as previously discussed appear to have resulted in fewer references.

The duty of arbitrators is to ascertain the substance of the dispute, to give directions as to proceeding, and to hear the parties as quickly as possible and make their decision known by way of an award. Arbitrators hear both sides of an argument, decide which they prefer and award accordingly. They cannot decide that they do not like either argument and substitute their own solution.

An arbitrator is not named within a contract, as it is usually thought better to wait until the dispute arises and then choose an appropriate person. The choice is made by the parties: each side exchanges suitable names and usually an acceptable choice emerges. For instances when parties are unable to agree, the form of contract will identify the appointer of the Arbitrator. This will usually be a president or vice-president from a body such as the Chartered Institute of Arbitrators, the Royal Institute of British Architects, or the Royal Institution of Chartered Surveyors. If a quantity surveyor is thought the most appropriate choice, then the President of the RICS would be the most likely to be approached.

When appointed, an arbitrator calls the parties together in a preliminary meeting. At this meeting the nature and extent of the dispute are made known to the arbitrator, the most suitable procedure will be agreed, and an order is sought for directions, fixing a timetable for submission of the pleadings (i.e. points of claim, points of defence etc.). The directions normally end by fixing a date and venue for a hearing if applicable to the procedure.

As the date for the hearing approaches, the parties will keep the arbitrator informed of the progress they are making in working their way through the timetable. If a compromise is achieved and the matter is settled, they will inform the arbitrator immediately. At the hearing, which maybe virtual, each party puts their case and then calls their witnesses, first the witnesses of the facts and then the expert witnesses. All witnesses normally give evidence under oath and are examined and cross-examined by the parties' advocates.

All arbitrators are bound by the terms of the Arbitration Act 1996. Their awards, once made, are final and binding. If an award is not honoured, then the aggrieved party can call on the courts to implement the award. The only exception to this is if the decision is found to have a serious irregularity as set out in section 68 of the Act.

Quantity surveyors are well suited to act as arbitrators in construction disputes, as the dispute frequently involves measurement, costs, and loss and/or expense and interpretation of documents. These are all matters falling within the expertise of quantity surveyors; guidance for this role is provided in the *RICS Surveyors Acting as Arbitrators in Construction Disputes*. When the matters in dispute concern quality of workmanship or design faults then arbitration is best left to architects or engineers. Equally, when matters of law are the prime consideration then the arbitrator should be a lawyer.

Non-adversarial alternative dispute resolution

Conflicts occur on a regular basis during the construction process, most are resolved between parties by negotiation without the need for more formal mechanisms. However, in the first instance when unresolved disputes arise a neutral third party can facilitate objective discussion, the narrowing of the area of dispute and hopefully a settlement. The facilitation by a third party or parties is a characteristic of non-adversarial ADR processes. Such processes preserve relationships and enhance communication between parties and therefore develop a working relationship that may mitigate further differences. ADR can take a variety of forms, and the most commonly used are:

- Conciliation
- Mediation
- Project mediation
- Early neutral evaluation
- Mini trial
- Dispute boards.
- Independent expert determination

Conciliation

Conciliation is a voluntary, consensual process where a neutral third party assists both parties in settling differences through structured negotiation. It is an 'evaluative' process; the conciliator's recommendations are not binding and their sole purpose is to assist the parties in achieving a settlement. The outcome, as for mediation, may not be one of reward or penalty; it might be a commitment to change practice, procedures or alter behaviour. An example of this approach is the RICS's Conflict Avoidance Process (CAP).

Mediation

Mediation is described as a facilitative process (RICS 2014) Increasingly there are clauses within building contracts that endorse 'where difference cannot be resolved by direct negotiation that each party give serious consideration to any request by the other to refer the matter to mediation' (Clause 9.1, JCT). The parties, select a neutral whose background reflects the matters in dispute, this maybe an RICS accredited mediator or a professional selected through the CIArb referral service.

The role of the mediator is '*to be the guardian and guide of the mediation process, to facilitate the exchange of information, to develop creative solutions, to help parties reality check their position and to reach an agreement.*' (CIC 2019)

A planning meeting is arranged by the mediator to discover the substance of the dispute, each party's position, areas of agreement and to decide how best to proceed.

The mediation will normally take place over a day and will commence with each parties make formal presentations in a joint session. This is then followed by a series of private meetings, or 'caucuses' as they are termed, between the mediator and each of the parties on their own. The mediator moves from one caucus to the next, reporting, with agreement, the views of each party in turn. This should lead to the mediator's being able to suggest a formula for agreement between the parties at a concluding joint meeting, which in turn may lead to a settlement. The Settlement Agreement will be signed by both parties.

Project mediation

Project mediation can be used on long complex projects where there is an express term in the contract requiring the parties to engage in the process. The objective of project mediation is to embed the mediation process into the project. Whilst this requires employing a project mediator, this cost could be offset by a reduction in disputes due to the enhanced non-adversarial nature of the relationships between the parties and the benefit that the mediator has a deeper understanding of the project and the parties involved.

Early neutral evaluation

Early neutral evaluation (ENE) can be used when disputes are not complex. It is an effective process when both parties have differing interpretations of how the law may affect the outcome of the dispute or the influence of the evidence which they hold. ENE involves a neutral party with appropriate specialist knowledge assessing the merits of the case. The evaluation gives parties and understanding of the strengths and weaknesses to inform and enable more constructive negotiation. ENE is also supported by the courts, as in the case of *Telecom Centre (UK) Ltd v Thomas Sanderson Ltd (Early Neutral Evaluation) (2020) EWHC 368 (QB)* (Jackson 2020).

Mini trial

Mini trials are most likely to be used for major disputes involving complex legal questions. Unlike what is suggested by the name, this process is not a 'formal trial' but a settlement procedure. Generally, but not necessarily, each of the parties is represented by a lawyer. First, there is a limited disclosure of documentation, or discovery, as it is known. The purpose of this limited discovery is to ensure that each party is aware of the opposite side's case and is not taken by surprise by the lawyer's presentation. Next, the lawyer for each side makes a short presentation of their client's case to the tribunal.

The tribunal usually is made up of a senior managerial representative from each party and an independent advisor referred to as a neutral. It is important to the success of the mini trial that each of the parties' representatives should not have been directly involved in the project. There is then less emotional involvement than with someone who has lived with the dispute who might have difficulty in taking a detached view. It is also essential that both parties' representatives have full authority to settle the matter. There is nothing worse than arriving at what appears to be consensus for one party then to disclose that they can only agree it subject to authorisation from a chairman, board, council, chief officer or some other third party.

The neutral must be someone with a knowledge of the industry but no knowledge or interest in the dispute. In this respect, they need to have the same requirements as those of a good arbitrator. However, unlike an arbitrator, whose task it is to hear the arguments and decide which is preferable, a neutral becomes much more involved, listening, suggesting and giving advice on matters of fact and sometimes on law as well.

After the initial presentation, experts and witnesses of fact may be called, following which the managers enter into negotiation with a view to coming to a consensus. The length of these negotiations will depend very much on the complexity of the matters in dispute. They will be assisted by the neutral who, if the parties remain deadlocked, may give a non-binding opinion. This may lead to a settlement after further negotiation.

Once a negotiated settlement is reached, the neutral will there and then draft the heads of a statement of agreement, which the parties will each initial. This will then be followed by a formal agreement ending the matter.

Dispute boards

Specific forms of contract enable the appointment and operation of a dispute resolution board; such as the FIDIC Dispute Avoidance/Adjudication Board and the JCT Dispute Adjudication Board Documentation 2021 that can be used with JCT Major Project Construction Contracts. These boards comprise three individuals appointed by the parties to the contract. The profile of the board will reflect the nature of the project and are likely to have a range of professional expertise. The boards' role being to identify disagreements as the project progresses and offer informal assistance in the first instance to aid resolution. If unresolved, in the case of a Dispute Review Board recommendations will be made that become legally binding after a stipulated time period. A Dispute Adjudication Board provides binding decisions. Whilst dispute boards have been used on international projects, the provision of statuary adjudication within the Construction Act means that there has been limited use in the UK.

Independent expert determination

Independent expert determination is suitable when the dispute is of a technical nature. Prior to the start of the project, the expert or mechanisms to select an expert should be agreed by the parties within the contract. If a dispute arises the expert, assisted by the parties, will carry out an investigation to inform the decision, which is final and binding. However, it is important to appreciate that the expert is not acting in a judicial capacity and therefore is not bound to apply legal principles or able to award costs.

Choice of dispute resolution

Ideally, contentious issues will be avoided as much as possible or can be resolved through timely discussions between the parties, but there will be those which are not settled and for which a more formal and systematic approach must be adopted.

Where matters do call for a formal process of dispute resolution, the recommended course of action will depend upon the procurement route involved, as this will usually have predetermined the form of contract in operation. Notwithstanding the common law right in all situations to take one's opposite number to litigation, the form of contract, for its part, will clearly state the options available and the route to be taken in seeking to resolve any dispute

In advising your client, you may find it helpful to summarise the key features of the differing dispute resolution techniques in a similar manner to that shown in Table 15.2, highlighting those which are in fact available to them, depending upon their chosen procurement route/form of contract. To some extent, one trades cost and speed against the quality or certainty of any decision. You must point out that, despite the increasing support given to adjudicators' decisions upon appeal, they are not as yet automatically binding on

Table 15.2 Principal features of techniques for dispute resolution.

Technique	Cost	Time	Quality
ADR – such as conciliation, mediation, ENE etc.	Economic – cost of preparation, nomination fees and neutral expert fees	Quickest process –dependent only on the parties finding earliest opportunity to meet and discuss the issues	A private decision based on consent, not binding on either party. Only of lasting value if the problem is fully resolved and not returned to by either party
Adjudication	Cost of preparation. The adjudicator's fee is shared between the parties	An on-site process which should be commenced as soon as possible upon notification of any dispute. The adjudicator has a limited time to decide, causing minimum disruption to the progress of the project	Effective process for disputes requiring quick determination by third party A private decision unless challenged and referred to the TCC
Arbitration	Costs to be paid for arbitrator's time and premises for hearing. Also, the parties may employ legal counsel etc. at additional cost	May be protracted over many months, as the arbitrator may give the parties several months to produce their respective evidence. The arbitrator may then take some time over his/her decision	The arbitrator's decision will be private, as above, but expert and binding unless challenged in the courts on a matter of legal process, not fact
Litigation	Court and judge provided free, but costs of specialist legal and technical advice, if used, and of barristers in court may make costs very high. However, legal costs recoverable from losing party.	Proceedings can be complex and time consuming.	The court's decision will be reported publicly, expert and binding and can only be challenged in a higher court (usually at great expense to the appellant). Only challenged on matters of legal process, not fact.

the parties as are those of the arbitrator or judge. You should inform your client of the procedural requirements of each of the three main techniques: adjudication, arbitration and litigation. Although notices may not need to be given by the client in person, there are certain timescales etc. involved of which they will wish to be advised, as these may affect their own cashflows and other arrangements.

The client should also be advised that the differing dispute resolution techniques carry with them varying degrees of privacy, ranging from total confidentiality in the case of on-site mediation, through to total disclosure in the case of litigation, where results are in the public domain (via Law Reports and the like). Both parties may seek to avoid publicity for the sake of their reputation, and for this reason alone may attempt resolution of their dispute through ADR.

Expert witness

Quantity surveyors, particularly those with more experience, may be briefed to give expert witness in a variety of circumstances:

- Litigation in the High Court or in the County Court
- In arbitration
- Before a tribunal.

While proceedings in court tend to be more formal than in arbitration, the same rules apply.

Expert evidence is evidence of opinion to assist with the technical assertions of each party. It is therefore essential that the expert is working within their subject expertise and experience. The expert must be seen at all times to be independent and to have no financial interest in the outcome. Expert witnesses are there to assist the court, the arbitrator or the tribunal and must not be seen to be solely advocating their client's case. Under the Civil Procedure Rules, an expert witness's duty is to the court, overriding any duty they owe to the party that has appointed them (Kavanagh 2017).

An expert opinion must be based on facts, and these facts have to be proved unless they can be agreed.

If instructed by a judge or arbitrator to agree facts and figures and to narrow the issues in dispute, the process of agreement is achieved by meetings of the experts. These meetings are usually described as being 'without prejudice', that is, nothing discussed can be used in evidence, enabling experts to discuss matters openly. Such meetings are followed by a joint statement of what has been agreed, and of those matters that are still outstanding. In fact, such meetings can often lead to a settlement of the dispute without the tribunal being troubled further.

The Civil Procedure Rules, which now govern procedure in the courts following the far-reaching reforms of Lord Woolf, strictly regulate the procedure relating to expert evidence and should be studied by all involved in the process of adducing such evidence. The Civil Procedures encourage the use of a single expert agreeable to both parties.

Acting as an expert witness is time-consuming, hence expert witnesses must ensure they have the resources to meet the required timescales. There needs to be careful reading of all relevant matters, some research where appropriate and a studiously prepared proof of evidence. As has been stated above, such proofs of evidence must be seen as being in the sole authorship of the expert. While they may listen to advice, the final document, on which they stand to be cross-examined, is theirs and theirs alone.

Following the UK Supreme Court ruling on *Jones* v. *Kaney (2011)*, expert witnesses have now lost the right to immunity from civil action for breach of duty for evidence given in court and any supporting reports. It is therefore important that surveyors follow the Practice Statement of *Surveyors Acting as Expert Witnesses* (RICS 2020).

Claims

Contractual claims arise where contractors assess that they are entitled to additional payments over and above that paid within the general terms and conditions of the contract for payment of work done. For example, the contractors may seek reimbursement for some alleged loss that has been suffered for reasons beyond their control. Claims may arise for several different reasons, such as:

- Extensions of time
- Changes to the nature of the project
- Disruption to the regular progress of the works by the client or designer
- Variations to the contract.

Claims arise in respect of additional payments that cannot be recouped in the normal way, through measurement and valuation. They are based on the assumption that the works, or part of the works, are considerably different from or executed under different conditions than those envisaged at the time of tender. The differences may have revised the contractor's intended and preferred method of working, and this in turn may have altered or influenced the costs involved. The rates inserted by the contractor in the contract bills do not now represent fair recompense for the work that has been executed.

Where a standard form of contract is used, attempts may be made by contractors to invoke some of the compensatory provisions of the contract in order to secure further payment to cover the losses involved. As with many issues in life, contractual claims are rarely the fault of one side only. A claim only becomes a dispute when it cannot be resolved, then one of the methods of dispute resolution referred to earlier needs to be invoked. Special care, therefore, needs to be properly exercised in the conduct of the negotiations since they may influence the outcome of any subsequent legal proceedings. Claims usually reflect an actual loss and expense to a contractor. Once the contract is made, the contractor is responsible for carrying out and completing the works in the prescribed time and for the agreed sum. Entitlement to additional payments occurs if a breach of contract is committed by the employer or a party for whom the employer is responsible, or if circumstances arise that are dealt with in express terms of the contract providing for additional payment, such as Clause 4.20 (JCT 2016). It should be noted that the costs incurred by contractors because of their own mistakes in pricing or whilst executing the works cannot form the basis of a successful claim.

Contractual claims – extensions of time

The two factors, time, and money, are often inextricably linked. Hence a claim for loss and expense may include some element relating to the cost to the contractor of an extension to preliminaries costs and the like (Table 15.1). However, a claim for an extension of time does not automatically give a right for a claim for loss and expense. Therefore, most forms of contract also allow for certain situations where the contractor has been delayed and may claim an extension of time but for which they will not receive any additional payment. Proving an entitlement to an extension of time in such cases is significance to the contractor as it removes the liability to pay the client liquidated and ascertained damages.

The quantity surveyor

The details of claims submitted by the contractor are usually investigated by quantity surveyors, since claims invariably have a financial consequence. A report will subsequently be made to the architect, engineer, or other lead consultant. The report should summarise the arguments involved and set out the possible financial effect of each claim. Quantity surveyors frequently end up negotiating with contractors over such issues, in an attempt to solve the financial problems and agree an amicable solution wherever possible. This is preferable to lengthy legal disputes described above.

The quantity surveyor may well be requested to negotiate with the contractor on specific issues relating to a claim before there is resort to adjudication, arbitration, or litigation. To guard against commitment, correspondence in the period before action is often marked 'without prejudice'. This will usually preclude reference to such correspondence in subsequent litigation but it is not, as is sometimes mistakenly believed, a safeguard against any form of binding obligation. In fact, should an offer made in such correspondence be accepted, a binding agreement will usually result. This may be so even if the matter agreed relates only to a detail and the negotiations as a whole eventually collapse.

The decision on the outcome of a claim usually rests with the architect (or engineer on civil engineering works). This is except in such matters of valuation as the parties to the contract have entrusted to the quantity surveyor. The quantity surveyor, in a preliminary consideration of them, should remember the principles that must guide the architect or engineer in a decision. The following thoughts are suggested as a guide to a decision on claims:

- What did the parties contemplate on the point at the time of signing the contract? If there is specific reference to it, what does it mean?
- Can any wording of the contract, though not specifically mentioning it, be *reasonably* applied to the point? In other words, if the parties had known of the point at the time of signing the contract, would they have reckoned that it was fairly covered by the wording?
- If the parties did not contemplate the particular matter, what would they have agreed if they had?
- If the claim is based on the contract, does it so alter it as to make its scope and nature different from what was contemplated by the parties signing it? Or is it such an

extension of the contract as would be beyond the contemplation of the parties at the time of signing it? In either case, the question arises whether the matter should not be treated as a separate contract, and a fair valuation made irrespective of any contract conditions.
- The value of the claim in monetary terms should not affect a decision on the principle.
- If the claim is very small, however, whichever party is concerned might be persuaded to waive it, or it may be eliminated by a little 'give and take'.

After receiving a contractual claim from the contractor, the following should be considered in assessing the value of the claim:

- Is the contractor's claim reasonable?
- What are the costs involved?
- What basis is the contractor using to justify a claim?

The RICS *Evaluation of Claims Checklist* provides further questions under Heads of Claim, Effect on Preliminaries and Financial Evaluation to support the quantity surveyor in ascertainment of the claim (RICS 2022).

It is also important to give due consideration to review any action of the client that may have been a contributory cause. If the claim is based on unanticipated misfortune, consideration of what the parties would have done, if they had anticipated the possibility, will often indicate whether it would be reasonable to ask the client to meet the claim to a greater or lesser extent.

The architect or engineer may make a recommendation to the client. If the contractor does not accept the client's offer, then it may be necessary to invoke a third party to help resolve the difficulty, such as in mediation, although if the architect or engineer is the certifier under the contract, they must make a decision on the validity of the claim and certify accordingly.

Contractors

Many contractors have well-organised systems for dealing with claims on construction projects and the recovery of monies that are rightly due under the terms of the contract. They are likely to maintain good records of most events, but particularly those where difficulties have occurred in the execution of the work. However, some of the difficulties may be due to the manner in which the contractor has sought to carry out the work and they thus remain the entire responsibility of the contractor.

Claims that are notified or submitted late will inevitably create problems in their approval. In these circumstances, the architect or engineer might not have the opportunity to check the details of the contractor's submission, and suitable records might not have been maintained.

The contractor must prepare a report on why a particular aspect of the work has cost more than expected, substantiate this with appropriate calculations and support it with reference to Architect Instructions, drawings, details, specifications, project communication, etc. The contractor must also be able to show that an experienced contractor could not have foreseen the difficulties that occurred. They will also need to satisfy themselves that the work was carried out in an efficient, effective and economic manner.

Example

The construction of a major new business park on a greenfield site requires a large earthmoving contract. The quantities of excavation and its subsequent disposal have been included in contract bills and priced by the contractor. During construction, due to variations to the contract and the unforeseen nature of some of the ground conditions, the quantities of excavated materials increase by 25% by volume, all of which needs to be removed from the site.

To continue to apply the contract bill rates is unfair. The rates no longer reflect the work to be carried out and the contractor's changed method of working. Other questions will also need to be asked. Is the type of material being excavated similar to that described in the contract bills? Does this material bulk at the same rate? Is it more difficult to handle? Does it necessitate the same type of mechanical plant?

Some other factors that the contractor may also consider including:

- The increase in the amount of excavated materials, on this scale, may also have other repercussions, such as an extension of the contract time.
- The method of carrying out the works might also now be different from that originally envisaged by the contractor.
- Different types of mechanical excavators may have been more efficient than the plant originally selected to do the work.
- The mechanical plant on site may no longer be the most appropriate to do the job. This is especially so where cut and fill excavations are considered, where motorised scrapers may need to be substituted for excavators and lorries.
- The contractor may also be involved in hiring additional plant at higher charges and employing workpeople at overtime rates, in order to keep the project on schedule.

The preparation of the claim will include two aspects:

- A report outlining the reasons why additional payments should be made to the contractor.
- An analysis showing how the additional costs have been calculated.

References

Arcadis. *2021 Global Construction Disputes Report*. Arcadis. 2021. Available at https://www.arcadis.com/en-gb/knowledge-hub/perspectives/global/global-construction-disputes-report (accessed 27/03/2022).

Bailii. *England and Wales High Court (Technical and Construction Court) Decisions*. Bailii. 2022. Available at https://www.bailii.org/ew/cases/EWHC/TCC/ (accessed 27/03/2022).

CIArB. *The Guidance Note: The Scheme for Construction Contracts*. 2nd Edition. Adjudication Society. 2020.

CIC. *CIC Model Mediation Agreement and Procedure*. 1st Edition. Construction Industry Council. 2019.

Crown. *Annual Report of Technology and Construction Court 2019–2020*. Crown. 2021. Available at https://www.judiciary.uk/wp-content/uploads/2021/06/6.7511_JO_TCC_Annual_Report_2019_20_WEB-2.pdf (accessed 27/03/2022).

Gould, N., King, C., and Britton P. Mediating Construction Disputes: An Evaluation of Existing Practice. King's College London. 2010.

HM Courts & Tribunal Service. *The Technology and Construction Court Guide*. Crown. 2015. Available at https://assets.publishing.service.gov.uk/government/uploads/system/uploads/attachment_data/file/819807/technology-and-construction-court-guide.pdf (accessed 27/03/2022).

Hughes, W., Champion, R. and Murdoch, J. *Construction Contracts Law and Management*. 5th Edition. Routledge. Oxon. 2015.

Hutt, J. and May, T. *Arbitration*. RICS. 2022. Available at https://www.isurv.com/site/scripts/documents.php?categoryID=162 (accessed 02/04/2022).

Jackson, S. Act early to avoid disputes. *Construction Journal*. 2020. Available at https://ww3.rics.org/uk/en/journals/construction-journal/act-early-to-avoid-disputes.html (accessed 03/07/2022).

JCT/CIMAR. *Construction Industry Model Arbitration Rules 2016*. London. Sweet and Maxwell. 2016.

Joint Contracts Tribunal (JCT). *Standard Building Contract with Quantities (SBC/Q)*. Sweet and Maxwell Thomson Reuters. 2016.

Kavanagh, B. *Avoiding & Resolving Disputes*. RIBA Publishing. 2017.

Ministry of Justice. *Pre-Action Protocol for Construction and Engineering Disputes*. 2nd Edition. Ministry of Justice. 2017. Available at https://www.justice.gov.uk/courts/procedure-rules/civil/protocol/prot_ced (accessed 27/03/2022).

Milligan, J.L and Jackson, A.L. *Report No. 18*. 2019. Available at https://www.adjudication.org/sites/default/files/Adjudication%20Report%2018%20-%20December%202019%20Rev%20A.pdf (accessed 01/04/22).

NBS. *National Construction Contracts and Law Report 2018*. RIBA Enterprises Ltd. 2018.

RICS. *Mediation*. 1st Edition. The Royal Institution of Chartered Surveyors. 2014.

RICS. *Surveyors Acting as Adjudicators in the Construction Industry*. 4th Edition. The Royal Institution of Chartered Surveyors. 2017a.

RICS. *Surveyors Acting as Arbitrators Construction Disputes*. 2nd Edition. The Royal Institution of Chartered Surveyors. 2017b.

RICS. *Surveyors Acting as Expert Witnesses*. 4th Edition. The Royal Institution of Chartered Surveyors. 2020.

RICS. *Pre-action Protocols*. RICS. 2022. Available at https://www.isurv.com/info/163/litigation/953/pre-action_protocols (accessed 27/03/2022).

16

Project Management

KEY CONCEPTS

- Project management
- Skills of the project manager (PM)
- Duties of the project manager
- Responsibilities of the project manager during the design and construction process.

LEARNING OUTCOMES

After reading this chapter, you should be able to:

- Distinguish between the competencies of a quantity surveyor and that of a project manager.
- Understand the importance and benefits of effective project management.
- Understand the responsibilities of the project manager at each stage of the RIBA Plan of Works from stage 0 (strategic definition) to stage 6 (handover).

COMPETENCIES

Competencies covered in this chapter:

- Managing projects (on Project Management pathway).

Introduction

Increasingly, clients are adopting a project culture in all aspects of their business. Project management is not new or specific to construction contracts.

The organisation and management of construction projects has existed in practice since buildings were first constructed. Long ago, the process was much simpler but, as knowledge increased and societies became more complex, so the principles and

Willis's Practice and Procedure for the Quantity Surveyor, Fourteenth Edition.
Allan Ashworth and Catherine Higgs.

procedures in management evolved. In some countries, notably the USA, the management of construction works began to emerge as a separate and identifiable professional discipline some years ago alongside architecture and engineering. Because of the differences in the way the construction industry is structured in the UK, the professions have not developed in the same way, nor to the same extent. It is now, however, being accepted by more and more clients that, to succeed in construction, someone needs to take the responsibility for the overall management of the construction project. The development of the standing of the project manager is shown further by contractual recognition, for example, within the New Engineering Contract (NEC) family of contracts where the *Project Manager* is the independent contract administrator (Eggleston 2019).

Project management is a very different function from either design or construction management and requires different knowledge and skills, which are not necessarily inherent in the more traditional disciplines and is regarded as a distinct discipline within the construction sector. To function well as a project manager, it is necessary to have a wide range of competencies. These are categorised in *BS ISO 215021* (2020) as *technical*, such as managing, planning and delivering projects, *behavioural*, such as leadership and people management and *business*, such as managing the project within the external environment. Table 1 within *BS 6079:*2019 provides further information on typical skills and competencies for project management roles. Some of these competencies are shared by QS practitioners as demonstrated by a comparison of the RICS core competencies for the two disciplines of quantity surveying and project management (Fig. 16.1).

However, it is important to distinguish between an extended quantity surveying involvement and formal project management. The quantity surveyor must take a substantial step to provide the comprehensive project management service required by clients. The RICS *Project Manager Services* provides a detailed list of both the core and supplementary services that might be offered (RICS 2019).

In addition to the RICS, professional recognition and chartered status of the project management function can be obtained via the Association of Project Managers (APM).

RICS Competency	PM Pathway	QS Pathway
Contract practice	Core Level 3	Core Level 3
Development/ project briefs	Core Level 3	Not required
Leading projects, people and teams	Core Level 3	Not required
Managing projects	Core Level 3	Not required
Programming and planning	Core Level 3	Optional Level 2
Construction Technology and Environmental Service	Core Level 2	Core Level 3
Procurement and tendering	Core Level 2	Core Level 3
Project finance	Core Level 2	Core Level 3

Fig. 16.1 Comparison of project management and quantity surveying and construction (*Source:* Adapted from Pathway Guides 2018a, b).

Project management

The success of the design team in achieving the client's objectives is greatly influenced by their ability to recognise each other's activities and to integrate them to the full. However, it is the nature of most professionals to see the project objectives in terms of their own discipline and, to a large extent, to operate from within their specialist perspective. Project management provides the important management function of bringing the project team together and may be defined as:

> 'The overall planning, coordination and control of a project from inception to completion, aimed at meeting a client's requirements in order to produce a functionally viable and sustainable project that will be completed safely on time, within authorised cost and to the required quality standards'. (CIOB 2014)

This definition reveals the essence of project management as that of managing and leading the project team toward the successful completion of the client's project objectives. Within most projects, there are several key aspects that are of importance to the client, and each will fall within the remit of the project management function. These include the fundamental considerations of time, cost, quality and sustainability decisions and their interrelationship and, within this framework, the management of procurement, risk and value, each of which is considered in detail elsewhere in this book. To justify the appointment of a project manager, it is important to be able to demonstrate to clients the added value that the function brings in respect to these key considerations.

The benefits of project management are:

- 'The increased likelihood of delivering the project outputs, achieving project outcomes and hence organisational objectives,
- An improved predictability and consistency of the project delivery,
- An improved certainty of delivery of the project within budget,
- Ability to achieve value for money,
- More efficient use of resources, and
- The development of the competence and capability to deliver change' (4.2.2, BS6079:2019+C1:2019)

Project management, therefore, adds value by fulfilling the management role within the context of a modern and increasingly complex construction industry and in recognition of client demand.

Terminology

Whatever name is given to the role of the project manager, and alternatives may include project controller, project administrator or project coordinator, the general intent is usually the same. The idea is that one person or organisation should take overall control and responsibility for coordinating the activities of the various consultants, contractors, subcontractors, processes and procedures for the full duration of the project. The project duration in this context starts at inception and ends on the completion of the rectification period. The management process may also extend into the time when the building is in use and thus link with the facilities management role. This whole-life view will be beneficial in bringing the design and development function more closely together with that of occupation and use.

Whilst the general intent of the project management role is understood, irrespective of designation, the title given to this function may denote differing levels of service. For example, the term 'project coordinator' is likely to indicate the exclusion of the responsibility for the appointment of the other consultants. In any event, the terminology used will always require precise definition in terms of the service to be provided that is stated in the terms of engagement. Clarification is certainly required to ensure that there is no misunderstanding as to the level of professional indemnity cover required. Figure 16.2 indicates a comprehensive list of project management duties differentiating between the project coordinator and project manager designations regarding both external and in-house situations (CIOB 2014).

When acting in the dedicated role of project manager, there should be no attempt to perform any of the functions normally undertaken by the design team, including the traditional duties of the quantity surveyor. These should always be separate to avoid having to make any compromised decisions that might otherwise occur.

The term 'project manager' may be seen within several contexts and is commonly used throughout industry. Within construction, it is important to distinguish between a contractor's project manager, who will primarily manage the construction process, and the client's project manager to whom this chapter is dedicated.

A further development of the management function may be seen in management-based contracts, for instance construction management. Such a procurement route necessitates the appointment of a construction manager whose role involves the management of the design, procurement and construction. No main contractor exists; instead, the construction manager manages the trade contractors and, as reimbursement is by way of a fee, this role can be carried out in a more objective manner with due recognition of the client's and project's interests. Although the duties of the construction manager differ from those of the project manager, with the former having a detailed responsibility for the management of the construction process, there is some commonality in the roles.

Duties of the project manager

The duties of a project manager in the construction industry will vary from project to project. Different countries around the world will also expect a different response to the situation, depending on the contractual systems that are in operation. The mistake, and perhaps the reason for the failure, of the traditional system in certain instances is that an attempt is made to use a single system to suit all circumstances. Any contractual arrangement, however good, must be adapted to suit the needs and experience of the client and

Duties[1]	Client's requirements			
	In-house project management		Independent project management	
	Project management	Project coordination	Project management	Project coordination
Be named in the contract	●		○	
Assist in preparing the project brief	●		●	
Develop project manager's brief	●		●	
Advise on budget/funding / programme / risk management arrangements	●		○	
Advise on site acquisition, grants and planning	●		●	
Arrange feasibility study and report	●	○	●	○
Develop project strategy	●	○	●	○
Prepare project handbook	●	○	●	○
Develop consultant's briefs	●	○	●	○
Devise project programme	●	○	●	○
Select project team members	●	○	●	○
Establish management structure	●	○	●	○
Coordinate design processes	●	○	●	○
Appoint consultants	●	●	●	○
Arrange insurance and warranties	●	●	●	○
Select procurement system	●	●	●	○
Arrange tender documentation	●	●	●	○
Organise contractor prequalification	●	●	●	○
Evaluate tenders	●	●	●	○
Participate in contractor selection	●	●	●	○
Participate in contractor appointment	●	●	●	○
Organise control systems including reporting procedures	●	●	●	●
Monitor progress	●	●	●	●
Manage and monitor meetings	●	●	●	●
Authorise payments	●	●	●	○

Fig. 16.2 A comprehensive list of project management duties (*Source:* Table 1.1, CIOB 2014/John Wiley & Sons).

Organise communication/reporting systems	●	●	●	●
Provide project coordination	●	●	●	●
Issue health and safety procedures	●	●	●	●
Address environmental aspects	●	●	●	●
Coordinate statutory authorities	●	●	●	●
Monitor budget and changes	●	●	●	●
Develop final account	●	●	●	●
Arrange precommissioning/commissioning	●	●	●	●
Organise handover/occupation	●	●	●	●
Advise on marketing/disposal	●	○	●	○
Organise maintenance manuals	●	●	●	○
Plan for maintenance period	●	●	●	●
Develop maintenance programme/staff training	●	●	●	○
Plan facilities management and coordinate BIM	●	●	●	○
Arrange for feedback monitoring and post-completion review	●	●	●	○
Investigate BIM implementation	●	●	●	●
Liaise with funding institutions	●	●	●	○
Liaise with ground landlord	●	○	●	○
Liaise on acquisition, valuation, disposal of land	●	○	●	○
Liaise with agents over leasing tenants' queries, etc.	●	○	●	○
Liaise with clients over move to new premises	●	○	●	○
Liaise coordination with legal agents	●	○	●	○
Advise and manage client's changes	●	●	●	●

[1]Duties vary by project and relevant responsibility and authority
Symbols ● = suggested duties; ○ = possible additional duties

Fig. 16.2 (Continued)

the project, and not vice versa. The project manager will need to employ a wide variety of skills and options for a whole range of different solutions. The key duties of the project manager are to:

- Communicate to the consultants the requirements of the client's brief
- Monitor the progress of design work, and the achievement of function and sustainability requirements by reference to the client's brief

- Monitor and regulate programme and progress
- Monitor and use reasonable endeavours to coordinate the efforts of all consultants, advisors, contractors and suppliers directly connected with the project
- Monitor the cost and financial rewards of the project by reference to the client's brief.

Although not stated in the above outline of duties, it is implicit that the project manager will also lead the project. The list above is not exhaustive, as demonstrated by Fig. 16.2, and the *Project Manager Services* (RICS 2019) provide a checklist to specific services offered.

The following sections provide more detailed consideration of some, but not necessarily all, of the duties of the project manager. In considering these duties, please note the significance of other sections of this book that are of major importance in project management, for instance procurement, value management and risk management.

BIM and the project manager

The increasing use of Building Information Modelling (BIM) within the industry has impacted on the role and responsibilities of the project manager. They have an important role in advising the client of the benefits of BIM implementation on the project and influencing the collaborative environment needed both in the context of the project and client's organisation. The project manager must ensure that the project benefits from the digitally supported and enabled processes that underpin the delivery and operation of the client's asset as detailed within *BS EN ISO 1650:*2018. Table 1 in the *RICS Building Information Modelling for Project Managers* illustrates how BIM adoption significantly impacts on the role and responsibilities of the project manager and the importance of having adequate BIM awareness to effectively implement BIM through the project lifecycle (RICS 2017).

Responsibilities of the project manager

The following sections provide more detailed consideration of some, but not necessarily all, of the responsibilities of the project manager. Stage references are to the RIBA Plan of Works 2020. In considering these responsibilities, please note the significance of other sections of this book that are of major importance in project management, for instance procurement, value management and risk management. In relation to the provision of cost advice from the quantity surveyor throughout the design stages, Appendix C of the NRM*1 Order of Cost Estimating and Cost Planning for Capital Building Works* (2021) provides a comprehensive list of the key information required from the project manager.

Stage 0 – strategic definition

Client's objectives

The starting point of the project manager's commission is to establish the client's objectives in detail. The success of any construction project can be measured by the degree to which it achieves these objectives. The client's need for a building or engineering structure may have arisen for several reasons: to meet the needs of a manufacturing industry, as part of an investment function, or for social or political demands. These requirements must be high-level strategic decisions and relate to the wider business objectives to ensure that the project objectives set align to client needs, such as the sustainability outcomes. In an attempt to provide satisfaction for the client, three major areas of concern will need to be considered. The weighting given to these factors will vary depending on the perception of the client's objectives:

- *Performance.* The performance of the building or structure in use will be of paramount importance to the client. This priority covers the use of space, the correct choice of materials, adequate design and detailing, and the aesthetics of the structure. Attention will also need to be paid to future maintenance requirements once the building is in use and the incorporation of sustainable technologies.
- *Cost.* All clients will have to consider the cost implications of the building's desired performance. The price that they are prepared to pay will temper, to some extent, the differences between their needs and wants. Clients today are also more likely to evaluate costs not solely in terms of initial capital expenditure, but rather on a basis of whole life cost management.
- *Time.* Once clients decide to build, they are generally in a hurry for their completed building. Although they may spend a great deal of time deliberating over a scheme, once a decision to build has been reached they often require the project to be completed as quickly as possible. In any event, in order to achieve some measure of satisfaction, and to prevent escalating costs, commissioning must be achieved by the due date.

The project manager's strategy for balancing the above three factors will depend on an interpretation of the client's objectives. It would appear, however, that there is some room for improvement in all three areas. The improvement of the design's completeness, particularly, should reduce the contract time and hence the constructor's costs. The correct application of project management should be able to realise benefits in these areas.

The client's objectives should be used as the key parameters in which decision making for the broader issues involved in the design and construction of the project are set. The discernment of these objectives will assist the project manager to decide which alternative construction strategies to adopt. It is very important that an adequate amount of time is allowed for a proper evaluation of the client's needs and desires. Failure to identify these properly at the outset will make it difficult for the project to reach a successful conclusion upon completion and in the long term.

It is also important for the project manager to develop a BIM strategy, which aligns with the client's asset management strategy. The BIM strategy should, within the constraints of available time and resource, aim to maximise the value of the technology for both the client and the project team (Shepherd 2015).

Stage 1 preparation and briefing

Client's brief

This involves the evaluation of the user requirements in terms of space, design, function, performance, sustainability, time and cost. The whole scheme is likely to be limited one way or another by cost, and this in turn will be affected by the availability of finance or the profits achieved upon some form of sale at completion. It is therefore necessary for the project manager to be able to offer sound professional advice on a large range of questions, or to be able to secure such information from one of the professional consultants who are likely to be involved with the scheme. This will include the coordination of all necessary legal advice required by the client. It is most important that the client's objectives are properly interpreted, as at this stage ideas, however vague, will begin to emerge, and these will often then determine the course of the project in terms of both design and cost. Consideration will also be given to the client's information requirements that will need to be incorporated to support the post occupancy stage of the project. A Digital Execution Plan will be developed, setting out how the design team will be expected to use digital tools collaboratively and effectively on the project (RIBA 2020). It is the project manager's responsibility to ensure that the client's brief is clearly transmitted to the various members of the design team, and also that they properly understand the client's aims and aspirations.

An increasingly used technique at this stage of a project is value management (see Chapter 7). This will assist in establishing clearly what the client's objectives are and is an excellent approach to the achievement of a consensus view of the brief, incorporating an evaluation of the needs and desires of all key stakeholders. Similarly risk management (see Chapter 8) is also an important consideration at this stage of the project.

Contractor involvement

The client will probably require some initial advice on the methods available for involving the contractor in the project. The necessity for such advice will depend on the familiarity of the client with capital works projects, although the growth and complexity of procurement options are such that some advice will almost certainly be of benefit. The correct evaluation of the client's objectives will enable the project manager to recommend a particular method of contractor selection. It may be desirable, for example, to have the contractor involved at the outset or to use some hybrid system of contractor involvement (see Chapter 9). The project manager will be able to exercise expert judgement in this respect by analysing the potential benefits and disadvantages of each procurement option in the context of the project concerned. Those where there is early contractor involvement, such as two-stage tendering will support integration of BIM, a collaborative approach and effective management of information flows at an early stage. This decision will need to be made reasonably quickly, as it can influence the entire design process and the necessity of appointing the various consultants.

Design team selection

The project manager may be responsible for the selection of the design team. The team will reflect the nature of the project, its complexity, procurement route and the discipline-specific expertise required for successful completion of the project. Although some situations will demand prompt negotiation with a proven team, if circumstances allow, proposals should be sought from at least three consultants. In addition to demonstrating the appropriate experience and expertise to undertake the project, if BIM is to be used, the consultants will need to show that their organisations have the appropriate technology, skills and experience to use the digital tools and produce the information requirements set out in the project's Digital Execution plan. Within each of the project team's contracts, the *CIC Building Information Modelling Protocol* maybe included as an 'incorporated' clause. The Protocol sets out how each team member must produce information at specified project stages as defined in a responsibility matrix (CIC 2018).

If the client has been involved in capital works projects previously, they may already have designated consultants within their framework agreements with whom the project manager will need to work. If this is the case, the project manager should make clear, at the outset, relevant concerns relating to any of the client-appointed consultants, including past performance, location and resources. Where the consultants are appointed on a regular basis by the client, the project manager might experience problems due to existing relationships and lines of authority. This will be particularly difficult in situations where the client is using project management for the first time.

The project manager is likely to be responsible for agreeing fees and terms of appointment of all consultants on behalf of the client. The relationships between the contributions from each consultant must be clear at the outset to avoid any misunderstandings that could occur later.

Feasibility and viability reports

During the early stages of the design process, it will be necessary for the project manager to examine both the feasibility and the viability of the project. Sound professional advice is very important at this stage, as it will determine whether or not the project should proceed. A feasible solution is one that is capable of technical execution and might be found only after some site investigation and discussion with the designers. A feasible solution may, however, prove not to be viable in terms of the project budget. Unless the project is viable in every respect, it will probably not proceed. The investigation work should be sufficiently thorough while taking note of the fees involved, particularly if the project should later be abandoned.

Once the project is established as both viable and deliverable, feasibility studies will be undertaken to further define and explore design solutions that best suit the client requirements. The project manager will manage the team's compliance with the BIM implementation plan, so that integrated information can support decision making.

At this stage of the project, it will be beneficial to carry out a detailed risk analysis of the proposed scheme or various options that are under consideration. Risk is an inherent part of every project and whilst a development may be determined as feasible, such a decision is reliant on assumptions and predictions relating to uncertainties that exist. Project management should incorporate a professional approach to the identification, assessment and management of risk (see Chapter 8).

Planning and programming

Once the project has been given the go-ahead it will then become necessary to prepare a programme for the overall project, incorporating both design and construction. The programme should represent a realistic coordinated plan up to the commissioning of the scheme. The project manager must carefully monitor, control and revise where necessary. Project management tools can be used to support the planning and programming of the project.

The construction industry has a poor reputation with regard to achieving project completion dates. It is a difficult task to predict the completion date of a proposed development at inception. There are likely to be many unknown factors, not least those relating to site conditions, design solution and construction method. Notwithstanding the difficulties, project management will only be considered a success if project deadlines are met.

With regard to the programming of the construction phase, subject to the method of procurement, additional difficulties exist throughout the project's duration. This is due to the method of construction being largely unknown until tenders are received and, in any event, forecasts are reliant on the contractor's expertise and cooperation.

Stage 2 concept design to stage 4 technical design

Design process management

Project information when uncoordinated leads to inefficiency, a breakdown in communications between the design team, frequent misunderstandings and an unsatisfied client. For example, the delayed early involvement of service engineers often results in changes to the design of the structure both to accommodate the engineering work and also incorporate good engineering ideas. An important task therefore for the project manager is to ensure that the various consultants are appointed at appropriate times and that they easily and frequently liaise with each other while maintaining their own individual goals. Collaboration and teamwork is vital for the success of the project. The project manager will therefore need to exercise both tact and firmness to ensure that the client's objectives remain paramount and that the design team follow the Digital Execution Plan for the project.

The project manager, although not directly involved in the process of designing in its widest sense, must nevertheless have some understanding of design in order to appreciate the problems and complexities of the procedures involved. Responsibility for the integration and control of the work from various consultants rests with the project manager who, in the first instance, will be directly answerable to the client for all facets of the project. This will include ensuring quality control systems of all aspects of the design (and construction) process are followed with regular reporting to enable audits on the developed design solutions. These will ensure that the design continues to satisfy the project objectives. The project manager must also ensure that the project is compliant with all statutory processes, such as planning, building regulations and health and safety (CDM).

The project manager must also be kept informed of the cost implications as the design develops, usually the responsibility of a quantity surveyor. The design team must be informed of what can or cannot be spent and promptly advised when problems are envisaged. In this respect, the control of the costs should be more effective than when relying on the efforts of the architect alone. The project manager must, of course, have a very clear understanding of the client's intentions and will also need to advise the client in those circumstances where the original requirements cannot be met in terms of design, cost or time. The project manager will always have an eye on the future state of the project and must keep at least one step ahead of the design team.

During this stage, unless the contractor has been appointed earlier, the project manager will need to consider a possible list of firms who are capable of carrying out the work, and to ensure that the proper timely action is taken to obtain all statutory approvals.

Stage 5 Manufacturing and Construction

Supervision and control during construction

During the contract period, the project manager will need to have regular meetings with the consultants and contractor and his subcontractors. A focus of the role is change management; progress of the works must be monitored and controlled and any potential delays identified. The project manager will manage, monitor and report on risks; *Table 8 in the RICS Management of Risk (2015)* provides the project manager's responsibilities in relation to other stakeholders. Actual costs will be compared with planned costs and the effects on the programme and the budget of any variations will also need to be monitored. The project manager must be satisfied that the project is finished to the client's original requirements; although one of the consultants may be responsible for the quality control, the project manager will need to be careful about accepting substandard or unfinished work. *BS6079: 2019+C1:2019* recommends the use of Earned Value Management (EVM) as a proven approach to monitor and control a project. Project management functions such as quality, change, safety and supply chain management can be further supported by BIM and other digital technologies.

As discussed in Chapter 15 if issues arise during the course of the project, the role of the project manager is key in mitigating and resolving conflict. Some problems may need to be discussed with the client, but early decisions should be sought to bring the project to a successful conclusion.

The project manager might have an ongoing role after the main construction contract to administer fitting-out work for occupiers and tenants.

Stage 6 handover

Evaluation and feedback

Evaluation and feedback represent the final stage of the project manager's duties. It should be ascertained that commissioning and performance checks of all services and installations have been carried out and systems are fully operational. The contractors final account needs to be agreed and the necessary 'as built' drawings and specifications have been supplied to the client. This might also include digital models and structured information to support the operation and maintenance of the finished building. To ensure the safety of any future maintenance, refurbishment or construction work associated with the building, the project manager will confirm the completeness of the Health and Safety File prior to handover to the client. The project manager will also need to advise on any current legislation affecting the running of the project, on grants, taxation changes and allowances.

To formalise the end of the project, the client should be issued with a 'close-out' report that states the responsibilities of all parties have been satisfactorily discharged and includes a comprehensive review of the project with recommendations to assist in any future capital works that the client might undertake.

References

BS 6079:2019+C1:2019. *Project Management – Principals and Guidelines for Management of Projects*. BSI Standards Institution. 2019.

BS EN ISO 1650:2018. *Organization and Digitization of Information About Buildings and Civil Engineering Works, Including Building Information Modelling (BIM) – Information Management Using Building Information Modelling*. ISO. 2018.

BS ISO 21502:2020. *Project, Programme and Portfolio Management – Guidance on Project Management*. ISO. 2020.

CIOB. *Code of Practice for Project Management for Construction and Development*. 5th Edition. Wiley Blackwell/Chartered Institute of Building. 2014.

CIC. *Building Information Modelling (BIM) Protocol: Standard Protocol for use in projects using Building Information Models*. 2nd Edition. Construction Industry Council. 2018.

Eggleston, B. *The NEC4 Engineering and Construction Contract: A Commentary*. 3rd Edition. Wiley Blackwell. 2019.

RIBA. *Plan of Work Overview 2020*. RIBA. 2020.

RICS. *Management of Risk*. The Royal Institution of Chartered Surveyors. 2015.

RICS. *Building Information Modelling for Project Managers*. The Royal Institution of Chartered Surveyors. 2017.

RICS. *Pathway Guide: Quantity Surveying and Construction*. The Royal Institution of Chartered Surveyors. 2018a.

RICS. *Pathway Guide: Project Management*. The Royal Institution of Chartered Surveyors. 2018b

RICS. *Project Manager Services*. The Royal Institution of Chartered Surveyors. 2019.

RICS. *Order of Cost Estimating and Cost Planning for Capital Building Works: NRM1*. RICS. 2021.

Shepherd, D. *BIM Management Handbook*. RIBA Publishing. 2015.

17

Facilities Management

KEY CONCEPTS
• Facilities management and services • Measures of building performance • Procurement of FM Services • Maintenance management • Cost management of maintenance works • Sustainability and FM

LEARNING OUTCOMES
After reading this chapter, you should be able to: • Understand the purpose of facilities management and its contribution to business success. • Recognise and explain the different types of building maintenance and how each may be procured. • Describe the use of cost management processes to support maintenance plans. • Appreciate the contribution the role FM has in meeting the UK 2050 net zero target. • Identify the potential use of technology in supporting FM.

COMPETENCIES
Competencies covered in this chapter: • Data management • Construction technology and environmental services • Contract practice • Design economics and cost planning • Procurement and tendering • Risk management • Sustainability

Willis's Practice and Procedure for the Quantity Surveyor, Fourteenth Edition.
Allan Ashworth and Catherine Higgs.

Introduction

The emergence of facilities management (FM) was in response to the realisation that a company's property assets are of vital importance to business success. This may be considered in two ways:

- In most companies, the costs associated with real estate are a major overhead, often representing the largest cost after wages and salaries. The control of this overhead, via cost management and use optimisation, is therefore likely to make a significant contribution to profitability.
- The impact that real estate may have upon income generation could be significant, indeed, more significant than the direct costs of provision. For instance, a factory with an ageing and inefficient layout might not be able to compete with a modern facility that accommodates new technologies, for example, AI and robotics.

In addition to the needs of the business, FM must also satisfy the needs of the workforce. *Raising the Bar III* states that the value of FM is the provision of high-quality, cost-effective facilities and related support services, that has a demonstrable impact on organisational effectiveness, such as workplace attraction, retention, and business productivity (IFMA and RICS 2017). Facility or facilities management can therefore be defined as an 'organisational function which integrates people, place and process within the built environment with the purpose of improving the quality of life of the people and the productivity of the core business' (BSI 2018).

Social, economic and environmental drivers, together with the post-COVID impact on working practices and a greater focus on well-being, have emphasised the need for a more human-centric approach to both the management of facilities and the measurement of its performance.

In a book of this type, it is by no means possible to adequately cover the entire subject matter that falls under the umbrella of facilities management. As shown below, the range of activities is potentially diverse and the knowledge and skills necessary for the fulfilment of each require specialist attention. The purpose of this chapter, therefore, is to introduce the reader to facilities management and highlight its significance in terms of the QS role.

Measuring building performance

The RICS *International Building Operation Standard (IBOS)* is a multidimensional tool for measuring building performance. The Standard approaches performance from different perspectives, referred to as the five pillars, with a single unified thread of user experience (RICS 2022). Whilst its focus is on performance metrics, the five pillars demonstrate the multi-faceted needs of facilities management:

Compliance. The provision of a building, which is compliant with all the relevant statutes and regulations.

Functional. Suitability of the building to meet both business and user needs in terms of layout, flexibility etc.

Economic. Cost efficient operating and other life cycle costs.

Sustainable. The social and environmental impact of the building and how it operates. This will be discussed later in the chapter.

Performing. The ability of the building and how it is managed to effectively support the user (RICS 2022).

Building services has a significant impact on the last pillar, CIBSE identify the following as impacting on the health and well-being of buildings:

- Thermal
- Humidity
- Air quality and ventilation
- Visual environment (CIBSE 2021)
- Water Quality
- Electromagnetic effects
- Noise and vibration

The Healthy Building movement also includes safety and security to this list (Wilding 2022). IBOS also include workspace optimisation and connectivity (RICS 2022).

The role of the facilities manager

The role of the facilities manager is not solely during the occupation stage of the project, it is important that, if the end user, the client's facility management team are involved early in the design process to ensure operational readiness and performance in use. This can be achieved by adopting a Soft Landings approach. This framework facilitates project team decision making during the design and the construction stages focusing on the operational outcome of the project (BSRIA 2018b). The FM's knowledge and past experience of the client's maintenance strategy will support the quantity surveyor in the provision of whole life costs. Their understanding of the client's business and operational strategies means that they are an important stakeholder at value management workshops (see Chapter 7).

Atkins and Brooks provide a comprehensive description of the range of functions covered by facility management, including 'real estate management, financial management, HR management, health, safety, security and the environment, change management and contract management, in addition to maintenance, facility services (e.g. cleaning, security and catering, business support services and utility services' (Atkin and Brook 2021). In practice, the scope of facilities management services may range from the provision of a single service at operational level (for example, maintenance management), to that in which a comprehensive range of services is provided including those at a strategic level. A major influence to the current changes in the facilities manager's role is supporting the client's sustainability agenda and its impact on the management and operation of the building. Sustainable facility management will also include energy management, water management, material and consumable management, waste management and the promotion of a sustainable community.

Outsourcing

In recent years, there has been an increasing trend to contract out or outsource certain support functions, usually regarded as non-core to the main raison d'être of the business, rather than provide such services via the employment of in-house personnel. This is consistent with the downsizing philosophy that many companies follow and is said to offer an efficient and cost-effective alternative to large, internally staffed empires.

The policy of outsourcing can be applied to many support services including catering, security, fleet management, IT management, maintenance and repairs to buildings and services, waste management, landscape management, travel, recruitment...the list is extensive. Although outsourcing may be applicable to a wide range of functions, it should not be seen as synonymous with facilities management, merely one approach to be considered in obtaining necessary services. There may be strong reasons for a service to remain under direct staff control, for example where commercial security is vital and would be endangered by office cleaning contracts being outsourced. Some of the issues surrounding the outsourcing philosophy include:

- *Competition.* In-house provision is prone to become less efficient due to the absence of competition. Outsourcing will promote economies of scale, right-sizing and is generally seen to result in a reduction in costs. Where services are provided internally, benchmarking and service-level agreements may assist in identifying inefficiencies and provide a means of comparison when competition is absent.
- *Specialisation.* In contracting out services, companies will have access to experts they would otherwise be unable to adequately justify as full-time personnel on their payroll. For example, a small company will be unable to sustain the employment of a quantity surveyor. In practice, the range of facilities management-related skills required are such that some outsourcing is inevitable.
- *Limited experience.* In-house personnel are generally restricted in exposure to one set of systems and are likely to work in isolation from other organisations and professional associates, thus limiting technical and managerial development. Alternatively, the in-house team is immersed in the business operations of its company which may be advantageous compared to an external organisation that is poorly acquainted with the client organisation.
- *Quality.* An external contractor is motivated toward profit maximisation, albeit via the provision of an adequate service, and can achieve this by providing the minimum acceptable service at the lowest possible cost. It is reasonable to assert that in-house teams are motivated simply to provide a service to the employer and that the quality objective cannot be compromised by the profit motive. Alternatively, in-house staff may be in sheltered employment positions and, as a consequence, be poorly motivated generally.
- *Confidentiality/security.* Some aspects of an organisation may be vulnerable to poor security/confidentiality. These would be better served by the provision of in-house services.
- *Community culture.* Personnel employed on a contract basis may be very transient and uninvolved with anything other than the service they are contracted to provide. Rapport with other staff and extended knowledge of the organisation will be absent. As a consequence, social benefits enhancing the service provided may be absent.
- *Flexibility.* In-house staff will be more widely involved within an organisation, which may result in more flexibility of service, for example cleaners could be requested to move furniture.

- *Employment law.* In deciding to outsource a particular service, attention should be given to relevant employment law.

Once the decision to outsource particular services has been taken, consideration as to how to organise and procure each particular service needs to be made. The organisation of contracted-out services may be dealt with in a variety of ways:

- *Individual contracts.* Each service required to be outsourced may be done on an individual contract basis, for example to supplement an otherwise comprehensive in-house facilities management provision. This may be managed either centrally or by several appropriate management or cost centres within an organisation. For instance, photocopying costs, furniture costs and cleaning costs could be handled by a sectional office manager, whilst IT provision is managed centrally. Centralising the outsourcing of service contracts promotes more efficient administration and performance monitoring.
- *Bundling.* Bundling involves the collection of particular facilities management services that could be grouped on the basis of efficiency or convenience and placed with a single supplier, for instance, cleaning, security and catering. This approach reduces the points of responsibility and could lead to economies of scale.
- *Total outsourcing.* With this approach, which is an extension to 'bundling', all aspects of facilities management are outsourced to a single facilities management company, including those that are operational as well as managerial. This approach allows a business organisation to benefit from a single point of contact, although the facilities management company appointed will be unable to provide all of the services required by the client and will in turn subcontract some of these out. There are further advantages to the appointment of single contractors. Administration should be reduced, economies of scale are likely to be achieved and, due to the increased commercial significance to the contractor, a larger commitment and better service should result.

Whichever outsourcing method is adopted partnerships are a key driver to the successful delivery of FM services. BS ISO 44001:2017 Collaborative Business Relationship Management provides both a framework and means of benchmarking to develop sustainable relationships, harnessing collaboration through the adoption of the standard principles.

The quantity surveyor has considerable knowledge and skills that can be applied to the procurement of services and is well placed to manage the outsourcing requirements of a business. An understanding of the principles surrounding effective procurement, including selection procedures and the preparation of relevant documentation, is important and within the traditional remit of the profession. Matters relating to the specification of particular services are likely to need the support of in-house advisors and other professional consultants.

Maintenance management

The focus of the rest of this chapter will be maintenance management. BS 8210 defines maintenance management as the 'process of ensuring that the most attractive and efficient maintenance programme is formulated and delivered to ensure that assets continue to perform their intended function' (BSI 2020). Assets refer to the building structure, its fabric and components and the engineering installations. Repair and maintenance is a significant sector within the market, with a total expenditure at £65.6 bn (2.96 of GDP) in 2019 and represents 32.5% of total construction output (BCIS 2021). In addition to the size of the

Preventive maintenance	Corrective maintenance
Replace timing chain	Replace indicator bulb
Oil change	Renew illegal tyre
Replace air filter	Touch up paintwork
Replace oil filter	Replenish screen wash
Replace brake pads	Adjust handbrake

Fig. 17.1 80,000-mile service on a car.

market, the workload activity is less vulnerable to market fluctuations in that maintenance is required to existing building stock and is not reliant on investment in new buildings.

There are two broad categories of building maintenance: preventative maintenance and corrective maintenance. In terms of any piece of equipment, there are some items that will be attended to following 'failure'; however, there are distinct advantages in adhering to a policy of preventative maintenance. For example, consider the example of a service to a car. Figure 17.1 highlights typical items of work that might be carried out and shows the category of service provided with reference to preventative and corrective maintenance. The benefits of carrying out the preventative maintenance work are clear. For instance, the timing chain may cost, say, £200 to replace, but could lead to a repair bill in excess of £1000 if it failed during road use. Likewise, the oil change, air filter and oil filter are relatively cheap measures that will protect the engine and extend its life. In addition to the long-term benefits, preventative maintenance provides peace of mind and improves reliability.

This philosophy works equally well in buildings; however, the complex nature of buildings and the business organisations that use them might result in a less well-organised approach.

The maintenance of buildings is an important element of facilities management in that it is essential to the efficient use and costs of operation. The production of a maintenance policy, similar in concept to the list of checks at various service intervals during the life of a car, will provide the direction necessary to achieve good standards of maintenance. The policy should consider the various categories of maintenance work, the standards to be attained, health and safety requirements, security factors and building access. In addition, an action plan with regard to each category of work, including response times and lines of communication, needs to be established, as well as methods of budgeting and payment. In conclusion, maintenance should not be considered as a pure overhead, but strategically as a means of adding value to the business. Once occupied it is essential that assets are maintained to ensure their effectiveness and efficiency to support the business as well as occupants' needs and well-being (BSRIA 2018a). Such maintenance will also avoid the risk of asset failure, maintain efficient energy consumption and extend product life span.

Problems of working in existing buildings

There are unique problems to be overcome when undertaking maintenance work, in existing buildings. The difficulties may be extended to the design, planning, costing and execution of the work and incorporate the following considerations:

- *Working conditions.* Work in existing buildings is likely to be hindered by continued occupation. Whilst this might not impact on some types of work, the execution of any major repair or maintenance will need to accommodate the needs of existing users. Health and safety provisions (see Chapter 4) must be strictly enforced and general nuisance factors such as noise and dust can cause serious disruption to the main activities of a business. Similarly, the image of a company could be adversely affected due to building operations. To compound the problems, working around occupants can be very time consuming and thus costly. Careful planning of maintenance work is essential if disruption and costs are to be minimised.
- *Abnormal hours.* Some aspects of a business cannot be interrupted during normal business hours; for example, any work to an existing IT installation may necessitate the closure of a networked system preventing internal and external communications, business transactions, etc. as well as the general use by all members of staff of the IT facility. This is major disruption that could result in a substantial loss to a business. Therefore, work of this type should be well planned and carried out beyond normal working hours where possible. Similarly, power supplies are vital to any organisation and must be maintained to allow continued production. Consider, for example, the cost implications of a loss in power in a large manufacturing organisation resulting in standing time of one hour.
- *Layout constraints.* The installation of new floor layouts, services and specialist installations could be dramatically constrained by the existing floor heights, internal walls and structures. Many older buildings are unsuited to adaptation and this factor may ultimately result in their obsolescence and disposal.
- *Legal constraints.* Planning restrictions and building regulations may restrict refurbishment potential. In particular, listed buildings, whilst possibly providing a unique atmosphere and enhancing company image, pose particular problems by preventing alteration and generating additional expense for simple repairs and maintenance.
- *Unique building structures.* Depending on the age and status of a building, repairs and maintenance may be problematic due to external controls, the availability of suitable materials and the inability to accurately assess and plan repairs. Some typical problems encountered are considered in the example in Fig. 17.2.

Due to the uncertainties that may exist, procurement of work to existing structures needs careful consideration. Without the opportunity to prepare adequate detailed designs, which as outlined above is not always achievable, it is likely that the client will be required to accept more risk than with new works.

Procurement

The approach to the execution of maintenance work will be influenced by the size and type of client. Large client organisations may have a comprehensive in-house team fulfilling the professional, technical and construction roles of building maintenance, whilst smaller businesses will be unable to sustain such an internalised operation and need to outsource all of the maintenance requirements. The issues surrounding the merits and concerns relating to outsourcing were discussed earlier in this chapter.

Premises: *Large landmark Victorian building being used as main administration headquarters*

Site problems:

- Repairs to existing façade required following storm damage; availability of matching imperial bricks caused delay and expense
- Outbreak of dry rot on upper floor detected during the works. Extensive access required to check for dry rot throughout the building, causing disruption to the occupiers due to remedial work and adherence to health and safety provisions
- Installation of secondary glazing throughout the west elevation of the building to exclude external noise. Irregularity of the window sizes and types involved much customisation and delay
- Large ornate window requires replacement. Due to listed building status, the replacement window must comply with the requirements of the planning authority and be an exact replica – a very expensive purpose-made unit taking several weeks to manufacture
- All external stonework needs to be cleaned and repaired. The organisation and budgeting of this work is difficult due to the inability to accurately assess the scope of the works. Full details will not be known until scaffolding is erected and cleaning occurs.

Fig. 17.2 Problem of unique building structures: example.

With regard to maintenance work, the employer may choose to contract with several contractors to carry out individual projects or maintenance contracts or appoint a single management contractor responsible for the entire maintenance requirement.

The types of maintenance work required by the employer will vary and, as stated above, broadly include work of either a preventative or corrective nature. Although there is varying terminology in use, the outline shown in Fig. 17.3 provides a clear indication of the range of maintenance work encountered. Any decisions relating to maintenance work undertaken should be considered from a whole life costing perspective (see Chapter 6).

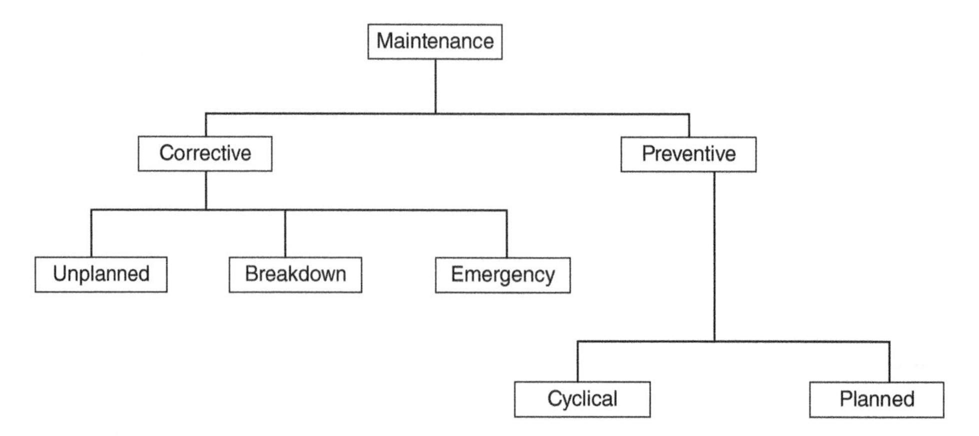

Fig. 17.3 Types of maintenance (*Source:* Adapted from Atkin and Brook 2021).

The procurement of maintenance work will depend on the type of work required, as follows.

- Corrective maintenance
  Unplanned maintenance This is unforeseen maintenance work to prevent failure and is normally in response to the unexpected deterioration of a component or element.
  Emergency maintenance Where sudden breakdown or accident occurs, prompt action may be necessary, for example in order to maintain a service or make a building safe, secure or watertight. This could extend to a 24 hour/365 day service.
  Breakdown response It would be inefficient and costly to categorise all corrective maintenance as 'emergency' work. Some items of work may be deferred until a regular visit occurs, for example the replacement of a light bulb or ceiling tile, or attendance to a dripping tap.
  The organisation for the execution of corrective maintenance work will require the predefinition of items of emergency work, financial limits, response times and clear lines of action. There may be a desire for local client facilities representatives to pursue emergency action in non-emergency situations. Surveyors involved in the maintenance process may be required to give appropriate direction in accordance with corporate policy and arrangements when a possible emergency situation is referred.
  The most appropriate contractual response for corrective maintenance work is likely to be some sort of term arrangement (see later in this chapter) with a general building contractor and/or specialist contractors.
- Preventive maintenance
  Planned maintenance This work will be identified in the organisation's planned maintenance schedule. Such a schedule will detail maintenance requirements over the long term and will accommodate user needs and business continuity. Such work might include flat roof replacement, repointing and smaller building alterations. Individual contracts are likely to be more suitable.
  Cyclical maintenance This work is essential to the ongoing safety and reliability of key components, for example a boiler or air conditioning plant. Such maintenance is carried out to prevent or reduce the changes of component failure. The suppliers of the equipment will usually recommend a programme of maintenance and also offer the maintenance service itself. Some care should be taken to ensure that neither under- nor over-maintenance occurs. In some instances, statutory requirements must also be adhered to, for example with lift installations. The most appropriate method of procurement is likely to be via a term contract, probably with the supplier of the equipment or other suitable specialist.
  It is also usual to carry out some areas of maintenance work at regular intervals, for example painting and decorating, irrespective of the actual condition of the building fabric. This type of work may also be influenced by cash-flow factors, e.g. leaving the repainting work until the following summer when a better financial position is anticipated. To some extent, this approach may defeat the purpose of the planned interval of work, which is the long-term protection of the premises. The choice of procurement for this type of work will be situation dependent. Term contracting may be appropriate but individual contracts are likely to be more suitable.

Tendering and contractual arrangements

Methods of selecting the contractor and tendering procedures have been discussed in Chapter 9, Procurement, the contents of which are equally applicable to maintenance work.

With maintenance work, particularly that relating to emergency situations, the location of the contracting organisation, relative to the properties to be included within the maintenance contract, should be carefully considered to ensure that logistics are consistent with an adequate response.

The choice of contract will relate to the nature of the work outlined above. The employer may appoint a contractor on a single project basis or on a term basis whereby the agreement will stand for a given period of time.

Project contracts

Individual project work may be let via one of the many arrangements that are discussed elsewhere in this book, depending on the procurement objectives relating to the specific project. These will include both measurement and cost reimbursement-based contracts.

Measured term contracts

A measured term contract may be awarded to cover a number of different buildings. It will usually apply for a specific period of time, depending on the necessity of maintenance standards and the acceptability of the contractor's performance. The contractor will at the outset be offered the maintenance work for various trades. The work when completed will then be paid for using rates from an agreed schedule. This schedule may have been prepared specifically for the project concerned or based on a standard document such as the NSR (National Schedule of Rates) or the BCIS Building Maintenance Price Book. Following the liquidation of Carillion in 2018, the PSA Schedule of Rates is no longer used for public measured term contracts.

Where the client supplies the rates for the work, the contractor is given the opportunity of quoting a percentage addition to or deduction from these rates. The contractor offering the client the most advantageous percentage will usually be awarded the contract. An indication of the amount of work involved over a defined period would therefore seem appropriate for the contractor's assessment of the prices quoted. The JCT publishes a Measured Term Contract (2016) suitable for use when work is to be measured and valued on the basis of an agreed schedule of rates. This contract is also applicable where maintenance work is to be carried out by a single contractor over a specific time and under a single contract (JCT 2016).

Managed contracts

A single management contractor may be appointed to manage the entire maintenance requirements of a client. Some clients might prefer this strategy since it reduces their input significantly and allows the benefits of outsourcing to be more fully achieved. With this approach, individual sections of maintenance work will be suitably packaged, procured and managed on behalf of the client by the management contractor.

Facilities management contract

BSRIA identifies the following facilities management contracts, which can also be used for maintenance services (BSRIA 2016). The CIOB Facilities Management Contract (2008) is a purpose designed standard form for the facilities management services. This contract is intended to focus on facilities management issues and can be used for a range of private and public sector facilities management work including maintenance, cleaning, security, etc. The contract allows the parties to agree the services to be provided and the specification to be met. The NEC4 Term service contract can also be used to appoint the contractor for a period of time to provide a maintenance service. Such a service can be reimbursed using a priced contract with a price list, a target cost with a price list or as a cost reimbursement contract (NEC 2017).

Budget and cost control

To most clients, the single most important justification of maintenance management is one of economics whereby value is improved through the desire to provide an optimum service at the minimum cost. This requires sound cost management skills.

This is a particular strength of the quantity surveying profession incorporating traditional cost management skills, including those of whole life costing considerations, extended to accommodate the needs of maintenance management. Cost control is considered in detail in Chapter 5 of this book; however, it is important to consider its application within the context of the facilities manager. The process of cost management is outlined as follows.

Setting budgets

As with any form of cost prediction, budget setting in facilities management is a difficult task and is accompanied by attendant risks. Many budgets will be established by reference to historical data, i.e. 'How much did it cost last year or last time'? Provided these estimates are adjusted to incorporate changes in circumstance and policy and also account for inflation, this is a sensible approach for many parts of the total maintenance budget. However, it will be of little value where budgets are to include items of major expenditure, for example purchase of new property or equipment.

Budgeting for a large business organisation will require sound organisation and the identification of cost centres, heads of expense and a clear understanding of the channels for the collection and communication of necessary data. The main cost headings will be case specific but will generally include the following.

- Maintenance costs: incorporating buildings' mechanical and electrical components and external works
- Repair costs: although clearly random in occurrence, larger organisations will be able to forecast overall budget allowances from previous data
- Running costs: energy; water; rent and rates; cleaning and security; waste disposal

When researching costs from previous data, more accuracy will be achieved by considering costs over a period, say the preceding four to five years, than merely examining last year's accounts. A one-year view could conceal exceptional circumstances which are likely to distort future budgets if wrongly applied.

As discussed in Chapter 6, NRM3 provide the basis for effective and accurate cost advice, as well as facilitating better control of the combined construction renewal and maintenance costs. For example, it provides information for advising clients on appropriate cashflow requirements for annual budgeting and informing the forecasting of the forward maintenance and renewal plans, which in turn informs the implementation of the maintenance strategy and procurement (Green 2012).

Risk should be accommodated within the overall budget provision determined in response to planned activities and levels of certainty relating to them. NRM3 recommends an allowance is included for 'maintenance risks'. These risks are associated with premature failure of components resulting in corrective maintenance, see Fig. 17.3 (RICS 2021). Element 13.2 of NRM3 provides examples of maintenance risks related to maintenance delivery and lifecycle replacement works. In addition to cost risks, there is a need to consider both schedule risks and risks to the quality of the facility maintained, which may interfere with the output of the core business of the client. The principles and practice of risk management that are discussed in Chapter 8 should be considered in the context of the facilities management service.

Whole life costing

Whole life costing is discussed in Chapter 6 and is of key importance in maintenance management. Although there is less opportunity to fully apply whole life costing principles to existing premises stock, there are many situations where it is a valuable and important technique, for example when the choice between repair and replacement of a major piece of equipment has to be made. The scenario in Fig. 17.4 illustrates the significance of whole life costing in this context.

When applying life cycle costing, selecting an appropriate life span can be problematic as an equipment manufacturer may provide a 2-year guarantee, stating that the equipment will last 8 years, but, if well maintained, perhaps much longer, say 12 to 15 years. Sensitivity analysis should therefore be applied to inform any decision.

Cost control

Costs relating to each of the cost headings referred to above should be monitored at regular intervals, or with single projects as and when the work is being carried out. Once a budget has been prepared, the programme of planned works should be carried out in accordance with the forward maintenance and renewal plans. Actual expenditure needs to be related

Scenario: The refrigeration plant in an existing food processing factory is underperforming and requires attention. Following the visit of the service engineer, the following two options become clear:

- *Option A* Repair the existing plant, now 10 years old, at a cost of £20 000. The engineer is able to guarantee the repair for 5 years; however, the expected life of the remainder of the plant is less certain. 15 years is a normal life expectancy.
- *Option B* Replace the entire plant at a total cost of £150 000. The new equipment, also with an expected life of 15 years, will provide several operating advantages and relative to the existing plant will reduce energy costs by approximately 15%.

The cost considerations to be made are clearly more than capital cost considerations. With reference to Chapter 6 the costs must be discounted to present day values, including all future maintenance and running costs, to allow a fair comparison of the options. In making the decision, advantages and disadvantages which cannot be costed also need to be taken into consideration, for example, possible disruption, reliability, improvement in operation and image.

Fig. 17.4 Example of significance of whole life costing.

to forecast and, where necessary, adjustments in planned expenditure should be made to avoid over-spend or problems, which often occur due to end-of-year budget surfeit.

An annual audit should be carried out to firstly ascertain any changes needed to improve value for money in relation to maintenance expenditure. Secondly to determine the extent maintenance has provided operational benefit. Such an audit will therefore inform budgetary decisions for subsequent maintenance plans (BSI 2020).

Taxation considerations

Since investment in facilities is likely to have a major impact on the tax liability of a business organisation, it is important to highlight the need to understand the potential taxation implications within this context (see also Chapter 6).

For example, with regard to capital allowances, issues that may be a concern to the facilities manager include:

- *The distinction between capital expenditure and revenue expenditure.* Revenue expenditure, for example expenditure on salaries, rent, cleaning costs, repairs, etc. may be deducted from revenue in the assessment of taxable profit. Alternatively, capital expenditure, for example the purchase of new equipment, is subject to a different set of rules established within the prevailing capital allowances arrangements. In terms of tax allowances these are generally much less advantageous.

  The distinction between these two types of expenditure may sometimes be difficult. For instance, repairs to an existing stonework façade will be considered as revenue expenditure, while the cost of its total renewal may become a capital item. In practice, the point at which a repair becomes replacement may often be unclear.
- *The capital allowances system.* This shows recognition of depreciation by prescribing, by statute, a range of varied allowances against expenditure on fixed assets. For example, items of 'plant and machinery' receive a preferential tax allowance and, as such, must be identified and valued separately, incorporating allowances for associated builders' work and fees.

The distinction, for taxation purposes, between what may or may not be allowed as plant and machinery is thus very significant; however, in practice, it is often less than clear.

The facilities manager should also be aware of the implications with regard to both capital gains tax (CGT) and value added tax (VAT). It is recommended that the facilities manager should always consult a tax consultant as taxation regulations can be complex.

Benchmarking to improve value

It may be important and/or necessary to demonstrate that the facilities strategy and organisation is achieving good value. Benchmarking techniques may be used in this respect to recognise areas of poor service and assist in establishing measures required to improve performance. An important element of benchmarking is the identification and agreement of relevant and measurable key performance indicators. With reference to maintenance work, strategic measures could include:

- Total cost of maintenance work per m_2 of building maintained.
- Cost of maintenance work per unit of occupancy (e.g. bed spaces)
- Cost per head of occupiers.

Service quality measures include:

- Number of complaints
- Percentage of emergency responses within the stated response time.

This approach to monitoring value is particularly helpful within large organisations that have sufficiently large portfolios to enable internal benchmarking comparisons to be made. The *RICS Strategic FM Framework* recognises external benchmarking is difficult but indicates that the European Standard BS EN 1522-7 establishes a standard procedure for benchmarking facilities (IFMA and RICS 2018). Benchmarking can be applicable to all areas of facilities management to assist in this measurement the facilities manager could use a computer-aided facilities management system (CAFM).

Sustainability

The business sector is estimated to be responsible for 18% of all greenhouse gases in the UK with carbon dioxide being the most prominent gas (Dept for Business, Energy and Industry Strategy 2022). Non-domestic electricity consumption accounts for 58% of all electricity and 36% of total gas consumption in Great Britain in 2020 (Dept for Business, Energy and Industry Strategy 2021). Therefore, the position of facilities managers, in that they are central to the management of the majority of energy use in buildings, is of great importance in both developing and implementing sustainable policies that support the UK in achieving its 2050 targets to reduce carbon emissions and use less energy to combat the impacts of climate change. Facilities managers are key due to the vital role the management of the facility has on its sustainable performance. The separation of economic management and environmental management is no longer a long-term sustainable option.

In addition to the contribution to environmental policy, the pursuit of sustainability can bring a range of other benefits, some of which support the more immediate business objectives of organisations. These include:

- A reduction in operating costs, for example through the introduction of low energy design. This is also likely to result in an increase in rental incomes as tenants recognise the benefits of energy efficient buildings.
- An improvement in the image of an organisation via adherence to green policies. Depending upon the nature of the core activity of the organisation concerned, green policy is seen to give a positive message to consumers and can be used to improve market position.
- An increase in the quality of the working environment, resulting in an improvement in occupant well-being and improved performance. For example, there are some well-known buildings where the use of natural ventilation and lighting has not only reduced the costs associated with heating, ventilation and cooling systems but is considered to have resulted in an improvement in productivity.

Although there are no valid arguments against the pursuit of green principles, the facilities manager is likely to be faced with the dilemma of balancing the traditional metrics of business success, i.e. cost and revenue, with the need or desire to achieve a high standard of environmental quality. Since the FM is not in an independent position, it is important to recognise the significance of the commitment of top management in implementing environmentally friendly policy.

In practice, the facilities manager will need to oversee the implementation of the corporate environmental strategy at a range of different levels by key actions, including:

- *Identifying the key aspects that might influence environmental performance.* These may seem obvious to most organisations, but it is important to take a wide perspective in determining a strategy that will be most effective. For example, switching the production of manufactured goods to a region of low labour costs may improve short-term profitability but increase the associated transport costs substantially with detrimental effects on the environment.
- *Establishing and communicating environmental goals.* At a macro level, this could be the reduction in associated transport costs referred to in the example above. At the micro level, it may relate to employee use of the private car. For example, one worthwhile goal for an organisation might be the reduction in car journeys when commuting to the workplace. This could then be translated into actions by both positive and negative communication, for example by simultaneously promoting the benefit of car sharing schemes whilst increasing car parking charges. At a more general level, it is critical that individuals adopt an appropriate sustainable lifestyle in the workplace and thus need to be educated in environmentally friendly principles.
- Embracing the philosophy of environmentally friendly design and construction. In terms of building design, it is important to remember that what we are planning today is likely to be with us in 40 to 60 years time and, given the need for conservation, perhaps much beyond. With this in mind, design needs to reflect concerns such as predicted temperature rises, rises in sea level, increased rainfall and more frequent severe weather occurrences. In designing buildings, clients and designers need to look at how long-term energy demand can be reduced, for example through more efficient climate control and power management. Likewise, the use of buildings and urban planning is a major factor in reducing energy waste through achieving reduced transportation costs and emissions. The environmental problems caused by construction are also significant, for example: manufacturing processes and use of plastics cause toxic waste; energy expended in transportation is often unnecessary; waste materials find their way through landfill into rivers; quarrying and mining lead to a degradation of the landscape; the need for timber leads to deforestation and loss of habitat.
- *Ensuring that whole life costs are an intrinsic element of all value judgements.* As discussed elsewhere, there is a need to consider whole life costs in designing buildings, in the knowledge that in the life of a building these will be much greater than the capital cost of construction. Whole life costs should also inform all maintenance decisions.
- *Eliminating waste from an organisation's processes.* Examples of excessive waste within any organisation abound and are well acknowledged but normally unresolved: lighting 20-person rooms occupied by two people; printing documents with less than a 1-day life; wearing jackets to compensate for the chill of an air-conditioning system. In terms of the FM, there is much that can be done to reduce and even eliminate such waste through

reviewing relevant systems throughout an organisation and building incentives into the management structure, e.g. benchmarking interdepartmental energy use.
- *Minimise carbon emissions.* Being aware of the activities within the facility that contribute to the organisation's carbon footprint and taking steps to minimise such emissions to meet operational carbon emissions performance targets.
- *Complying with environmental legislation.* The scope of environmental law is wide and includes all regulations that are designed to control the use of natural resources by humans. Adhering to environmental legislation is, of course, not an option and the FM has a clear responsibility to ensure that relevant law is not transgressed. Whilst such legislation may be driven and derived globally, it is more visible in action locally, for example via policy implemented by planning authorities with regard to land use, or in compliance with building regulations relating to double glazing.

There is a compelling case for the facilities manager to be involved in environmental management. The activities of the organisations they represent need to be managed to ensure that environmental objectives are achieved. Consequently, successful policy and implementation will contribute to an improvement in the well-being of a sustainable built environment.

Facilities management and BIM

Traditionally spreadsheets have been used to store and manipulate data to support facility management functions. Software systems such as Computer Aided Facility Management (CAFM) and Integrated Workplace Management Systems (IWMS) are also used widely in the sector to store, analyse and report date to inform decision making relating to space planning, workplace services, maintenance management, energy management and the organisation's sustainability policies.

The potential integration between 7D BIM and FM technologies enables a data-rich environment, which enhances the efficiencies and effectiveness of FM services. As identified by Wiggins (2021), the use of BIM offers the following benefits:

- Better handover of premises
- Fully populated asset database set for CAFM and IWMS systems
- More accurate space and asset data
- Enhanced post occupancy evaluation
- Data capture to inform future projects

Nicol and Wodynski's research (2016) also identified the application of BIM to support renovation and retrofit planning, maintainability studies, energy analysis and control and safety/emergency management. The ability of the model to store life cycle data (see Chapter 6) together with data such as energy consumption, carbon emissions has the potential to enable performance analysis to support preventive maintenance costing and long-term maintenance plans. Using an up-to-date BIM model equipment and machinery can be classified according to location to better support information storage relating to service agreements and manuals.

The use of industry wide Construction Operation Building Information Exchange (COBie) datasets means information can be exported into FM software systems. COBie provides a data format which will deliver consistent and structured asset information to support post-occupancy decision making and improve the measurement and management of assets (Hamil 2018). The RICS research *Big Data: A new revolution in the UK facilities management sector* (2018a) identified the future potential of the use of applications of Big Data (BDA), Internet of Things (IoT), cloud technology, artificial intelligence (AI) and augmented reality (AR) in transforming maintenance engineering, reducing business risk, maximising equipment performance and enhancing customer-oriented services.

Education and training for the facilities manager

There are several institutions with an interest in facilities management, which not only promote its professional standing but also provide vehicles for formal training and qualification. The International Facility Management Association (IFMA) and RICS have collaborated to both raise the profile of facilities management and provide a suite of credentials for FM professional recognition and development. The RICS recognises certain qualifications from the following organisations to gain AssocRICS status:

- International Facility Management Association (IFMA)
- Institute of Workplace and Facilities Management (IWFM) formerly BIFM
- Institute of Asset Management
- Building Futures Group.

Chartered status can be further gained via the RICS Facilities Management pathway. The RICS also recognised the credentials of both certified facility managers (CFM) and BIFM level 6/7 for direct eligibility status to MRICS.

Facilities management opportunities for the quantity surveyor

Many quantity surveyors are now employed in the sector of facilities management which, relative to the traditional quantity surveyor market, offers great opportunity in that it provides the prospect of being involved beyond the initial procurement of a building. The facilities management market has spread from the USA throughout Europe and Japan and therefore on an international scale is immense.

The quantity surveyor has many of the skills required in the provision of a facilities management service although, when considering the discipline in the wider context, there are clear limitations in existing quantity surveying expertise. This position is also true of other construction professionals, as a considered review of the work of the facilities manager might reveal. There are some aspects of the role that will be better served by designers or contractors, electrical engineers and building surveyors. Likewise, management consultants and accountants are well placed with regard to strategic management consultancy and are beginning to enter the FM market. It seems that the facilities management function may naturally divide into operational and managerial affairs. Without investment in extensive additional human resources, which is likely to be unavailable to all but the larger practices, the majority of opportunities for the quantity surveyor will remain at the operational level.

When considering the service of facilities management, emphasis should be placed on the term management. There is no assertion that the FM role demands the knowledge and ability to apply all of the required skills, any more than a project manager is able to carry out all of the design aspects of a proposed project. However, knowledge of the range of FM functions is important to the adequate fulfilment of the management role.

Consideration of the management and specialist skills required in facilities management as listed below (RICS 2018b) will demonstrate the breath of FM knowledge:

- Asset management
- Big data
- BIM management
- Business alignment
- Client care
- Commercial management
- Contract administration and practice
- Health and safety
- Management (CAFM)
- Maintenance management
- Legal and regulatory compliance
- Lifecycle costing/Occupancy cost analysis
- Outsourcing performance measurement
- Procurement and tendering
- Project finance
- Risk management
- Supply management
- Strategic advice and planning
- Workplace strategy

It is important to note that, whilst the quantity surveying profession may possess some of the skills necessary to provide the facilities management service, the discipline of facilities management should be seen apart. The skills relating to cost management can be outsourced in the same way as can design or engineering requirements. In appointing a facilities manager, clients are seeking managers, not technicians.

References

Atkin, B. and Brooks, A. *Total Facility Management.* 5th Edition. Wiley Blackwell. 2021.

BCIS. *Economic Significance of Maintenance: 2021 Report.* BCIS. 2021.

BSI. *BS ISO 44001:2017 Collaborative Business Relationship Management – Requirements and Frameworks.* British Standards Institution. 2017.

BSI. *BS EN 41011:2018 Facility Management – Vocabulary*. British Standards Institution. 2018.
BSI. *BS 8210:2020 Facilities Maintenance Management – Code of Practice*. British Standards Institution. 2020.
BSRIA. *Maintenance Contracts: A Guide to best Practice for Procurement*. BSRIA. 2016.
BSRIA. *Soft Landings and Business Focussed Maintenance: TG19/2018*. BSRIA. 2018a.
BSRIA. *BG54/2018 Soft Landings Framework 2018: Six Phases for Better Building*. BSIRA. 2018b.
CIBSE. *Environmental Design: CINSE Guide A*. 8th Edition. CIBSE. 2015. Incorporating Jan 2021 amendments.
Dept for Business, Energy and Industry Strategy. *Subnational Electricity and Gas Consumption Statistics*. National Statistics. 2021. Available at https://assets.publishing.service.gov.uk/government/uploads/system/uploads/attachment_data/file/1079141/subnational_electricity_and_gas_consumption_summary_report_2020.pdf (accessed 25/06/2022).
Dept for Business, Energy and Industry Strategy. *UK Greenhouse Gas, Emissions, Final Figures*. National Statistics. 2022. Available at https://assets.publishing.service.gov.uk/government/uploads/system/uploads/attachment_data/file/1051408/2020-final-greenhouse-gas-emissions-statistical-release.pdf (accessed 25/06/2022).
Green, A. A measure of success. *Construction Journal*. RICS. June–July 2012.
Hamil, S. What is COBie? 2018. Available at https://www.thenbs.com/knowledge/what-is-cobie (accessed 26/06/2022).
IFMA and RICS. *Raising the Bar: From Operational Excellence to Strategic Impact in FM*. RICS. 2017.
IFMA and RICS. *Defining FM Excellence*. RICS. 2018. Available at https://www.rics.org/globalassets/rics-website/media/qualify/accreditations/defining-fm-excellence-2018 (accessed 26/06/2022).
Joint Contracts Tribunal (JCT). *Measured Term Contract 2016*. Sweet and Maxwell Thomson Reuters. 2016.
NEC. *NEC4 Term Service Contract*. Thomas Telford Ltd. 2017.
Nicał, A and Wodyński, W. *Enhancing Facility Management through BIM 6D, Procedia Engineering*, 164, 299–306. 2016. Available at https://www.sciencedirect.com/science/article/pii/S1877705816339649 (accessed 26/06/2022).
RICS. *Big Data: A New Revolution in the UK Facilities Management Sector*. RICS. 2018a.
RICS. *Facilities Management Assessment of Professional Competence Pathway*. RICS. 2018b.
RICS. *NRM3- Order of cost Estimating and cost Planning for Building Maintenance Works*. 2nd Edition. RICS Books. 2021.
RICS. *International Building Operation Standard (IBOS): A Framework for Assessing Building Performance*.1st Edition. RICS. 2022.
Wiggins, J. *Facilities Manager's Desk Reference*. 3rd Edition. Wiley Blackwell. 2021.
Wilding, J. *Future-Proof Facilities: Optimizing Indoor Environments*. FMJ. 2022. P18–20. Available at http://fmj.ifma.org/publication/?m=30261&i=745973&p=1&ver=html5 (accessed 12/06/2022).

18

Sustainability in the Built Environment

KEY CONCEPTS
• Environmental and sustainability targets • UN Sustainable Development Goals (SDGs) • Sustainability legislation • Sustainability assessment methods • Whole life carbon assessments

LEARNING OUTCOMES
After reading this chapter you should be able to: • Appreciate the significance of the built environment on climate change. • Appreciate how legislation and regulations relating to sustainability affects construction. • Describe how UN Sustainability Development Goals (SDGs) relate to the construction sector and the role of the quantity surveyor. • Understand, in relation to construction, embodied carbon emissions, operational carbon emissions and whole life carbon emissions and their assessment.

COMPETENCIES
Competencies covered in this chapter: • Sustainability

Introduction

Scientists have identified that human activity has been the dominant cause of increases in climate warming compared to pre-industrial levels of the 1850s–1900 period. The increased levels of greenhouse gas emissions, such as water vapour, carbon dioxide (CO_2) and

Willis's Practice and Procedure for the Quantity Surveyor, Fourteenth Edition.
Allan Ashworth and Catherine Higgs.

methane, derived by the demand for energy, has led to a rise in the average surface temperature of the planet. This has resulted in rising sea levels and acidification of the oceans, a changing cryosphere and extreme weather events (Climate Change Committee 2022). In the United Kingdom, in 2020 net territorial greenhouse gas emissions were estimated to be 405.5 m tonnes carbon dioxide equivalent ($MtCO_2e$) with carbon dioxide making up around 79% of the total (Department for Business, Energy and Industry Strategy 2022a). In 2021 there was an increase of 4.7% in greenhouse gas emissions and 6.3% in CO_2 emissions reflecting post pandemic increases in travel and economic activity (Department for Business, Energy and Industry Strategy 2022b).

As a sector the built environment is a significant contributor to climate change, demanding a more sustainable approach to decision making in relation to the development of land and operation of building assets. Construction and property accounts for globally 40% of energy-related global CO_2 submissions and 46% of all raw materials extracted and in the UK for 60% of all the sawn softwood consumed (Bioregional 2018).

This textbook has, so far, comprehensively explained the role of the quantity surveyor, however, it would not be complete without considering how 'sustainability development' impacts on the services quantity surveyors provide and the method in which they practice.

Sustainable development was defined within the *Our Common Future* (or commonly referred to as the Brundtland Report) as development that meets the needs of the present without compromising the ability of future generations to meet their own needs. (World Commission on Environment and Development 1987). This report was a catalyst for the government's desire to improve the efficiency of the industry and meet their global obligations in addressing global warming.

This chapter provides an overview of sustainability and how quantity surveyors need to be aware of sustainability drivers and how they relate to the client's key sustainability objectives. Consequently, there is a need to have an understanding of a project's sustainability outcomes and whole life carbon impact and how this knowledge is embedded within quantity surveying practice.

Sustainable development

A global perspective

The following provides an overview of the chronological steps that have led to the adoption of a sustainable development agenda within the United Kingdom. In response to global environmental concerns, the UK government was one of first countries to develop and publish a national strategy on the environment and development, *This Common Inheritance* (HMSO 1990). In June 1992, the United Kingdom signed the United Nations Framework Convention Climate Change (UNFCCC), commonly known as the Earth Summit, committing to reduce atmospheric concentrations of greenhouse gases. In 1994, after the annual Earth Summit, the government published *Sustainable Development: The UK Strategy* (HMSO 1994) and in 1997 signed the Kyoto Protocol. This committed the United Kingdom, with other Annex 1 countries, to reduce greenhouse gases, measured as carbon dioxide equivalents, by 5.2% between 1998 and 2002 compared to the base year of 1990.

The Paris Agreement, came into force in 2016, replacing the Kyoto Protocol. It set long-term sustainable goals which included reducing global greenhouse gas emissions to limit global warming to below 2 °C with an aim for 1.5 °C (United Nations Climate Change 2022).

Since 1995 global events have been held to review and maintain the ambitions set out in both the Kyoto Protocol and the Paris Agreement. These global sessions are referred to as COPs, Conference of the Parties of the UNFCCC. The 27th session, COP27 was held in the Arab Republic of Egypt in November 2022.

UN sustainability development goals

In 2015, the United Nations agreed 17 Sustainable Development Goals (SDGs); the details of which were set out in the *2030 Agenda for Sustainable Development*, 'The Sustainable Development Goals are the blueprint to achieve a better and more sustainable future for all. They address the global challenges we face, including poverty, inequality, climate change, environmental degradation, peace, and justice'. (United Nations 2022a). Figure 18.1 provides a list of the 17 SDGs. Within each of these goals are goal-specific strategic objects. Information on the 17 goals and 169 targets can be found on the United Nations website (www.sdgs.un.org/goals).

The United Nations produces an annual report reviewing, analysing and evaluating the global progress against each of the SDGs. The *Sustainability Development Report 2022*

United Nations Sustainability Development Goals
SDG 1 End poverty
SDG 2 Zero Hunger
SDG 3 Good health and wellbeing
SDG 4 Quality education
SDG 5 Gender equality
SDG 6 Clean water and sanitation
SDG 7 Affordable and clean energy
SDG 8 Decent work and economic growth
SDG 9 Industry, innovation and infrastructure
SDG 10 Reduced inequality
SDG 11 Sustainable cities and communities
SDG 12 Responsible consumption and production
SDG 13 Climate action
SDG 14 Life below water
SDG 15 Life on land
SDG 16 Peace, justice and strong institutions
SDG 17 Partnerships for the goals

Fig. 18.1 United nations sustainable development goals.

provides evidence of the negative impact of the COVID-19 pandemic and the Ukraine war on such progress (United Nations 2022b).

The construction sector perspective

The adoption of sustainable development and implementation of both national and local strategies led to the production of strategies for individual sectors of industry. Recognising that the construction industry has a significant impact on the natural environment in relation to the use of natural resources, and informed by Egan's *Rethinking Construction* (Egan 1998), the government published its first sustainable strategy addressed to the industry, *Building a Better Quality of Life: A Strategy for Sustainable Construction* (Department of the Environment, Transport and Regions 2000). Though it is not the intention to outline all the strategies, the key theme areas of this document outlined the practical ways by which the industry, at the time, could contribute to the reduction of greenhouse gas emissions as follows:

- *Reuse existing built assets.* Refurbishment and renovation should be considered as viable alternatives to new construction in meeting clients' functional needs.
- *Design for minimum waste.* The industry, which contributes 17% of the total waste stream, should seek to eliminate waste through the whole project lifecycle from extraction to disposal. Consideration should be given to lean design and improved construction processes that allow more efficient use of material resources and recycled materials.
- *Lean construction.* The industry should seek continuous improvement, with a strong client focus and consideration of value for money.
- *Minimise energy in construction.* Adoption of energy-efficient operational practices.
- *Minimise energy in use.* Recognising that energy use within the house contributes to 27% of all carbon emissions, design teams should consider incorporating energy-efficient solutions into design as well as involving energy produced from renewable sources such as solar panels, wind turbines, wave power and ground heat pumps.
- *Reducing pollution.* The industry should seek to reduce waste materials, vehicle emissions, noise and release of contaminants into the atmosphere, ground and water.
- *Preserving and enhancing biodiversity.* Consideration of opportunities both to protect existing and provide new habitats, from extracting of raw materials to landscaping of completed developments.
- *Conservation of water resources.* Project teams to design building service systems that promote increased water efficiency and general water conservation within the built environment.

To measure the response of the industry an eleventh issue was added to the construction industry's key performance indicators to benchmark sustainability at project level, addressing waste, energy, water, ecology, transport and recycling.

Securing the Future (HM Government 2005) aligned the environment, social progress and economy into policy-making procedures. Based on a shared recognition of the need to deliver a radical change in the sustainability of the construction industry, the government and industry published a joint initiative called *Strategy for Sustainable Construction* (HM Government 2008). The aim of the strategy was to provide clarity around the policy framework at that time, as well

Means	Ends
Procurement	Climate change mitigation
Design	Climate change adaption
Innovation	Water
People	Biodiversity
Better regulation	Waste
	Materials

Fig. 18.2 Overarching themes (*Source:* Adapted from HM Government 2008).

as give a signal to future policy direction. The overarching themes of the policy are identified in Fig. 18.2. The 'means' identify the setting of objectives and targets based on the practical themes identified above, whilst the 'ends' identify the potential environmental and conservation benefits as an overall reduction of carbon emissions, waste and water consumption. In June 2011, the government published the *Low Carbon Construction Action Plan* in response to the Low Carbon Construction Innovation and Growth Team Report, in which a series of recommendations were made to support the transition to a low carbon built environment. Included within the recommendations was that industry and government should jointly showcase the deliverance of London 2012 as an example of how to implement plans for a low carbon development, embracing design and engineering and lean construction through an integrated supply chain. Indeed, the use of sustainable procurement ensured value for money and lower operational costs of the project whilst protecting the environment and bringing wider societal benefits (DEFRA 2013). Published in the same year and developed jointly by government and industry the *Construction 2025 Strategy* sets out the vision for success in the sector (HM Government 2013). Together with smart technologies and overseas trade, it identified green construction as a growth area within the sector. It set a target of a 50% reduction in greenhouse gases in the build environment and aspirations that the industry would lead the world in low-carbon and green construction exports.

The *Construction Sector Deal* reinforced and provided a framework to deliver these targets, identifying three strategic areas: the use of digital techniques, offsite manufacturing and whole life asset performance (HM Government 2018).

Data, technology, and improved delivery models are perceived as key in supporting a sustainable built environment. The *RoadMap to 2030* introduces the 'Built Environment Model' and the concept of systems thinking. Decisions, about the built environment, therefore, must recognise the interconnectedness of the services the built environment seeks to support together with the natural environment. Decisions should also be made not just against project outcomes but on societal outcomes as articulated by the 17 SDGs (IPA 2021). The *Construction Playbook* builds on the *RoadMap* by focusing on strategies that promotes both a reduction in carbon emissions linked to buildings, infrastructure and within the construction supply chain, as well as promoting social value (HM Government 2022).

SDGs: the sector perspective

As outlined in the Introduction to this chapter, the built environment sector makes a significant impact on the use of the world's resources and is a major contributor to

Goals	SDGs
Advanced:Key SDGs for the built environment	3,7,11,12,13,17
Progressive:SDGs impacted by the built environment	6,8,9,10,15
Growth:with contributory role for the built environment	1,2,4,5,14,16

Fig. 18.3 Tiered SDGs (*Source:* Adapted from (World Green Building Council 2020).

greenhouse gas emissions. The United Nations SDGs as detailed in Fig. 18.1 provides a roadmap to sustainable development and is increasing being used as a framework to align design, construction and maintenance decisions.

The WorldGBC global network has co-created three targets for the Built Environment for 2050:

- *Climate Action* (SDGs 7,9,13): Total decarbonisation of the built environment.
- *Health and Wellbeing* (SDGs 3,6,11): A built environment that delivers healthy, equitable and resilient buildings, communities and cities.
- *Resources and Circularity* (SDGs 8,12,15): A built environment that supports the regeneration of resources and natural systems, providing socio-economic benefit through a thriving circular economy. (World Green Building Council 2020)

In addition, they have recognised that the built environment has differences in influence in meeting each of the SDGs and have identified a tiered goal approach to prioritise focus (Fig. 18.3).

Goubran (2019) in an analysis of sustainable literature also proposed both a direct and indirect dependence on construction and property activities in meeting the targets set in the *2030 Agenda*. Of the 169 targets, 29 were identified as having direct impact and a further 43 indirectly. The research identified the sector having the most significant impact on achieving SDG goals 7 (Affordable and clean energy), 11 (Sustainable cities and communities) and 6 (Clean water and sanitation). In reflecting the volume of raw material the industry extracts for the supply of materials and products, *Build a better future with the Sustainable Development Goals* highlights the additional significance of SDG 12 – Responsible consumption and production (Bioregional 2018). This document also further identifies potential responses organisations working in the sector can make to address 56 targets of the 2030 Agenda. The responses suggested relating to a company's downstream and upstream operations as well as its internal business operations. As discussed in Chapter 3, businesses within the quantity surveying sector have committed to *The Ten Principles* of the United Nations Global Compact. These principles address human rights, labour, the environment, and anti-corruption.

At project level, in terms of the design, construction and operational decisions made by built environment professionals in considering an individual building's impact on both the environment and the quality of life of occupants, a building with 'green' features will have positive impacts on SDGs – 3,7,8,9,11,12,13,15 and 17 (World Green Building Council 2022).

The *RIBA Plan of Work 2020* incorporates a sustainability strategy that provides a framework to support the project team to deliver buildings that both meet clients' requirements and address climate needs (RIBA 2020). The sustainability strategy references

RIBA Sustainable Outcome	Mapped to UN SDG
Good health and wellbeing	3 Good health and well-being
Sustainable water cycle	6 Clean water and sanitation
Net zero operational carbon emissions	7 Affordable and clean energy
Sustainable life cycle cost	8 Decent work and economic growth
Sustainable connectivity and transport	9 Industry, innovation and infrastructure
Sustainable communities and social value	11 Sustainable cities and communities
Net zero embodied carbon emissions	12 Responsible consumption and production
Whole life carbon emissions	13 Climate action
Sustainable land use and ecology	15 Life on land

Fig. 18.4 RIBA sustainable outcomes and the UN SDGs (*Source:* Adapted from Table 1 RIBA 2019).

the RIBA 'eight sustainable outcomes' that align with the UN SDGs (Fig. 18.4). In their sustainability survey, the NBS identified that 14% of projects used sustainable outcomes all of the time and 25% used such outcomes 25% of the time (NBS 2022). The role of the quantity surveyor in supporting these sustainable outcomes with reference to the RIBA Plan of Work is outlined later in the chapter.

Legislation

Against this background, it is clear to see the significant impact which a sustainable construction industry will have on the UK's sustainability baseline. The government has recognised this and has enacted several fiscal policies, regulations and voluntary compliance schemes; those that quantity surveyors should be aware of are Building Regulations Part L, the Energy Act 2011, the Climate Change Act 2008 and the Environment Act 2021. Such legislation makes the targets set within the previously mentioned strategies legally binding.

Part L Conservation of fuel and power, revised in June 2022 sets building performance targets in relation to CO_2 emissions, measures to promote energy efficiencies and the requirements for SAP (Standard Assessment Procedures) calculations to assess energy performance. The carbon emission targets set within Part L are an interim measure and will align with the *Future Home Standard* and *Future Building Standard* due in 2025. The Energy Act 2011 sets minimum performance standards for buildings with reference to Minimum Energy Efficiency Standard (MEES).

Within the Climate Change Act 2008, the government has set targets for a 100% reduction of greenhouse gases measured in carbon dioxide (CO_2) equivalent emissions by 2050 relative to 1990. The Climate Change Committee is a statutory body that advises the government on progress in meeting the targets. On its recommendations, the Act provides a framework in which to do this by setting out legally binding carbon budgets; a cap on the amount of greenhouse gases emitted in the United Kingdom over a five-year period. In 2021, the UK Government published its *Net Zero Strategy*, outlining plans for the Fourth, Fifth and Sixth Carbon Budgets. (Devolved administrations, Wales and Scotland have the autonomy to produce their own programmes and policies.) Strategies outlined include the decarbonisation of the way we heat and power our buildings. The Environment Act 2021 is the UK's framework of environmental protection. It provides legislation to improve air quality, restore natural habitats, increase biodiversity, reduce waste and make more effective and efficient use of resources (Gov.uk 2022).

Assessment methods

Quantity surveyors are increasingly required to refer to industry assessment tools in terms of the cost significance of ratings and their influence in value engineering exercises, as well as supporting certification applications in relation to life cycle costing. Whilst these tools are largely environmentally biased, it is important that the project team consider other aspects of sustainability in their decision making, such as the wider social and economic context. The following are the tools most used. Diagram 6 in the *RIBA Sustainable Outcome Guide* provides an overview of others, such as Greenstar and NABERS. Certification may be required as part of the planning process, however where voluntary they provide a benchmark against similar building assets and demonstrate performance against approved sustainability metrics.

BREEAM

The aim of BREEAM 2018 is to mitigate the lifecycle impacts of buildings on the environment in a robust and auditable manner. This requires project teams to view environmental concerns as an integral component of design and construction decision making. Figure 18.5 details the BREEAM rating system. Currently all government major projects are required to meet a minimum rating of outstanding for new build and excellent for refurbishment projects. The BREEAM scheme is linked to the RIBA Plan of Work in which the interim certification application is made at Stage 3, Spatial Consideration and an Interim Design Stage BREEAM certificate is awarded prior to Stage 5, Manufacturing and Construction. The Final (post construction) Certificate is awarded during handover (Stage 6). Globally more than 500,000 projects have achieved a BREEAM rating, with over 12,945 projects in the United Kingdom (Gulacsy 2020).

With particular reference to the quantity surveyor, *Management Section 02* of the BREEAM 2018 document recognises and encourages lifecycle costing and service planning

BREEAM Rating	Credit Score	Benchmark	Descriptor
Outstanding	≥85%	Less than the top 1 %	Innovator
Excellent	≥70%	Top 10%	Best practice
Very Good	≥55%	Top 25%	Advanced good practice
Good	≥45%	Top 50%	Intermediate good practice
Pass	≥30%	Top 75%	Standard good practice
Unclassified	≤30%	Non-compliant	

Fig. 18.5 BREEAM ratings benchmarks (*Source:* Adapted from BRE 2018).

to improve the design and specification of materials and components through the whole life of the building. More detailed discussion of compliance with Section 02 can be found in Chapter 6 (Life Cycle Costing).

Materials Section 01 recognises and encourages the use of low environmental impact assessment, including embodied carbon, over the life of the building. Embodied carbon emissions will be discussed in more detail later in this chapter.

As part of their sustainability departments, several quantity surveying organisations offer BREEAM assessor services.

LEED

LEED (Leadership in Energy and Environmental Design) is, like BREEAM, an environmental assessment rating system. Established in 1998 by the US Green Building Council, LEED has been used on 73,000 projects worldwide and 95 in the UK (Gulacsy 2020). Due to its US roots, it is most likely to be used by US global organisations operating in the United Kingdom.

The main aim of LEED is to encourage and accelerate global adoption of sustainable buildings, operations and neighbourhood developments. The main differences between the two rating systems are the process of certification and methods of assessments; LEED uses a percentage threshold compared to BREEAM's quantitative method and awards a rating of platinum, gold, silver or certified.

Health and well-being certification

The WELL Building Standard version 2 is exclusively concerned with occupier health and well-being. WELL aims to advance human health through design interventions and operational protocols and awards a rating of platinum, gold, silver or bronze (Wellcertified 2022). WELL unlike BREEAM and LEED addresses social sustainability and for this reason, a WELL rating can be combined with both. The *Assessing Health and Wellbeing in Buildings* publication provides guidance for projects wishing to obtain both a certified BREEAM rating and WELL Certification. Fitwel is also a building certification system that sets standards for design and operational strategies to promote health and well-being.

CEEQUAL

CEEQUAL is a framework and sustainability assessment for civil engineering, infrastructure, refurbishment of assets and work in the public realm. It holistically considers the economic, environmental and social performance of a project's specification, design and construction. The whole life carbon, resilience and security, ecological enhancements and the circular economy are also considered within the rating award.

Net zero

As previously discussed within the Climate Change Act 2008 (amended in 2019), the government has legally set targets for an 100% reduction of greenhouse gases, measured in carbon dioxide (CO_2) equivalent emissions by 2050 relative to 1990. As part of the sixth carbon budget, a target of reducing emissions by 78% by 2035 has been set (UK Parliament 2021).

The government has also committed the UK under Article 4 of the Paris Agreement to reducing economy-wide greenhouse gas emissions by at least 68% by 2030 (Department for Building Energy & Industry Strategy 2022). The *Net Zero Strategy: Build Back Greener* sets out government's strategy to meet the carbon budgets and a vision for a decarbonised economy in 2050 (HM Government 2021) underpinned by the sector-specific *Heat and Buildings* strategy. With approximately 25% of the UK's total greenhouse gas emissions being attributable to the built environment, the carbon emissions associated with the sector must be significantly and rapidly reduced to meet the net zero targets. The Government has identified the following areas to improve sustainability within the sector:

- Developing and making mandatory standards, methodology and reporting frameworks for embodied and whole life carbon.
- Addressing barriers for greater use of low-carbon building materials and products. (House of Commons 2022)

The *Net Zero Carbon Buildings Framework Definition* sets out high-level principles for achieving net zero carbon for construction and for operational energy (UK Green Building Council 2019). In response to the need within the UK real estate sector for a robust means of verifying buildings as net zero carbon, a UK *Net Zero Carbon Standard* is in development (BBP 2022).

Carbon definitions

This shift from energy to carbon as a unit of measurement for embodied impact reflects the commitment to addressing global warming through a reduction in CO_2 emissions. Carbon emissions are not solely focused on carbon dioxide CO_2, but any greenhouse gas that has global warming potential, which includes methane (CH_4), nitrous oxide (N_2O), hydrofluorocarbons (HFCs), perfluorocarbons (PFCs) and sulphur hexafluoride (SF_6) (WLCN et al. 2021).

The Office for National Statistics defines net zero as meaning 'the UK's total greenhouse gas (GHG) emissions would be equal to or less than the emissions the UK removed from the environment. This can be achieved by a combination of emission reduction and emission removal' (ONS 2019). Emission removal refers to approaches such as planting trees or bioenergy with carbon capture or storage.

Carbon emissions are associated with the whole life of a building, from the inception to the end of a building's life, meaning its demolition and recycling and disposing of materials. Figure 18.6 shows the emissions during the whole life carbon cycle of a building. The *Improving Consistency in Whole Life Carbon Assessments and Reporting: Carbon Definitions for the Built Environment, Building and Infrastructure* provides sector-accepted definitions. These definitions refer to BS EN 15978: 2011 and uses the life cycle modular structure (A1-A5, B1-B5, C1-C5 and D) as shown in the diagram.

The RICS *Whole life carbon assessment for the built environment* professional statement cites typical breakdowns of whole life carbon emissions for different building types, highlighting the relative weight of operational and embodied carbon. Taken over a 60-year life, the embodied carbon emissions contribute between 67 and 78% of the whole life carbon assessment (RICS 2017).

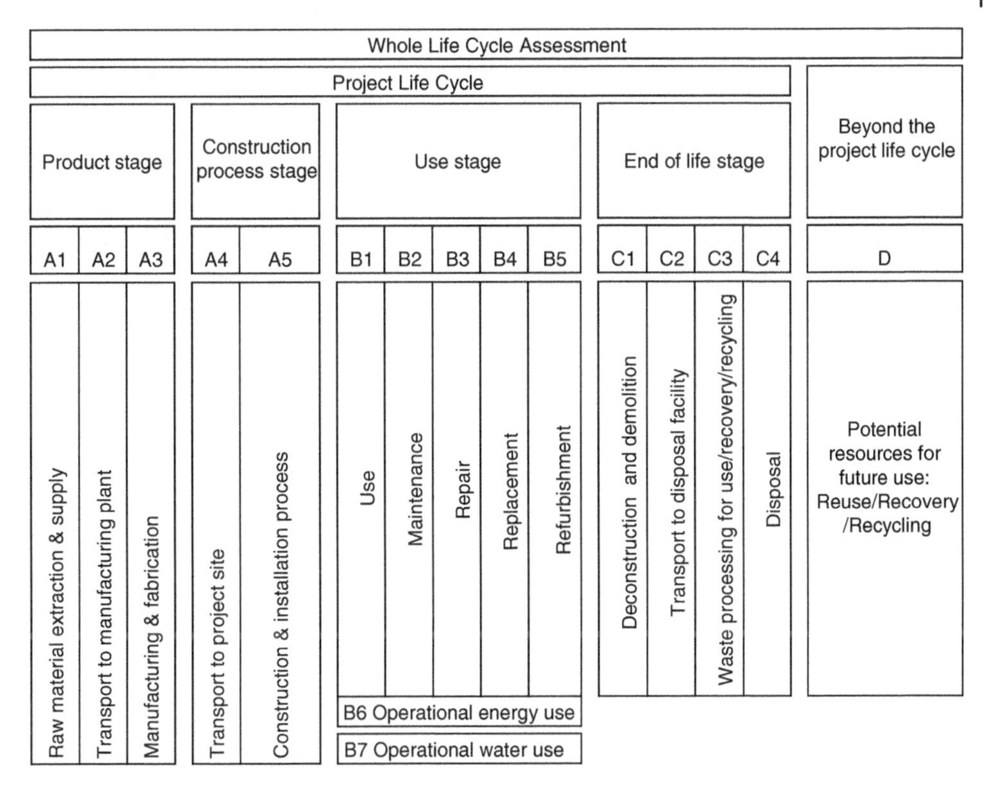

Fig. 18.6 Whole life carbon cycle of a building (*Source:* Adapted from British Standards Institution 2012).

Embodied carbon

Embodied carbon emissions are those emissions associated with the sourcing of materials, transportation, fabrication, construction of the building, its repair and maintenance and at the end of a building life, its demolition and disposal. Embodied carbon emissions up to practical completion are referred to as upfront carbon. Upfront carbon is a significant proportion of the whole life carbon cycle and will be impacted by material selection and building operations. Quantity surveyors therefore need to develop a good understanding of low carbon building materials, how to promote effective material usage, increased use of modern methods of construction, effective material transportation, consideration of onsite plant and reuse of reclaimed materials. This knowledge is important in optimising the relationship between embodied and operational carbon emissions; whilst the choice of a product may offer energy efficiencies, the embodied emissions associated with extraction, manufacturer and installation must not be overlooked.

Operational carbon

Operational carbon emissions refer to all emissions associated with energy use in the operation of a building. Whilst there has been a reduction in operational emissions since 2010, this is attributed to decarbonisation of the electricity grid rather than improvements in the energy efficiency of buildings (UK Green Building Council 2021). Operational carbon emissions are classified as either regulated or unregulated. 'Regulated' refers to carbon from energy uses regulated by legislation, such as Part L and the proposed Future Building Standard. Such energy uses include heating, cooling, ventilation, lighting, and hot water. Unregulated emissions are those associated with the use of appliances, computers, IT servers etc. Quantity surveyors therefore need to develop a good understanding of the cost influence of energy efficient design to meet the requirements of 2050 in value engineering exercises, to avoid the need to retrofit.

Sequestered carbon

Carbon sequestration is the natural or artificial process by which carbon dioxide is removed from the atmosphere and held in solid or liquid form (LETI 2020). Within the context of building, an example would be using timber components in construction as it will retain the carbon. At the end of the building life, the natural biodegrading of the timber will release the carbon and regenerate living systems.

Whole life carbon

Whole life carbon as shown in Fig. 18.6 is the combined total of embodied and operational emissions over the life cycle of a building. It is important that the quantity surveyor considers both, rather than in isolation to ensure decision making that will align with sustainability outcomes.

Circular economy

Elimination of waste at the end of a building life by either landfill or incineration leads to further carbon emissions. The circular economy concept proposes the consideration of the building component 'disposal' during the development of the project strategic brief to reduce resource consumption. The circular economy principles, driven by design, are to:

- Eliminate waste and pollution
- Circulate products and materials (at their highest value)
- Regenerate nature. (Ellen Macarthur Foundation 2022)

The UKGBC *Insights on how circular economy principles can impact carbon and value* provides guidance on the circular economy principles for construction (Fig. 18.7). Using circular principles will lead to lower waste production, lower use of virgin materials and increased material efficiency, resulting in reduced energy consumption and lower CO_2 emissions. UKGBC Solutions Library provides project case studies outlining sustainable objectives, approaches and solutions and a reflection on lessons learnt (www.ukgbc.org/solutions). For circular principles to be successful, it requires both a whole life building approach and the client's strategic brief to have a clear vision for applying circular economy principles (UKGBC 2019a).

Quantity surveyors therefore need to consider the use of circular principles in cost advice during the design stages of a project. In relation to minimising the impact of waste the RICS *Sustainability Report 2021* reported that 55% of respondents to their survey noted an increased demand for recycled and reusable materials compared to the previous year, with 13% deeming the demand to be significant (RICS 2021).

Circular Economy Principle	
Maximise reuse	Reuse the existing building asset for another business use Recover materials and products on site or from another site Share materials or products for onward reuse
Design for optimisation	When making design decisions consider: • Longevity • Flexibility • Adaptability assembly, disassembly • and recoverability
Use standardisation	Use standardised products and modular designs to maximise reuse
Products as a service	Purchase products as a service transaction.
Minimise impact and waste	Use low impact new materials Use recycled content or secondary materials Design out waste Reduce construction impacts

Fig. 18.7 Circular economy principles for construction (*Source:* Adapted from Fig. 2 UK Green Building Council 2022).

Measuring embodied carbon emissions drivers

At the time of authoring this textbook in October 2022, there is no mandatory requirement for assessing whole life carbon (WLC). Whilst regulated operational emissions are addressed through building regulations and planning requirements there is an increasing demand for the built environment industry to consider embodied carbon emissions. Indeed, within the government's *Building to net zero: costing carbon in construction* it is noted that the single most significant policy the Government could introduce is a mandatory requirement, set within the building regulations and planning system, to undertake whole life carbon assessments for buildings (House of Commons 2022). Such action would align the United Kingdom with precedents set within other country's national policies, such as The Netherlands, France, Denmark, and Sweden. In anticipation of the United Kingdom making WLC assessments mandatory, many local authorities, including the Greater London Authority's new London Plan, have introduced a pre-planning application stage, during stage 2/3 of the RIBA and a post-construction handover assessment.

Whole life carbon assessments

A better understanding of whole life carbon across the building life cycle will enable benchmarking across the sector and in doing so create a set of targets that has a trajectory to achieve carbon targets. The RICS Professional Standard *Whole life carbon assessment for the built environment* has wide support within the industry. Bodies such as the RIBA, LETI, ACAN and others have recommended that the RICS methodology be adopted as the UK industry approach, though it is noted that it does require updating to address issues associated with the lack of consistent data sources and measurement metrics (House of Commons 2022). This professional statement provides whole life carbon (WLC) assessment requirements and guidance in accordance with BS EN 15978 life cycle stages (Fig. 18.6).

Based on BCIS element categories, the RICS methodology specifies all items listed in the project's Bill of Quantities, cost plan or as identified in other design information (drawings, specifications etc.) should be included in WLC assessment (RICS 2017). These categories are identified in Fig. 18.8 and represent elements with potentially significant carbon values. Using these BCIS groups facilitates continuity, consistency and interoperability with other project information produced by the quantity surveyor.

The professional standard provides guidance on carbon measurement for each stage of the project life cycle with reference to the modules identified in BS EN 15978. Any quantities used within the assessment must be clearly stated. Due to the availability of project information during the design and project stages, the standard identifies an order of preference of sources to be used for quantities within the WLC assessment; these are (in priority order);

- Materials delivery records
- BIM model

	Building Part/ Element Group	Building Element
	Demolition	01 Toxic/Hazardous/Contaminated Material treatment 0.2 Major Demolition Works
0	Facilitating works	0.3 & 0.5 Temporary/Enabling Works 0.4 Specialist groundworks
1	Substructure	1.1 Substructure
2	Superstructure	2.1 Frame 2.2 Upper floors incl. balconies 2.3 Roof 2.4 Stairs and ramps 2.5 External Walls 2.6 Windows and External Doors 2.7 Internal Walls and Partitions 2.8 Internal Doors
3	Finishes	3.1 Wall finishes 3.2 Floor finishes 3.3 Ceiling finishes
4	Fittings, furnishings and equipment (FF&E)	4.1 Fittings, Furnishings & Equipment
5	Building services/MEP	5.1–5.14 Services
6	Prefabricated Buildings and Building Units	6.1 Prefabricated Buildings and Building Units
7	Work to Existing Building	7.1 Minor Demolition and Alteration Works
8	External works	8.1 Site preparation works 8.2 Roads, Paths, Pavings and Surfacings 8.3 Soft landscaping, Planting and Irrigation Systems 8.4 Fencing, Railings and Walls 8.5 External fixtures 8.6 External drainage 8.7 External Services 8.8 Minor Building Works and Ancillary Buildings

Fig. 18.8 Building element groups to be considered in a WLC assessment (*Source:* RICS 2017).

- Bill of quantities or cost plan
- Estimations from contractors' drawings.

The standard also provides guidance on acceptable sources of carbon data for materials and products. The environmental information for construction products and their application is stated within its Environmental Product Declaration (EPD). At the time of authoring this text, the BRE is developing the Built Environment Carbon Database (BECD), which will become the lead source for the construction industry to measure and reduce carbon. Just as a quantity surveyor makes use of experience and indices to make regional adjustments for costing advice, in assessing the environmental performance of a product the relevance of the local context – the nation, region and site in which the building is located – needs to be considered.

Carbon results should be reported in $kgCO_2e$ (or appropriate metric multiples thereof) and normalised (in $kgCO_2e$/unit of measurement, e.g. m^2). The RICS *Results reporting template* detailed in table 13 of the professional standard has become the standard for reporting the project carbon results. It is also the format adopted within the many carbon measurement software tools available in the market. The 2017 edition of the professional standard is due to be updated to align with the International Cost Management Standard (ICMS3). Whilst the ICMS 3 does not cover whole life costs and whole life carbon emissions, it is a high-level international standard providing the reporting format for carbon metrics and in doing so provides a mechanism for quantity surveyors to present both cost and carbon calculations.

Quantity surveying and sustainability

As a member of the project team, the quantity surveyor needs to have a deep understanding of sustainability drivers and how they relate to both the client's key sustainability objectives and the wider sector sustainability targets discussed in this chapter. Whilst in the past there might have been a tendency to relate solely cost advice to a building, it is now important that the quantity surveyor adopts a wider outlook and provides advice mindful of the impact of the development on the environment. The definition of sustainability is broad, so to participate fully in decision making the surveyor needs to be aware of aspects such as ecological concerns, corporate responsibility, waste management and other issues associated with sustainability development goals. The quantity surveyor must competently identify the cost significance of decisions on the whole life cost and whole life carbon cost of the project and impacts on other environmental measures during all stages of the project (Fig. 18.9). If the client is seeking BREEAM, LEED or WELL, the quantity surveyor must be conversant with the content of the reports required for the scheme, relating to their professional contribution and that of the contractor.

As industry databases improve and more benchmark data is provided, the quantity surveyor's greater understanding of the relationship of cost and carbon is key to ensure, in collaboration with other built environment professionals, that the design and construction of projects will meet future environmental targets. To support design decision making, quantity surveyors need to provide cost information on alternative sustainability technology; they therefore need a good understanding of alternative energy solutions, including solar technology, ground source heat pumps, wind power, biomass generators etc. An appreciation of the payback periods of alternative solutions will also be useful. Thc quantity surveyor should also understand cost-influencing factors and sustainable implications for low environmental impact materials, reuse and recycling of materials, whole life assessments of products and the

Stage	RIBA Plan of Work	Quantity Surveyors Role
0	Strategic Definition	Develop high level whole life cost plan incorporating the value of sustainability outcomes.
1	Preparation and Briefing	Produce order of cost estimates with additional emissions appraisals.
2	Concept Design	Develop formal cost plan 1 including sustainability outcomes.
3	Spatial Coordination	Assess relative impacts of design options on whole life cost and whole life carbon assessment.
4	Technical Design	Consider the choice of specification of sustainable materials and products on life cycle cost and carbon assessment. Produce pricing documents that define environmental constraints, requirements, and compliance. Ensure tendering process evaluates sustainability outcomes.
5	Manufacturing and Construction	Review impact of the construction process on carbon assessments. Evaluate variations in the context of whole life cost and carbon assessment and sustainability outcomes.

Fig. 18.9 QS role and sustainable project outcomes.

impact on carbon emissions. In any decision making that affects cost, whether during value engineering exercises or variations during the project the quantity surveyor needs to ensure with any change, the project still aligns to its sustainability outcomes.

References

BBP. *UK Net Zero Carbon Buildings Standard*. 2022. Available at https://www.nzcbuildings.co.uk/_files/ugd/6ea7ba_2bd05d6f5d484cc999108e475a9d8c9c.pdf (accessed 08/10/2022).

Bioregional. *Build a Better Future with the Sustainable Development Goals: A Practical Guide for Construction and Property Companies*. Bioregional. 2018. Available at https://www.bioregional.com/resources/build-a-better-future-the-built-environment-and-the-sustainable-development-goals (accessed 01/10/2022).

BRE. *BREEAM UK New Construction: Nondomestic Building (England)*. BRE Global Ltd. 2018.

British Standards Institution. BS EN 15978:2011 *Sustainability of Construction Works – Assessment of Environmental Performance of Buildings – Calculation Method*. The British Standards Institution Ltd. 2012

Climate Change Committee. *What is Causing Climate Change*. 2022. Available at https://www.theccc.org.uk/what-is-climate-change/what-is-causing-climate-change2/ (accessed 01/10/2022).

Department for Building Energy & Industry Strategy. *The UK's Nationally Determined Contribution Communicated to the UNFCCC (Updated September 2022)*. 2022. Available at https://www.gov.uk/government/publications/the-uks-nationally-determined-contribution-communication-to-the-unfccc (accessed 04/10/2022).

Department for Business Energy & Industry Strategy. *Final UK Greenhouse Gas Emissions National Statistics: 1990 to 2020*. GOV.UK. 2022a. Available at https://www.gov.uk/government/statistics/final-uk-greenhouse-gas-emissions-national-statistics-1990-to-2020 (accessed 01/10/2022).

Department for Business Energy & Industry Strategy. *2021 UK Greenhouse Gas Emissions: Provisional Figures – Statistical Release*. GOV.UK. 2022b. Available at https://assets.publishing.service.gov.uk/government/uploads/system/uploads/attachment_data/file/1064923/2021-provisional-emissions-statistics-report.pdf (accessed 01/10/2022).

Department for Environment Food & Rural Affairs. *London 2012 Olympic and Paralympic Games the Legacy: Sustainable Procurement for Construction Projects*. Crown. 2013. Available at https://assets.publishing.service.gov.uk/government/uploads/system/uploads/attachment_data/file/224038/pb13977-sustainable-procurement-construction.PDF (accessed 25/09/2022).

Department of the Environment, Transport and the Regions. *Building a Better Quality of Life: A Strategy for More Sustainable Construction*. 2000.

Egan, J. Rethinking construction: the report of the construction task force", Department of Trade and Industry, UK. 1998.

Ellen Macarthur Foundation. *What is a Circular Economy?* 2022. Available at https://ellenmacarthurfoundation.org/topics/circular-economy-introduction/overview (accessed 05/10/2022).

Goubran, S. On the role of construction in achieving the SDGs. *Journal of Sustainable Research*. Virtual issue. 2019. Available at https://sustainability.hapres.com/htmls/JSR_1126_Detail.html#sec11 (accessed 01/10/2022).

Gov.uk. *World Leading Environment Act Becomes Law*. 2022. Available at https://www.gov.uk/government/news/world-leading-environment-act-becomes-law (accessed 02/10/2022).

Gulacsy, E. *BREEAM, LEED or WELL: The Interest in Building Assessment Methods Keeps Growing*. BSRIA Ltd. 2020.

HM Government. *Securing the Future: Delivering UK Sustainable Development Strategy. Crown* 2005.

HM Government. *Strategy for Sustainable Development*. BIS. 2008.

HM Government. *Construction 202AQ5: Industrial Strategy: Government and Industry in Partnership*. Crown. 2013.

HM Government. *Industry Strategy: Construction Sector Deal*. Crown. 2018.

HM Government. *Net Zero Strategy: Build Back Greener*. Crown. 2021.

HM Government. *The Construction Playbook*. Version 1.1. 2022. Available at https://assets.publishing.service.gov.uk/government/uploads/system/uploads/attachment_data/file/1102386/14.116_CO_Construction_Playbook_Web.pdf (accessed 26/09/2022).

HMSO. This Common Inheritance: Britain's Environmental Strategy, Volume 1. Department of Environment. 1990.

HMSO. *Sustainable Development: The UK Strategy*. Dept. of Environment. 1994.

House of Commons. *Building to Net Zero: Costing Carbon in Construction*. 2022. Available at https://committees.parliament.uk/publications/30124/documents/174271/default/ (accessed 04/10/2022).

IPA. *Transforming Infrastructure Performance: Roadmap to 2030*. Crown. 2021.

LETI. *Climate Emergency Design Guide: How New Buildings can Meet UK Climate Change Targets*. 2020. Available at https://www.leti.uk/_files/ugd/252d09_3b0f2acf2bb24c019f5ed9173fc5d9f4.pdf (accessed 05/10/2022).

NBS. *Sustainable Futures Report 2022*. NBS. 2022.

ONS. *Net Zero and the Different Official Measures of the UK's Greenhouse Gas Emissions*. 2019. Available at https://www.ons.gov.uk/economy/environmentalaccounts/articles/netzeroandthedifferentofficialmeasuresoftheuksgreenhousegasemissions/2019-07-24 (accessed 03/10/2022).

RIBA. *RIBA Sustainable Outcomes Guide*. RIBA. 2019. Available at https://www.architecture.com/knowledge-and-resources/resources-landing-page/sustainable-outcomes-guide (accessed 02/10/2022).

RIBA. *RIBA Plan of Work Overview*. RIBA. 2020. Available at https://www.architecture.com/knowledge-and-resources/resources-landing-page/riba-plan-of-work (accessed 02/10/2022).

RICS. *Sustainability Report 2021*. 2021. Available at https://www.rics.org/uk/wbef/home/reports-and-research/sustainability-report-2021/ (accessed 06/10/2022).

United Nations. *Take Action for the Sustainable Development Goals*. 2022a. Available at https://www.un.org/sustainabledevelopment/sustainable-development-goals/ (accessed 25/09/2022).

United Nations. *Sustainability Development Report 2022*. 2022b. Available at https://unstats.un.org/sdgs/report/2022/ (accessed 25/09/2022).

United Nations Climate Change. *The Paris Agreement*. 2022. Available at https://unfccc.int/process-and-meetings/the-paris-agreement/the-paris-agreement (accessed 25/09/2022).

UK Green Building Council. *Circular Economy Guidance for Construction Clients: How to Practically Apply Circular Economy Principles at the Project Brief Stage*. 2019a. Available at https://www.ukgbc.org/ukgbc-work/circular-economy-guidance-for-construction-clients-how-to-practically-apply-circular-economy-principles-at-the-project-brief-stage/ (accessed 05/10/2022).

UK Green Building Council. *Net Zero Carbon Buildings: A Framework Definition*. 2019b. Available at https://ukgbc.s3.eu-west-2.amazonaws.com/wp-content/uploads/2019/04/08140941/Net-Zero-Carbon-Buildings-A-framework-definition.pdf (accessed 08/10/2022).

UK Green Building Council. *Net Zero Whole Life Carbon Roadmap A Pathway to Net Zero for the UK Built Environment*. 2021. Available at https://www.ukgbc.org/ukgbc-work/net-zero-whole-life-roadmap-for-the-built-environment/ (accessed 05/10/2022).

UK Green Building Council. *Insights on How Circular Economy Principles can Impact Carbon and Value*. 2022. Available at https://ukgbc.s3.eu-west-2.amazonaws.com/wp-content/uploads/2022/08/23174556/Whole-Life-Carbon-Circular-Economy-Report.pdf (accessed 05/10/2022).

UK Parliament. *Climate Change Targets: The Road to Net Zero?* 2021. Available at https://lordslibrary.parliament.uk/climate-change-targets-the-road-to-net-zero/ (accessed 03/10/2022).

Wellcertified. *Well Version 2 Introduction*. 2022. Available at https://v2.wellcertified.com/en/wellv2/overview (accessed 02/10/2022).

WLCN, Leti, and RIBA. *Improving Consistency in Whole Life Carbon Assessments and Reporting: Carbon Definitions for the Built Environment, Building and Infrastructure*. 2021. Available at https://asbp.org.uk/wp-content/uploads/2021/05/LETI-Carbon-Definitions-for-the-Built-Environment-Buildings-Infrastructure.pdf (accessed 04/10/2022).

World Commission on Environment and Development. *Report of the World Commission on Environment and Development: Our Common Future*. UN Documents: Gathering a Body of Global Agreements. 1987.

World Green Building Council. *Sustainable Buildings: For Everyone, Everywhere*. 2020. Available at https://worldgbc.org/sites/default/files/Sustainable%20Buildings%20for%20Everyone%2C%20Everywhere_FINAL.pdf (accessed 02/10/2022).

World Green Building Council. *Green Buildings and the Sustainable Development Goals*. 2022. Available at https://www.worldgbc.org/green-building-sustainable-development-goals (accessed 01/10/2022).

Index

Willis's Practice and Procedure for the Quantity Surveyor, Fourteenth Edition.
Allan Ashworth and Catherine Higgs.